Applications and Techniques for Experimental Stress Analysis

Karthik Selva Kumar Karuppasamy
Indian Institute of Technology, Guwahati, India

Balaji P.S.
National Institute of Technology, Rourkela, India

A volume in the Advances in Chemical and Materials Engineering (ACME) Book Series

IGI Global
DISSEMINATOR OF KNOWLEDGE

Published in the United States of America by
IGI Global
Engineering Science Reference (an imprint of IGI Global)
701 E. Chocolate Avenue
Hershey PA, USA 17033
Tel: 717-533-8845
Fax: 717-533-8661
E-mail: cust@igi-global.com
Web site: http://www.igi-global.com

Library of Congress Cataloging-in-Publication Data

Names: Karuppasamy, Karthik Selva Kumar, 1988- editor. | P.S, Balaji, 1986- editor.
Title: Applications and techniques for experimental stress analysis / Karthik Selva Kumar Karuppasamy and Balaji P.S., editors.
Description: Hershey, PA : Engineering Science Reference, 2020. | Includes bibliographical references. | Summary: “”This book examines how experimental stress analysis supports the development and validation of analytical and numerical models, the progress of phenomenological concepts, the measurement and control of system parameters under working conditions, and identification of sources of failure or malfunction”--Provided by publisher”-- Provided by publisher.
Identifiers: LCCN 2019032850 (print) | LCCN 2019032851 (ebook) | ISBN 9781799816904 (h/c) | ISBN 9781799816973 (s/c) | ISBN 9781799816911 (eISBN)
Subjects: LCSH: Strains and stresses--Analysis.
Classification: LCC TA407 .A635 2020 (print) | LCC TA407 (ebook) | DDC 620.1/1230287--dc23
LC record available at https://lccn.loc.gov/2019032850
LC ebook record available at https://lccn.loc.gov/2019032851

This book is published in the IGI Global book series Advances in Chemical and Materials Engineering (ACME) (ISSN: 2327-5448; eISSN: 2327-5456)

British Cataloguing in Publication Data
A Cataloguing in Publication record for this book is available from the British Library.

All work contributed to this book is new, previously-unpublished material. The views expressed in this book are those of the authors, but not necessarily of the publisher.

For electronic access to this publication, please contact: eresources@igi-global.com.

Advances in Chemical and Materials Engineering (ACME) Book Series

J. Paulo Davim
University of Aveiro, Portugal

ISSN:2327-5448
EISSN:2327-5456

Mission

The cross disciplinary approach of chemical and materials engineering is rapidly growing as it applies to the study of educational, scientific and industrial research activities by solving complex chemical problems using computational techniques and statistical methods.

The **Advances in Chemical and Materials Engineering (ACME) Book Series** provides research on the recent advances throughout computational and statistical methods of analysis and modeling. This series brings together collaboration between chemists, engineers, statisticians, and computer scientists and offers a wealth of knowledge and useful tools to academics, practitioners, and professionals through high quality publications.

Coverage

- Ductility and Crack-Resistance
- Industrial Chemistry
- Nanomaterials
- Wear of Materials
- Toxicity of Materials
- Sustainable Materials
- Electrochemical and Corrosion
- Ceramics and glasses
- Statistical methods
- Numerical Techniques

IGI Global is currently accepting manuscripts for publication within this series. To submit a proposal for a volume in this series, please contact our Acquisition Editors at Acquisitions@igi-global.com or visit: http://www.igi-global.com/publish/.

The Advances in Chemical and Materials Engineering (ACME) Book Series (ISSN 2327-5448) is published by IGI Global, 701 E. Chocolate Avenue, Hershey, PA 17033-1240, USA, www.igi-global.com. This series is composed of titles available for purchase individually; each title is edited to be contextually exclusive from any other title within the series. For pricing and ordering information please visit http://www.igi-global.com/book-series/advances-chemical-materials-engineering/73687. Postmaster: Send all address changes to above address. ©

Titles in this Series

For a list of additional titles in this series, please visit:
https://www.igi-global.com/book-series/advances-chemical-materials-engineering/73687

Design of Experiments for Chemical, Pharmaceutical, Food, and Industrial Applications
Eugenia Gabriela Carrillo-Cedillo (Universidad Autónoma de Baja California, Mexico) José Antonio Rodríguez-Avila (Universidad Autónoma del Estado de Hidalgo, Mexico) Karina Cecilia Arredondo-Soto (Universidad Autónoma de Baja California, Mexico) and José Manuel Cornejo-Bravo (Universidad Autónoma de Baja California, Mexico)
Engineering Science Reference • ©2020 • 350pp • H/C (ISBN: 9781799815181) • S $255.00

Handbook of Research on Developments and Trends in Industrial and Materials Engineering
Prasanta Sahoo (Jadavpur University, India)
Engineering Science Reference • © 2020 • 370pp • H/C (ISBN: 9781799818311) • US $295.00

Nanocomposites for the Desulfurization of Fuels
Tawfik Abdo Saleh (King Fahd University of Petroleum and Minerals, Saudi Arabia)
Engineering Science Reference • © 2020 • 300pp • H/C (ISBN: 9781799821465) • US $225.00

Enhanced Heat Transfer Mechanism of Nanofluid MQL Cooling Grinding
Changhe Li (Qingdao University of Technology, China) and Hafiz Muhammad Ali (University of Engineering and Technology, Taxila, Pakistan)
Engineering Science Reference • © 2020 • 350pp • H/C (ISBN: 9781799815464) • US $235.00

Advanced Catalysis Processes in Petrochemicals and Petroleum Refining Emerging Research and Opportunities
Mohammed C. Al-Kinany (King Abdulaziz City for Science and Technology, Saudi Arabia) and Saud A. Aldrees (King Abdulaziz City for Science and Technology, Saudi Arabia)
Engineering Science Reference • © 2020 • 257pp • H/C (ISBN: 9781522580331) • US $195.00

Nanotechnology in Aerospace and Structural Mechanics
Noureddine Ramdani (Research and Development Institute of Industry and Defense Technologies, Algeria & Harbin Engineering University, China)
Engineering Science Reference • © 2019 • 356pp • H/C (ISBN: 9781522579212) • US $265.00

Polymer Nanocomposites for Advanced Engineering and Military Applications
Noureddine Ramdani (Research and Development Institute of Industry and Defense Technologies, Algeria & Harbin Engineering University, China)
Engineering Science Reference • © 2019 • 435pp • H/C (ISBN: 9781522578383) • US $225.00

IGI Global
DISSEMINATOR OF KNOWLEDGE

701 East Chocolate Avenue, Hershey, PA 17033, USA
Tel: 717-533-8845 x100 • Fax: 717-533-8661
E-Mail: cust@igi-global.com • www.igi-global.com

Editorial Advisory Board

Table of Contents

Detailed Table of Contents

For the design and development of new machine components, the researchers and engineers must have an extreme understanding of the stress, strain, and the basic equations/laws relating the stress to the strain. In this chapter, the authors show the basic concepts of stress developed in a component concerning the external loading and the loading concerning the body force. In this chapter, the following aspects were proposed to be briefly discussed: type of stresses, introduction to stress at particular node, stress equation relates the equilibrium of body, laws related to transformation of stress, states of stress, and sample solved problems related to the simple state of the stress system.

In this chapter, the basic concepts of strain and the subject of components deformation and associated strain is proposed to be briefly discussed. The strain is a pure geometric quantity so there will be no limitation on the material of the component is required. The significant aspects related to strain and stress-strain relationship briefly discussed in this proposed chapter were as follows: displacement and strain, principal strains, compatibility equations, relationship between stress and strain, two and three dimensional state of stress with respect to strain, stress-strain relationship curve, and overview of various experimental techniques employed for the measurement of stress and strain respectively.

Stress and strain are mechanical behaviour of materials, subjected to mechanical or thermal loading. The detrimental effect of such loading is the ultimate failure of materials due to generation of high stress and strain. Therefore, measurement and prediction of stress and strain values help in proper design and maintenance of engineering equipment and structures. The present contents elaborate and summarize different methods adopted by researchers for mechanical stress and strain measurements. The content is focused to provide an overview regarding the measurement techniques adopted for strain measurement. The analysis holds information regarding working principle of different strain measuring technique along with a brief description about the history of strain measurement. Special attention has also been devoted for explanation of thermal stress and strain measurement techniques. The modern non-contact techniques have evolved as a potential tool for such measurements even at higher temperature conditions.

Strain gauge method is one of the essential and fundamental methods in experimental stress techniques that uses the resistance of the material to determine the stress at a point. The strain gauges can be used in a different combination called Rosette to obtain stress in various directions. This chapter intends to cover types of strain gauges, materials, and Rosette arrangements to provide the reader with an overview of the techniques. The chapter will discuss the basic physics behind the resistance measurement and take the reader into insights on how the developments were made to the application of strain gauges as experimental techniques.

The strain gauge system consists of a metallic foil supported in a carrier and bonded to the specimen by a suitable adhesive. Previous chapters discussed the construction, configuration, and the material of the strain gauge. The strain gauge has advantages over the other methods. A strain gauge can give directly the strain value as output. However, in optical methods, it is required to interpret the results. It is also required to be aware that the strain gauge technology is majorly used, and it can also be easily wrongly used. Hence, it is required to obtain the proper knowledge of the strain gauge to get the full benefit of the technology. This chapter covers the majorly on the performance of the strain gauge, its temperature effects, and strain selection. Further, this chapter also covers the brittle coating technique that is used to decide the position of the strain gauge in the applications.

Today, measurement of strain plays a crucial role in different areas of research such as manufacturing, aerospace, automotive industry, agriculture, and medical. Many researchers have used different types of strain transducers to measure strain in their relevant research fields. Strain can be measured using mainly two methods (i.e., electrical strain sensors and optical strain sensors). Electrical strain sensors consist basically of strain gauges, piezo film, etc. In electrical strain sensors, the strain gauge is one of the oldest and reliable strain sensors which are available in different types (i.e., wire strain gauge, foil strain gauge, and semiconductor strain gauge). Piezofilm is also playing an important role in the field of strain measurement due to easy availability and less cost.

Stress will be produced in most of the engineering components related to their manufacturing process or because of their loading condition. For some special cases, both types also combined together and produce stress. Manufacturing processes like casting, welding, machining, and hot forming are creating the stress in the components. This stress will produce due to alteration of the microstructure (size, shape, phase composition, and orientation). Loading conditions are also produced stresses in the engineering components. This stress may be classified into compression, shear, tension, and fatigue. These depend on the load. Measuring the stresses in the components is very important because it can save a lot in terms of money, material, and manpower. A lot of techniques are used in industries to measure the stresses. Based on that, measurement techniques are broadly classified into two category, namely destructive and non-destructive techniques. Each method has its own advantages and limitations too. In this chapter, the optical method of measuring stress is discussed briefly.

Lattice stress and strain was analysed with estimated crystalline size of the synthesised ZnFe2O4 nanoparticles from x-ray diffraction data using Williamson-Hall (W-H) method. This very peculiar method was used to analyse the other physical parameters such as strain, stress, and energy density. Values calculated from the W-H method include uniform deformation model, uniform deformation stress model, and uniform deformation energy density model. These are very useful methods to label each data point on the Williamson-Hall plot according to the index of its reflection. Particularly, the root mean square value of strain was calculated from the interplanar distance using these three models. The three models have given different strain values by reason of the anisotropic nature of the nanopartcles. The average grain size of ZnFe2O4 nanoparticles estimated from FESEM image, Scherrer's formula, and W-H analysis is relatively correlated.

modelling has found wide applicability for the same. This chapter is aimed at providing background knowledge, a comparison of BG approach with the FE approach, and a review of research progress of past two decades in this direction.

The stress generated due to the temperature difference is called thermal stress. Generally, the temperature gradients, thermal shocks, and thermal expansion or contraction are most effective contributors to thermal stress. The improper temperature profile of a body results in the formation of cracks, fractures, or plastic deformations at single or multiple spots depending upon two factors (i.e., the magnitude of temperature distribution or other variables of heating and material properties). So, this chapter analyses the causes of thermal stress and their measurement techniques. However, as most of the engineering problems, the thermal stress is due to the thermal expansion or sudden temperature changes happening in the body. Therefore, a brief analysis of temperature measurement devices with their proper data capturing methodology is also discussed.

Finite element analysis (FEA) is a numerical technique in which product behavior under various loading conditions is predicted for ease of manufacturing. Due to its flexibility, its receiving research attention across domain discipline. This chapter aims to provide numerical investigation on chip formation in micro-end milling of Ti-6Al-4V alloy. It is widely used for medical applications. The chip formation process is simulated by a 3D model of flat end mill cutter with an edge radius of 5 μm. Tungsten carbide is used as a tool material. ABACUS-based FEA package is used to simulate the chip formation in micro-milling operation. Appropriate input parameters are chosen from the published literature and industrial standards. 3-D orthogonal machining model is developed under symmetric proposition and assumptions in order to reveal the chip formation mechanism. It is inferred that the developed finite element model clearly shows stress development in the cutting region at the initial stage is higher. It reduces further due to tool wear along the cutting zone.

CFRP/Ti6Al4V stacks are widely used in aerospace and automobile industries as structural components. The parts are made to near net shape and are assembled together. Aerospace standards demand rigid tolerance for the holes. While drilling stacks, during the exit of drill from CFRP and entry into Ti6Al4V,

Foreword

All the Engineering components undergoes stresses and strains caused by various loads during the operation and by other minor factors such as loading of wind and vibrations. Engineers face challenges in deploying a method to evaluate these stresses in a real-time engineering component and they are hindered due to the limited availability of the required essential information on stress evaluation. Classically the stress evaluation is performed using analytical and by numerical methods such as finite element methods. This book, *Applications and Techniques for Experimental Stress Analysis,* presents a comprehensive and modern approach using experimental methods for stress and strain evaluations. Focusing on establishing formal methods and recent research works, this book helps to understand the various experimental and finite element techniques for the measurement of stress and strain analysis in real-time engineering problems. This book also briefing about:

- Stress evaluation of a variety of structures and components.
- Provides new ideas to researchers in Experimental and Computational mechanics.
- Discusses complex models.
- Compilation of a treasure of documented real-time research works.
- Enables stress analysts to extend the range of problems and applications they can address.

Further this book is a compilation of various significant approaches in the experimental stress analysis and observes their application to numerous states of the stress of foremost technical interest, prominence aspects which are not always covered in the common literature. It is enlightened how experimental stress analysis supports in the authentication and accomplishment of analytical and numerical models, the development of phenomenological concepts, the measurement and control of system parameters under various working conditions, and identification of the source of the fault. Cases addressed include measurement of the state of stress and strain in models, measurement of actual loads on structures, verification of stress states in circumstances of complex numerical modelling, assessment of stress-related material damage, and brief analysis of artefacts (e.g. dental bone) that interact with biological systems.

I am very pleased to provide a foreword for this book, offering real-time research examples, this book is considered to be an indispensable tool for all the undergraduate, postgraduate students and doctoral researchers, in the field of experimental mechanics and engineering analysis. The people include automobile industries, construction, mechanical, and aerospace engineers engaged in analysing the stress and strain of structures and components can be able to find this book as a valuable reference.

Muhammad Ekhlasur Rahman
Curtin University, Australia

Preface

The design of mechanical components for various engineering applications requires the understanding of stress distribution in the materials. The nature of stress distribution on the certain components can be achieved more precisely with various experimental techniques. This book is a timely research publication that examines how experimental stress analysis supports the development and validation of analytical and numerical models, the progress of phenomenological concepts, the measurement and control of system parameters under working conditions, and identification of sources of failure or malfunction. Highlighting a range of topics such as deformation, strain measurement, and element analysis, this book is essential for mechanical engineers, civil engineers, designers, aerospace engineer's researchers, industry professionals, academicians, and students.

ORGANIZATION OF THE BOOK

The book is organized into 15 chapters. A brief description of each of the chapters follows:

Chapter 1 introduces the basic concepts of stress developed in a component due to the external loading and the loads from the body forces. This chapter briefly discusses the type of stresses, stress at a particular node, stress equations related to the equilibrium of body, laws related to the transformation of stress, states of stress and sample solved problems related to the simple state of the stress system.

Chapter 2 covers the basic concepts of strain and the subject of components deformation and associated strain are discussed. The strain is a pure geometric quantity so there will be no limitation on the material of the component is required. The significant aspects related to strain and stress-strain relationship are briefly discussed, Further, it gives insight into displacement and strain, principal strains, compatibility equations, relationship between stress and strain, two and three dimensional state of stress with respect to strain, stress-strain relationship curve and overview of various experimental techniques employed for the measurement of stress and strain respectively.

Chapter 3 elaborates and summarizes the different methods adopted by researchers for mechanical stress and strain measurements. The content is focused to provide an overview regarding the measurement techniques adopted for strain measurement. The analysis holds information regarding the working principle of different strain measuring technique along with a brief description of the history of strain measurement. Special attention is also devoted to thermal stress and strain measurement techniques.

Chapter 4 introduces the strain gauge method which is one of the essential and fundamental methods in experimental stress techniques that uses the resistance of the material to determine the stress at a point. The strain gauges can be used in a different combination called Rosette to obtain stress in various directions. This chapter covers the types of strain gauges, materials and rosette arrangements to provide the reader with an overview of the techniques.

Chapter 5 discusses the performance of the strain gauge system. It is also required to be aware that the strain gauge technology is majorly used and it can also be easily wrongly used. This chapter covers the majorly on the performance of the strain gauge, its temperature effects and strain selection. Hence it is required to obtain the proper knowledge of the strain gauge to get the full benefit of the technology. this chapter also covers the brittle coating technique which is used to decide the position of the strain gauge in the applications.

Chapter 6 deals with the measurement of strain using strain gauge and piezoelectric sensors. As today measurement of strain plays a crucial role in a different area of research such as manufacturing, aerospace, automotive industry, agriculture and medical etc. Many researchers have used different types of strain transducer to measure strain in their relevant research field. Strain can be measured using mainly two methods i.e. Electrical strain sensors and Optical strain sensors. These techniques are discussed in this chapter.

Chapter 7 briefly describes the optical methods for the evaluation of stress. The stress can be also produced due to the alteration of the microstructure (size, shape, phase composition and orientation). This stress may be compression, shear, tension and fatigue. Measuring the stresses in the components are very important because it can save a lot in terms of money, material and manpower. Optical methods can be used to evaluate the stress to avoid failure in the components.

Chapter 8 analyses the crystalline size of the synthesized ZnFe2O4 nanoparticles from X-ray diffraction data using Williamson-Hall (W-H) method. This method is very peculiar to analyze the other physical parameters such as strain, stress, and energy density and the values that are calculated from the W-H method including uniform deformation model, uniform deformation stress model and uniform deformation energy density model. These are very useful method to label each data point on the Williamson-Hall plot according to the index of its reflection.

Chapter 9 assesses the deformation of stainless steel sheet using a shock tube. In this study, a high-velocity sheet metal forming experiment has been performed using a hemispherical end nylon striker inside the shock tube. The striker moves at a high velocity and impacts the sheet mounted at the end of the shock tube. Various forming parameters such as dome height, effective strain distribution, limiting strain, hardness and grain structure distribution are analyzed. The dome height of the material increases monotonically with the high-velocity. The effective-strain also follows a similar variation and a bi-axial stretching phenomenon is observed.

Chapter 10 discusses the micromotion analysis of a dental implant system. Micromotion of the novel implant was obtained using static structural FE analysis. Compared to the existing International team for implantology implant, the micromotion of the novel implant model was considerably decreased by static structural analysis. Compared to the original model, the micromotion is 0.01944 mm and micromotion of improved design version is 0.01244 mm.

Chapter 11 provides the modelling stress distribution in a flexible beam using bond graph approach. The Bond Graph (BG) approach of modelling provides a unified approach for modelling the systems having components belonging to multi-energy domains. Moreover, as evident by its name, it is a graphical approach. The graphical nature provides a tool for conceptual visualization of the model. It also provides some algorithmic tools because of its formal structure and syntax, thereby enable model consistency check such as checking algebraic loops, etc. This chapter is aimed at providing background knowledge, a comparison of BG approach with the FE approach, and a review of research progress of past two decades in this direction.

Chapter 12 presents a theoretical study of thermal stress for engineering applications. The stress generated due to the temperature difference is called thermal stress. Generally, the temperature gradients, thermal shocks and thermal expansion or contraction are most effective contributor to thermal stress. This chapter analyses the causes of thermal stress and measurement techniques. However, as most of the engineering problems, the thermal stress is due to the thermal expansion or in sudden temperature changes happen in the body. Therefore, a brief analysis of temperature measurement devices with their proper data capturing methodology is also discussed.

Chapter 13 deals with the finite element analysis of chip formation in micro-milling operation. Finite Element Analysis (FEA) is a numerical technique in which product behavior under various loading conditions predicted for easy of manufacturing. This chapter aims to provide numerical investigation on chip formation in micro-end milling of Ti-6Al-4V alloy. Chip formation process is simulated by 3D model of flat end mill cutter with edge radius of 5 µm. Tungsten carbide is used as tool material. ABACUS based FEA package is used to simulate the chip formation in micro-milling operation. Appropriate input parameters are chosen from the published literature and industrial standards. 3-D orthogonal machining model is developed under symmetric proposition and assumptions in order to reveal the chip formation mechanism.

Chapter 14 presents the understanding of the position of drill tool using acoustic emission signals during the drilling of CFRP/Ti6Al4V stacks. CFRP/ Ti6Al4V stacks are widely used in aerospace and automobile industries as structural components. The parts are made to near net shape and are assembled together. Aerospace standards demand rigid tolerance for the holes. While drilling stacks, during the exit of drill from CFRP and entry into Ti6Al4V there is a change in the overall behavior of the drilling process due to changes in the mechanical properties of the two materials. Hence stacks should be drilled under their optimal machining conditions in order to achieve better hole quality. This chapter proposes the use of acoustic emission (AE) signals to monitor the drill position while drilling of CFRP/Ti6Al4V stacks.

Chapter 15 presents the simulation of Mn2-x Fe1+x Al intermetallic Alloys microstructural formation and the stress-strain development in steel casting. In this simulation, the permeation of the n-phase precipitation to the Mn2 Fe Al crystallization is induced by the steel casting solidification process by JMatPro. Using the model, the morphological evolution of the Fe and Mn in different percentages was obtained, in which the heated data obtained by simulating casting and extreme heat treatment processes were adopted. This chapter describes a model of the computer model for calculating the phase transition and properties of materials required to predict the deviation during the heat treatment of steel. The current model has the advantage of using a variety of shape memory alloys including medium to high aluminum based Heusler alloys.

This book covers essential basics on stress measurement techniques and also provides research work carried out on the stress evaluation. The knowledge of experimental stress evaluation technique with the associated research work can serve as a resource with essential technique and applications.

Karthik Selva Kumar Karuppasamy
Indian Institute of Technology, Guwahati, India

Balaji P. S.
National Institute of Technology, Rourkela, India

Acknowledgment

The editors would like to acknowledge the support of all the individuals involved in this project. Further, the editors would like to thank each one of the authors for their contributions. Our sincere appreciation goes to the chapter authors who contributed their time and expertise to this project. Second, the editor wishes to acknowledge the valuable contributions of the reviewers regarding the improvement of quality, coherence, and content presentation of chapters. Most of the authors also served as referees; We extremely appreciate their contribution. The editors are very grateful to Dr. Ram Krishna Upadhyay for his timely response in the review process of the chapters. The editors would also like to acknowledge the support from the Department of Mechanical Engineering at Indian Institute of Technology Guwahati and National Institute of Technology Rourkela.

Karthik Selva Kumar Karuppasamy
Indian Institute of Technology, Guwahati, India

Balaji P. S.
National Institute of Technology, Rourkela, India

Chapter 1
Introduction to the Basics of Stress

Karthik Selva Kumar Karuppasamy
https://orcid.org/0000-0001-5431-2542
Indian Institute of Technology, Guwahati, India

Niranjan Sahoo
Indian Institute of Technology, Guwahati, India

Balaji Selvaraj
https://orcid.org/0000-0002-6364-4466
National Institute of Technology, Rourkela, India

ABSTRACT

For the design and development of new machine components, the researchers and engineers must have an extreme understanding of the stress, strain, and the basic equations/laws relating the stress to the strain. In this chapter, the authors show the basic concepts of stress developed in a component concerning the external loading and the loading concerning the body force. In this chapter, the following aspects were proposed to be briefly discussed: type of stresses, introduction to stress at particular node, stress equation relates the equilibrium of body, laws related to transformation of stress, states of stress, and sample solved problems related to the simple state of the stress system.

INTRODUCTION

Engineering structures are subjected to various forces that can result in instability and cause structural failures on the components. Similarly, the structural materials in the path of fluid flow are subjected to distortion due to the high current of fluids. This causes a concentration of stress and strain on the surface of engineering component that can lead to the failure of components (Karthik & Dhas, 2015, 2016, 2018). The excessive vibrations on the material also induces stress and cause failure (Balaji et al., 2015; Leblouba et al., 2015; Balaji et al., 2016a; 2016b). Furthermore, the temperature can cause stress on the

DOI: 10.4018/978-1-7998-1690-4.ch001

components (Balaji & Yadava, 2013). The present chapter covers the overview on the measurements of stress and strain elements in various structures. The authors determined the stress and strain measurements on the structural elements by the physical features of the resource employed and the external forces influencing the assemblies. To be resolute cautiously in the course of the design stage. The structure must be consistent as well as cost-effective in the course of the period of utilization. Researchers ensure the consistency once they consider the positive and fixed characteristics, such as durability, strength, stability, and stiffness, for the development of the complete structure. The cost invested in the required resource, concerning the introduction of newer technology and also on the application of cost-effective resources, influences the construction budget. Given the aforementioned information, it is understandable that the consistency and the budget are contradictory parameters. This is due to the fact that the strength of materials is dependent on the practical understanding over and above the conceptual knowledge.

General Theories

Researchers have to understand the materials characteristics before analyzing the forces on the structure and their aftereffects. For example, the strength is the capability of the component to resist the external forces applied to it, and the strength of materials is extensively dependent on the hypothetical theory of mechanics. Stiffness represents the capability of the component or element to resist the strains by the applied external forces. Stability is the special characteristic of any structure to retain its initial position of equilibrium (Ashby et al., 2007). Additionally, durability is the special characteristic of the component to protect its mechanical properties for the duration of the time that forces (external or internal) act on the surface of the structure. The fundamental issue in engineering is the identification of suitable techniques or methods for designing a machine component or structures by considering various essential factors within the expected resilience and cost-effectiveness. To analyze a real-time component of a machine/application, researchers must select a consistent, precise computational method. The computational techniques are tangible for use so that the unnecessary features can be easily identified and rectified in the design stage. To select the precise computational techniques, researchers have to establish the hypotheses concerning the material strength. The significant parameters have to be considers in the design stage to avoid structural failures on the test components were briefly listed as follows,

1) The materials used for the construction of the structure or the components:
    a) Continuity of the resource material:
        i) The resource material is consistently spread throughout the volume of the whole body.
    b) Homogeneity of the resource material:
        i) All the nodes of the component have identical properties (material).
    c) Isotropy of the resource material:
        i) The properties of the materials were identical in each direction.
    d) The ability of the structure to be deformed:
        i) The capability to transform the aforementioned original form and proportions under the influence of external forces.
    e) The elasticity of the resource material:
        i) The capability of the structure to reestablish its original form and size when the acting forces are removed.
2) Applied forces:

   a) The concentrated load on the component.
   b) Researchers understand that when an object subjected to an external force, the result is a minor effect of deformation on the structure.
3) Confined equilibrium:
   a) If the component is considered to be in a state of equilibrium, at that moment, each element of the components must be in the state of equilibrium as well.
4) Code of superposition:
   a) The ultimate scale of a mass deliberated and triggered by the set of peripheral forces can be acquired as an arithmetical summation of the number of enormities initiated by the specific forces put together conventionally.
5) Hardening:
   a) Researchers observe a component to have a definite profile and measurements before the application of the load.
6) Rigid body:
   a) A body is made up of particles, whereas researchers observe the distances between each of the particles to be in an indestructible state.
7) Deformable body:
   a) A body is made up of particles, whereas researchers observe the distances between each of the particles to be changing with the applied load.

Researchers can employ extremely insignificant stress elements to understand the normal stress on a component by characterizing how state of stress acts at a particular point. Further, starting the investigation by making an allowance for an element with the known stresses, which further lead the transformation relations for the estimation of the stresses acting on the edges of the component concerned in different direction. During the stress analysis of any component, researchers should always bear in mind that merely a single fundamental state of stress is existent at a location of a body subjected to stress. Nevertheless, the positioning of the component is cast-off to represent the state of stress. Whenever the elements with contradictory alignments develop on the test component, researchers observe the stresses acting on the surfaces of the two elements to be altered. On the other hand, the elemnts still epitomize the similar state of stress, specifically, the stress at the particular position under consideration. The perception of stress is more difficult than trajectories in arithmetic; this type of stresses was known as tensors. The authors observed the strain and the moment of inertia to be the additional tensor measurements in mechanics.

INTRODUCTION TO NORMAL STRESS

To understand the mechanical behavior of a component subjected to loading, it is necessary for engineers to develop a precise design, regardless of whether the component is an aircraft, automobiles, pressure vessels, buildings structures, machines, or a spacecraft. In-depth understanding the mechanics of materials can be obtained by examining the stresses and strains in real components. An experiment stress analyst must have a thorough understanding of stress, strain and the laws relating the stress and strain. For this reason, the authors discuss the basic concept of stress and how it is developed in a body subjected to external and body force in this chapter. There are two basic types of forces that cause the development of stresses on a body or structure. These forces are surface forces and body forces.

Surface Forces

Surface forces are forces that act on the surface of the structure or body, Surfaces forces are generally applied, when one structure comes into contact with other structures.

Body Forces

Body forces act on each element of the structure or body. Body forces are commonly developed due to gravitational, centrifugal, and other force fields. The most common body forces are gravitational. Body forces are significantly smaller when compared to the surface forces.

The authors characterized the forces applied on the structures as loads. The visibly applied forces may be a result of the conditions illustrated in Figure 1. According to the mechanics of deformable solids, the forces applied externally on a body lead to the deformation of structure. From the perspective of equilibrium, the aforementioned act would be resisted by the forces acting inside the constituent part of body material due to cohesion. This resistance from the internal forces against the applied forces leads to the concept of stress.

Stress

To understand the stress developed in a body, one can consider a rectangular bar exposed to a certain load or force (in Newton's), as illustrated in Figure 2. Further, the authors presume the same rectangle to be cut into two halves at section (**a** and **a'**). subsequently, each portion of this rectangular bar is considered to be in an equilibrium state under the action of load F and the internal forces acting at the section (a and a'), as shown in Figure.3. Now the stress developed in the body is defined by the force per unit area. The authors represent stress by the symbol σ .

Figure 1. Causes for the external forces on the structures

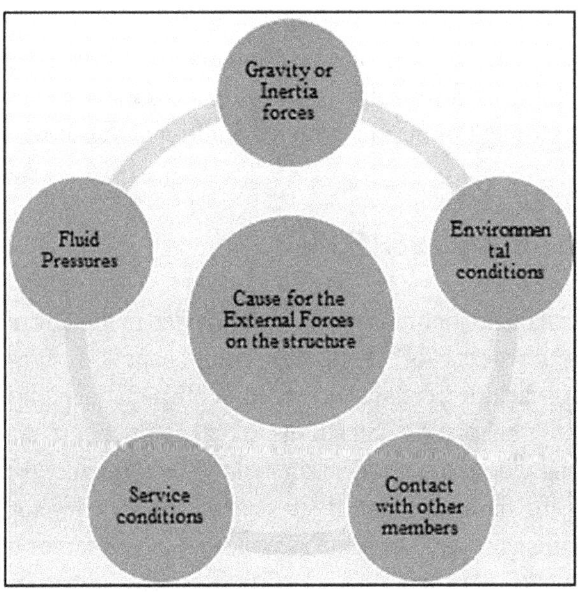

$$\sigma = \frac{P(or)F}{A} \tag{1}$$

While P or F represent the load acting on the body, A represents the cross-section area of the body (a and a'). To understand the nature of stress, must consider the overall force or the entire load applied to the rectangular bar, and the authors assume it to consistently spread throughout its cross-sections. However, the stress distribution on the body may differ in many aspects. If a component is subjected to a non-uniform force or load throughout its cross-sectional area, then high stress develops in a particular node or element, which is called stress concentrations. According to the load distribution on each area, the authors can investigate the stress development in a particular area (δA) due to the load applied (δP) on the component. Stress can be calculated using the following equation:

$$\sigma = \frac{\delta \mathrm{P}}{\delta \mathrm{A}} \tag{2}$$

Types of Stress

The stresses acting on a body depends upon various phenomenon, which the authors briefly explain below:

- Bending stress is a combination of compressive, shear, and tensile stresses.
- Torsional stress happens upon twisting a shaft component.

Figure 2. A rectangular bar subjected external forces

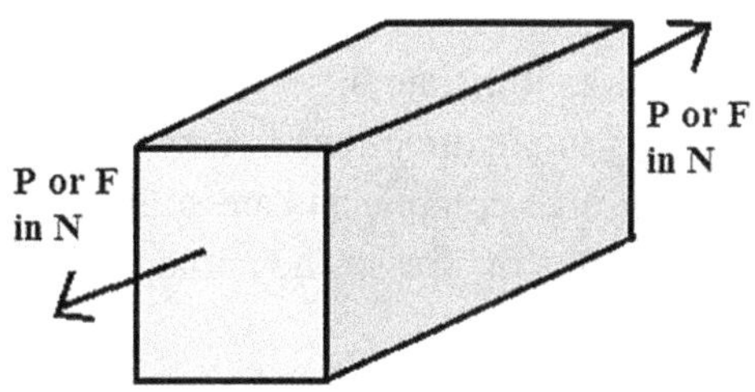

Figure 3. The rectangular bar is sliced into two splits at section (a and a')

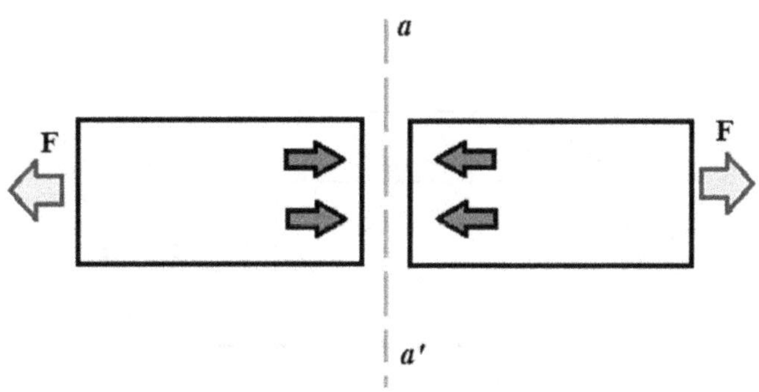

Figure 4. Basics stress acting on a body

The authors illustrate the basic stress acting on a structural component in Figure 4.

Normal Stress

Researchers characterize stress that is normal to the concerned areas as normal stress. Normal stress is commonly indicated by σ, and the authors illustrate this in Figure 5.

Researchers call the stress acting on single directions uniaxial stress. Commonly, this type of stress is a rare phenomenon because the real-time components are subjected to loads in multiple directions, and stress can be bi-axial or tri-axial states of stress. The authors illustrate a bi-axial state of stress in Figures 6 and 7.

The normal stress can be either tensile or compressive. The type of stress depends on how the forces are applied on the surface of the component, as illustrated in Figure 8.

Figure 5. Uniaxial stresses on a structure

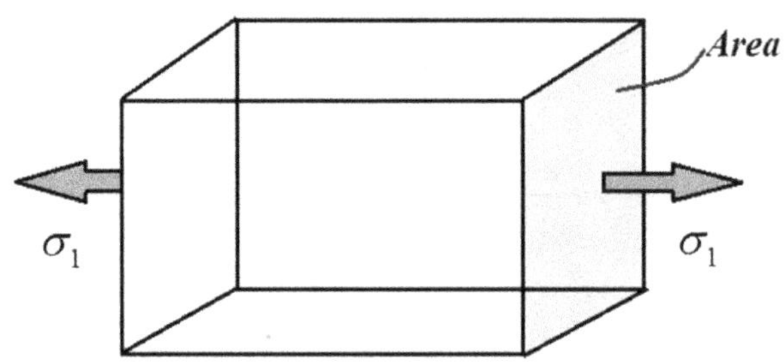

Figure 6. Bi-axial state of stress on a structure

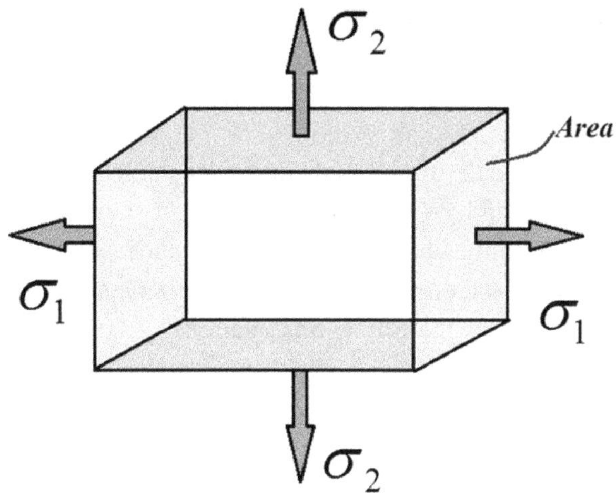

Figure 7. Tri-axial state of stress on a structure

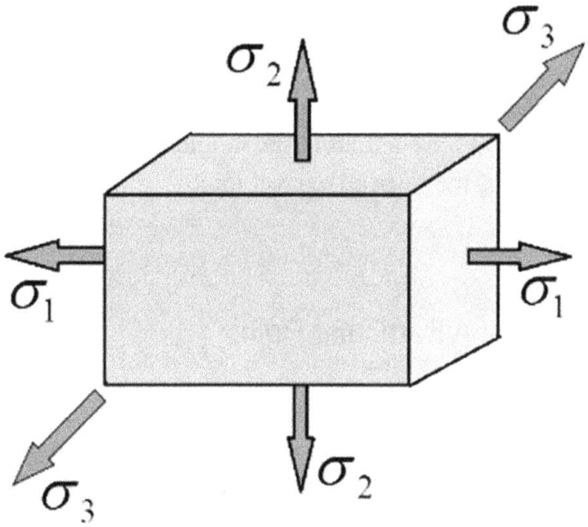

Figure 8. Tensile and compressive state of stress on the structure

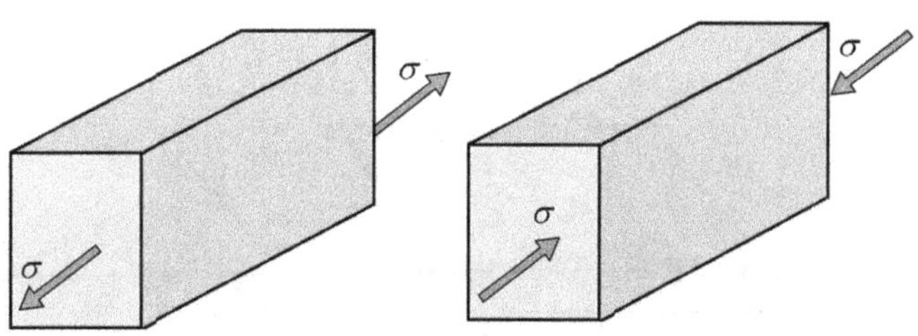

When one component is hard-pressed against the other, the bearing stress (compressive stress) is developed at the contact surface, as shown in Figure 9.

Shear Stress

When a cross-sectional area of the bulk of materials undergoes deformation by slippage along a plane parallel to the enforced stress, researchers consider the resultant shear to be significant in nature. Researchers call the forces allied with the shearing of materials shear forces. Shear stress is the ratio of applied force (F) to the cross-sectional area (A) of the structure. Generally, shear stress is represented by the Greek symbol τ, and is given by the following equation:

$$\tau = \frac{F}{A} \tag{3}$$

Generally, researchers determine the resultant stress at any point on a structure by two different components (i.e., σ and τ). The components, as mentioned above, act in a unique direction. For example, one component could act in the parallel direction, and the other one could act in the perpendicular direction, as illustrated in Figure 10. When investigating the shear phenomenon in any component, one must consider several features, such as the single shear rate that occurs on the single plane and the cross-sectional of the rivet that has to be considered as the shear area. When the authors study the shear phenomenon of rivets in butt joints, it is observed to have a double shear, and one must consider the shear area has as double the cross-sectional region of the respective rivet joint.

Assessment of Stress

Nature of the Stress Acting at A Particular Point

Researchers define stress acting at a particular position in a as a force per unit area, whereas the magnitude and resultant stress direction σ_R depends on the alignment of the plane passed over the particular point. Researchers can employ the number of resultant stress vectors on a number of planes that can be

Figure 9. Bearing a state of stress on the structure

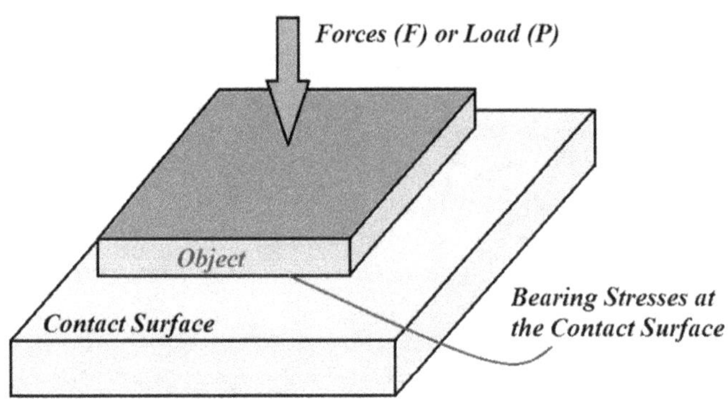

Figure 10. Shear failure in rivets

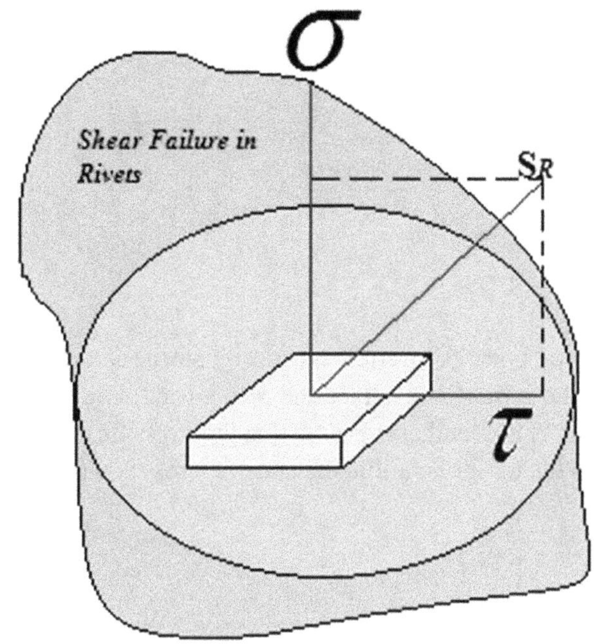

passed at each point to represent the resultant stress at each point of the component. Further, the magnitude and direction of the resultant stress vectors can be stated in the expressions of nine Cartesian stress components at the particular point (Boresi, 2000).

$$\sigma_{Rx} A - \sigma_{xx} \cos(R,x) - \breve{\tau}_{yx} A \cos(R,y) - \breve{\tau}_{zx} A \cos(R,z) + \breve{F}_{x} 0.3333hA = 0 \tag{4}$$

Once removing the common factor 'A' from the expression about, one can observe that without the body forces, the average stresses turn out to be exact stresses at the point P. Further, the expression will be written as equations concerning the Cartesian components are:

$$\sigma_{Rx} = \sigma_{xx} \cos^{*}(R,x) - \breve{\tau}_{zx} \cos^{*}(R,y) - \breve{\tau}_{zx} \cos^{*}(R,z) \tag{5}$$

$$\sigma_{Ry} = \sigma_{xy} \cos^{*}(R,x) - \breve{\tau}_{yy} \cos^{*}(R,y) - \breve{\tau}_{zy} \cos^{*}(R,z) \tag{6}$$

$$\sigma_{Rz} = \sigma_{xz} \cos^{*}(R,x) - \breve{\tau}_{yz} \cos^{*}(R,y) - \breve{\tau}_{zz} \cos^{*}(R,z) \tag{7}$$

Once the authors identified the resultant stress for the particular plane concerning the three Cartesian components, the resultant stress can be stated by the following expression:

$$\sigma_{R} = \sqrt{\sigma^{2}_{Rx} + \sigma^{2}_{Ry} + \sigma^{2}_{Rz}} \tag{8}$$

Researchers can identify the normal stress as well as the shear stress, which is acting on the particular plane using the following expressions:

$$\sigma_N = \left|\sigma_R\right| \cos^*(\sigma_{R,} R) \tag{9}$$

$$\tau_N = \left|\sigma_R\right| \sin^*(\sigma_R, R) \tag{10}$$

Tensor Representation of Stress

Researchers designed the "9 components" of stresses to designate the nature of stress at a point completely. As illustrated in Figure 11, the authors considered the σ, z, y (or τ, z, y) to be the stress generated by a shear force in the direction of Y. Researchers observe that the stress further acts in a plane normal to the direction of Z. According to the tensor notation, the state of stress at a point is expressed as:

$$\tau_{ij} = \begin{vmatrix} \sigma_{xx} & \sigma_{yx} & \sigma_{xx} \\ \sigma_{yx} & \sigma_{yy} & \sigma_{yz} \\ \sigma_{zx} & \sigma_{zy} & \sigma_{zz} \end{vmatrix} \tag{11}$$

where i and j are reiterated over x, y, and z. The shear stresses are proportioned, hence there are six autonomous variables in place of nine.

Figure 11. The average stresses are acting on the face of elemental tetrahedron at a point P

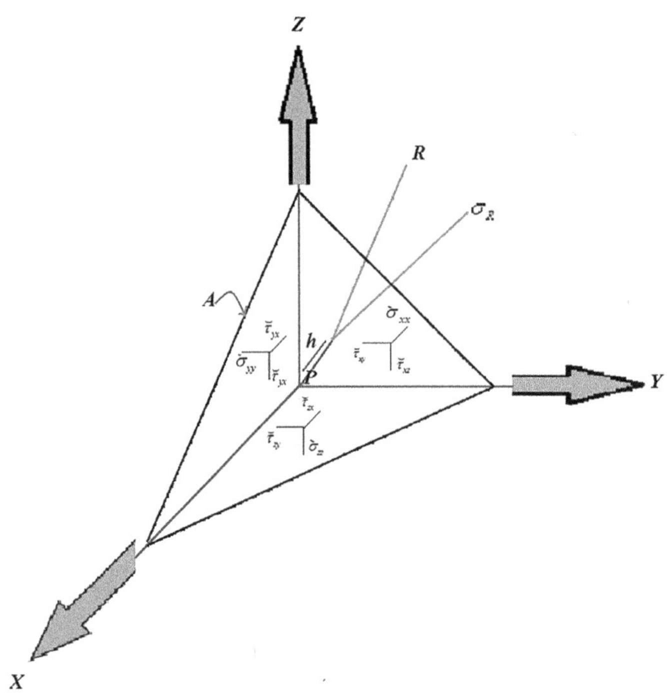

Stress Equation for the Equilibrium State

Stresses at variable magnitude and direction will develop throughout the structure when a body is subjected to forces. The distribution of the stresses throughout the structure has to preserve the overall equilibrium of the body at each element (Boyer, 1987).

Stress Transformation Laws

In the earlier statement, the authors mentioned that the resultant stress vector σ_R acts on an arbitrary plane defined by the outer normal ***"R"***. Researchers can determine this by replacing the independent six Cartesian components of stress into equations 5, 6, and 7. Researchers generally employ the transformation equation with reference to the coordinate system of the stress components as follows.

Normal Stress transformation equation on a plane with Cartesian components

$$\begin{aligned}\sigma_{x'x'} &= \sigma_{xx}\cos^{2*}(x',x)+\sigma_{yy}\cos^{2*}(x',y)+\sigma_{xx}\cos^{2*}(x',z)\\ &+2\tau_{xy}\cos^{*}(x',x)^{*}\cos^{*}(x',y)+2\tau_{yz}\cos^{*}(x',y)^{*}\cos^{*}(x',y)+2\tau_{zx}\cos^{*}(x',z)^{*}\cos^{*}(x',x)\end{aligned} \tag{12}$$

$$\begin{aligned}\sigma_{y'y'} &= \sigma_{yy}\cos^{2*}(y',y)+\sigma_{zz}\cos^{2*}(y',z)+\sigma_{xx}\cos^{2*}(y',x)\\ &+2\tau_{yz}\cos^{*}(y',y)\cos^{*}(y',z)+2\tau_{zx}\cos^{*}(y',z)\cos^{*}(y',x)+2\tau_{xy}\cos^{*}(y',x)\cos^{*}(y',y)\end{aligned} \tag{13}$$

$$\begin{aligned}\sigma_{z'z'} &= \sigma_{zz}\cos^{2*}(z',z)+\sigma_{xx}\cos^{2*}(z',x)+\sigma_{yy}\cos^{2*}(z',y)\\ &+2\tau_{zx}\cos^{*}(z',z)\cos^{*}(z',x)+2\tau_{xy}\cos^{*}(z',x)\cos^{*}(z',y)+2\tau_{yz}\cos^{2}(z',y)\cos^{*}(z',z)\end{aligned} \tag{14}$$

Shear Stress transformation equation on a plane with Cartesian components

$$\begin{aligned}\tau_{x'y'} &= \sigma_{xx}\cos^{*}(x',x)\cos^{*}(y',x)+\sigma_{yy}\cos^{*}(x',y)\cos^{*}(y',y)+\sigma_{zz}\cos^{*}(x',z)\cos(y',z)\\ &+\tau_{xy}\left[\cos^{*}(x',y)\cos^{*}(y',y)+\cos^{*}(x',y)\cos^{*}(y',x)\right]\\ &+\tau_{yz}\left[\cos^{*}(x',y)\cos^{*}(y',z)+\cos^{*}(x',z)\cos^{*}(y',y)\right]\\ &+\tau_{zx}\left[\cos^{*}(x',z)\cos^{*}(y'x)+\cos^{*}(x',x)\cos^{*}(y',z)\right]\end{aligned} \tag{15}$$

$$\begin{aligned}\tau_{y'z'} &= \sigma_{yy}\cos^{*}(y',y)\cos^{*}(z',y)+\sigma_{zz}\cos^{*}(y',z)\cos^{*}(z',z)+\sigma_{xx}\cos^{*}(y',x)\cos^{*}(z',x)\\ &+\tau_{yz}\cos^{*}\left[\cos(y',y)\cos^{*}(z',z)+\cos^{*}(y',z)\cos^{*}(z',y)\right]\\ &+\tau_{zx}\left[\cos^{*}(y',z)\cos^{*}(z',x)+\cos^{*}(y',x)\cos^{*}(z',z)\right]\\ &+\tau_{xy}\left[\cos^{*}(y',x)\cos^{*}(z',y)+\cos^{*}(y',y)\cos^{*}(z',x)\right]\end{aligned} \tag{16}$$

$$\begin{aligned}\tau_{z'x'} &= \sigma_{zz}\cos^*(z',z)\cos^*(x'z) + \sigma_{xx}\cos^*(z',x)\cos^*(x',x) + \sigma_{yy}\cos^*(z',y)\cos^*(x',y)\\ &+\tau_{zx}\left[\cos^*(z',z)\cos^*(x',x) + \cos^*(z',x)\cos^*(x',z)\right]\\ &+\tau_{xy}\left[\cos^*(z',x)\cos^*(x',y) + \cos^*(z',y)\cos^*(x',x)\right]\\ &+\tau_{yz}\left[\cos^*(z',y)\cos^*(x',z) + \cos^*(z',z)\cos^*(x',y)\right]\end{aligned} \tag{17}$$

The equations (12-17) allow the six Cartesian components of stress concerning the 0, x, y, x coordination to be converted into a variable set of **'6'** Cartesian constituents concerning a 0, x', y,' x' coordinate system.

PRINCIPAL STRESSES

Thus, if a plane is chosen such that σ_R coincides with the outside normal R, the resultant stress vector σ_R at a certain point depends upon the choice of the plane where the stress acted. Further, researchers understand that the shear stress vanishes when the σ_R, σ_N and R is in the state of coincidence. If the R is selected so that it coincides with the σ_R, a defined plane in relation to the outer normal considered as a principal plane, the path specified by the R represents the principal direction. On the other hand, researchers observe the normal stress acting on that particular plane to be known as principal stress. In general, that state of stress at any component can be understood with the presence of at least three principal planes. Further, each plane was conjointly positioned at perpendicular with each other planes with unique principal stress.

$$\begin{Bmatrix}\sigma_{Rx} = \sigma_N{}^*\cos^*(R,x)\\ \sigma_{Ry} = \sigma_N{}^*\cos^*(R,y)\\ \sigma_{Rz} = \sigma_N{}^*\cos^*(R,z)\end{Bmatrix} \tag{18}$$

Correlating the equations 5, 6, and 7 with equation 16 will leads to the following expressions:

$$\sigma_N\cos^*(R,x) = \sigma_{xx}\cos^*(R,x) - \breve{\tau}_{yx}\cos^*(R,y) - \breve{\tau}_{zx}\cos^*(R,z) \tag{19}$$

$$\sigma_N\cos^*(R,y) = \sigma_{xy}\cos^*(R,x) - \breve{\tau}_{yy}\cos^*(R,y) - \breve{\tau}_{zy}\cos^*(R,z) \tag{20}$$

$$\sigma_N\cos^*(R,z) = \sigma_{xz}\cos^*(R,x) - \breve{\tau}_{yz}\cos^*(R,y) - \breve{\tau}_{zz}\cos^*(R,z) \tag{21}$$

The significant cubic equations obtained after resolving equations 18, 19, and 20 for any of the directions of the cosines.

$$\left\{\sigma_N^3 - (\sigma_{xx} + \sigma_{yy} + \sigma_{zz})\sigma_N^2 + (\sigma_{xx}\sigma_{yy} + \sigma_{yy}\sigma_{zz} + \sigma_{zz\sigma_{xx}} - \tau_{xy}^2 - \tau_{yz}^2 - \tau_{zx}^2)\sigma_N \right. \\ \left. -(\sigma_{xx}\sigma_{yy}\sigma_{zz} - \sigma_{xx}\tau_{yz}^2 - \sigma_{zz}\tau_{xy}^2 + 2\tau_{xy}\tau_{yz}\tau_{zx})\right\} \tag{22}$$

Researchers can obtain the three principal stresses from the roots of the cubic equation written above. Once the relationship between the principal stress and its respective plane has been established, researchers can apply equations 19, 20, and 21, which further leads to three sets of simultaneous equations. In addition to the relationship as represented in the equation 23, researchers can resolve to provide a unique set of direction cosines describing the principal planes.

$$\cos^{2*}(R,x) + \cos^{2*}(R,y) + \cos^{2*}(R,z) = 1 \tag{23}$$

MAXIMUM STATE OF SHEAR STRESS

To obtain the maximum state of shear stress equation, the researchers should note that a special condition has considered. That is, the derivative of the $\tau=0$ with respect x, y, and z. leads to the following equation:

$$\tau_{xy} = \tau_{yz} = \tau_{zx} = 0 \tag{24}$$

When the authors consider this set of conditions, they observe it to have a significant output without any loss. Further, it deals with the re-alignment of the axis to coincide with the principal direction. Researchers may represent the principal direction with R_1, R_2, and R_3 respectively. Also, the resultant stress on an oblique plane will be represented by:

$$\sigma_R^2 = \sigma_{RX}^2 + \sigma_{Ry}^2 + \sigma_{Rz}^2 \tag{25}$$

Introducing the above parameters in equation 5, 6, and 7 with principal normal and zero shear yields the following equation:

$$\sigma_R^2 = \sigma_1^2\cos^2(R,R_1) + \sigma_2^2\cos^2(R,R_2) + \sigma_3^2\cos^2(R,R_3) \tag{26}$$

Practical assistance for visualizing the wide-ranging nature of stress at a position is the 3-Dimensional Mohr's circle, shown in Figure 12. This representation is more or less similar to usual 2-Dimensional Mohr circle. Additionally, this representation states the maximum shearing stresses, principal stresses, and the limits on the choice on normal state of stress and the stress components for the state of shear-stress.

Figure 12. Mohr's Circle representation for the 3-D nature of stress

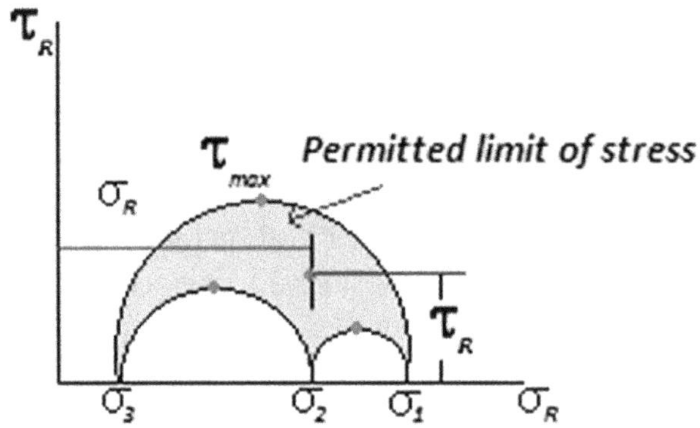

NATURE OF STRESS IN A 2-DIMENSIONAL PERSPECTIVE

For a 2-D state of stress fields, researchers can consider that $\tau_{zx}=\tau_{zy}=\tau_{zz}=0$, so equations 11-16 become:

$$\sigma_{x'x'} = \sigma_{xx} \cos^2\theta + \sigma_{yy} \sin^2\theta + 2\tau_{xy} \cos\theta \sin\theta = \frac{\sigma_{xx} + \sigma_{yy}}{2} + \frac{\sigma_{xx} - \sigma_{yy}}{2} \cos 2\theta + \tau_{xy} 2\theta \tag{27}$$

$$\sigma_{y'y'} = \sigma_{yy} \cos^2\theta + \sigma_{xx} \sin^2\theta - 2\tau_{xy} \cos\theta \sin\theta = \frac{\sigma_{xx} + \sigma_{yy}}{2} + \frac{\sigma_{yy} - \sigma_{xx}}{2} \cos 2\theta - \tau_{xy} \sin 2\theta \tag{28}$$

$$\tau_{x'y'} = \sigma_{yy} \cos\theta \sin\theta - \sigma_{xx} \cos\theta \sin\theta + \tau_{xy} \left[\cos^2\theta + \sin^2\theta\right] = \frac{\sigma_{yy} - \sigma_{xx}}{2} \sin 2\theta + \tau_{xy} \cos 2\theta \tag{29}$$

The relationship between the stress components in the expression above can be graphically represented using the Mohr's circle in which the normal stress components are plotted horizontally and the components of shear stresses are plotted vertically. Further, the tensile stress is plotted to the right side of the shear stress axis, and the compressive stress components is plotted to the left side of the shear axis (Courtney, 1990). Apart from the plotting/arrangement of stress, the authors note that the shear stress components have the nature of developing a clockwise rotation all around the surroundings of minor elements, which is plotted above the axis of the normal stress. Also, the authors note that those elements tend to develop a counterclockwise rotation are plotted in the downward direction. Whenever the authors observe this type of plotting arrangement, they denote the components of normal stress connected with each plane through the point by a point on the circle. Researchers employ the Mohr's circle and the equation (equations 27-29). Further during an experimental stress analysis when the components of the stress are transformed from one coordinate system to another coordinate system. Since the authors consider the 2-D stress system in more detail in the following discussions, it is considered to be more significant than keeping the principal stress in the loop due to its occurrence in the 2-D state of the stress system.

STRESSES RELATING TO THE PRINCIPAL COORDINATE SYSTEM

Suppose the considered system coordinates xyz is coincides with the three-principal direction R1, R2, R3, then $\sigma_1=\sigma_{xx}$, $\sigma_2=\sigma_{yy}$, $\sigma_3=\sigma_{zz}$ and the shear stress in the respective direction will be $\tau_{yz}=\tau_{zx}=\tau_{xy}=0$. From the above observation, that the authors note that certainly, it has reduced the six-component stress into a state of the three-component stress system.

$$\sigma_{x'x'} = \sigma_1 \cos^{2*}(x',x) + \sigma_2 \cos^{2*}(x',y) + \sigma_3 \cos^{2*}(x',z) \tag{30}$$

$$\sigma_{y'y'} = \sigma_2 \cos^{2*}(y',y) + \sigma_3 \cos^{2*}(y',z) + \sigma_1 \cos^{2*}(y',x) \tag{31}$$

$$\sigma_{z'z'} = \sigma_3 \cos^{2*}(z',z) + \sigma_1 \cos^{2*}(z',x) + \sigma_2 \cos^{2*}(z',y) \tag{32}$$

$$\tau_{x'y} = \sigma_1 \cos^{*}(x',x)\cos^{*}(y',x) + \sigma_2 \cos^{*}(x',y)\cos^{*}(y',y) + \sigma_3 \cos^{*}(x',z)\cos^{*}(y',z) \tag{33}$$

$$\tau_{y'z'} = \sigma_3 \cos^{*}(z',z)\cos^{*}(x',z) + \sigma_1 \cos^{*}(z',x)\cos^{*}(x',x) + \sigma_2 \cos^{*}(z',y)\cos^{*}(x',y) \tag{34}$$

$$\tau_{z'x'} = \sigma_3 \cos^{*}(z',z)\cos^{*}(x',z) + \sigma_1 \cos^{*}(z',x)\cos^{*}(x',x) + \sigma_2 \cos^{*}(z',y)\cos^{*}(x',y) \tag{35}$$

Hence, to understand the principal stresses acting on any particular planes using the experimental investigations, researchers can employ equations 30-35.

NATURE OF STRESS

Generally, the authors observe two states of stress to occur frequently in applications, which are classified as follows:

- Pure shearing nature of stress
- Hydrostatic nature of stress

The combination of the pure shearing nature of stress and the hydrostatic state of stress, which will result in the formation of general stress on a body (Ugural, 2010). The general nature of stress is represented as follows

The general nature of stress = Pure shearing nature of stress + Hydrostatic nature of stress

DISTORTIONS

When an external force acts on a structure or body, all the nodes or points in the structure can be dislocated into a new position. The dislocation of any position in a body may possibly be due to the result of a distortion in the structure, the motion of the rigid-body in transformation and revolution, or a near mixture of the aforementioned factors. Researchers observe experience of exchanging the positions of nodes or points of an element in a body to undergo the distortion phenomenon. The space between any two nodes in the structure remains unaltered; nevertheless, the dislocation is evident, which implicates the motion of the rigid-body (Jenkins & Khanna, 2005). In certain cases, the total quantity on account of all the loads acting altogether on a member may be obtained by defining the magnitude distinctly as a result of respective loads and at that moment of relating the results gained. The principle of superposition allows an intricate loading that may be substituted by two or more lighter loads.

CONCLUSION

In this chapter, the authors briefly discussed various characteristics of stresses on components. In continuation of this, the authors discuss some of the sample problems related to stress in different components at the end of this chapter.

EXAMPLE PROBLEMS AND SOLUTIONS

1. A hollow tube made of steel with an inside radius of 50 mm is subjected to transmit a tensile load of 500 kN. Find out the diameter in the outside of the tube with respect to a stress limited to 120 MN/m^2.

Given

σ =120MP

P = 500 kN =500*10^{3}N

Internal Diameter d=100 mm

$$A= \frac{1}{4}\pi D^2 - \frac{1}{4}\pi(100^2) = \frac{1}{4}\pi\left(D^2 - \pi(100^2)\right)$$

Solution

$$\sigma = \frac{P(or)F}{A},\ 120 = \frac{500^{*}10^{3}}{\frac{1}{4}\pi\left(D^2 - \pi(100^2)\right)}$$

$$D^2 = \frac{500^*10^3 + 300000\pi}{30\pi}, D^2=15307.855$$

Answer

D=123.72mm

EXERCISE PROBLEMS

1. A uniform 900 kg bar AB is held at both ends with a cable, as displayed in Figure 13. Determine the minimum area of each cable if the stress is not to surpass 95 MPa in bronze and 125 MPa in steel.
2. The uniform bar, as shown in Figure 14, is held by a smooth pin at C and a cable that runs from A to B around the smooth peg at D. Find the stress in the cable if its diameter is 15.24 mm and the bar weighs 2721 kg.
3. A bar is composed of an aluminum section firmly sandwiched between the steel and bronze sections, as shown in Figure 15. Axial loads are applied at the positions indicated. If F = 1360 kg and the cross-sectional area of the rod is 0.6 in^2, determine the stress in each section.

Figure 13.

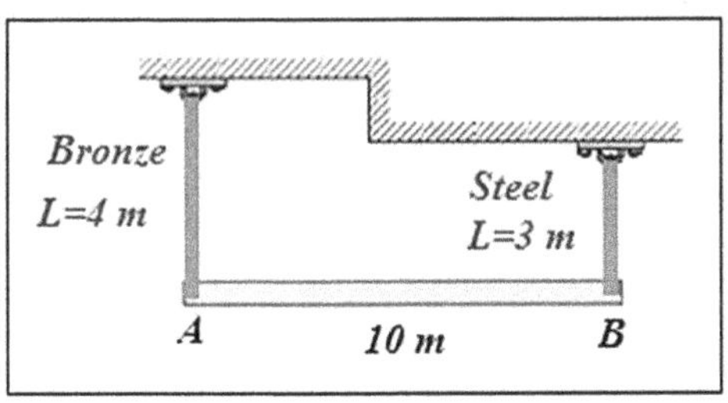

Figure 14.

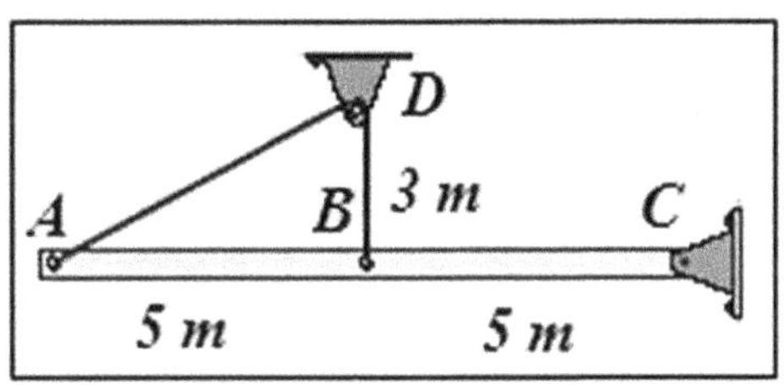

Figure 15.

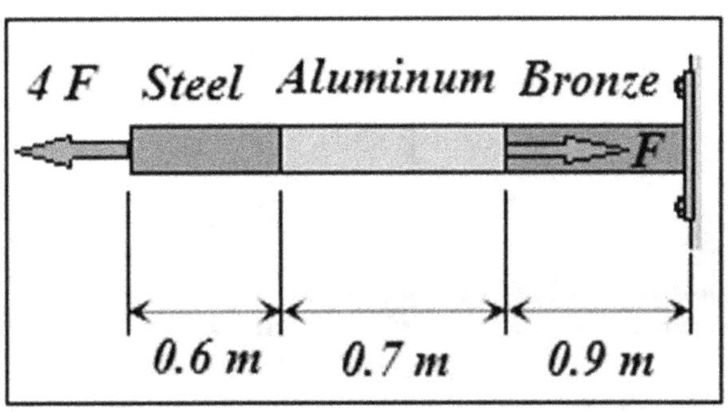

Figure 16.

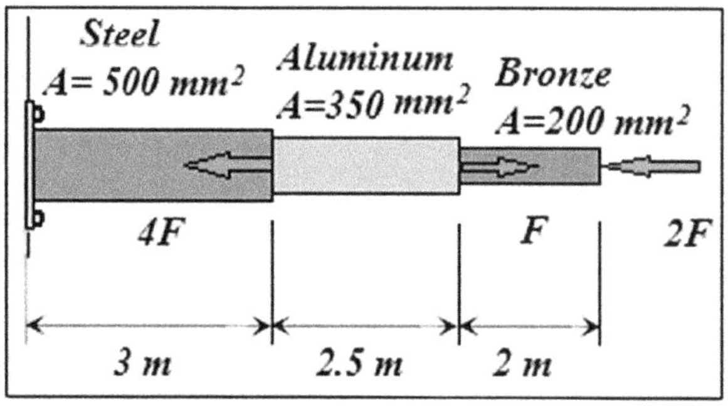

Figure 17.

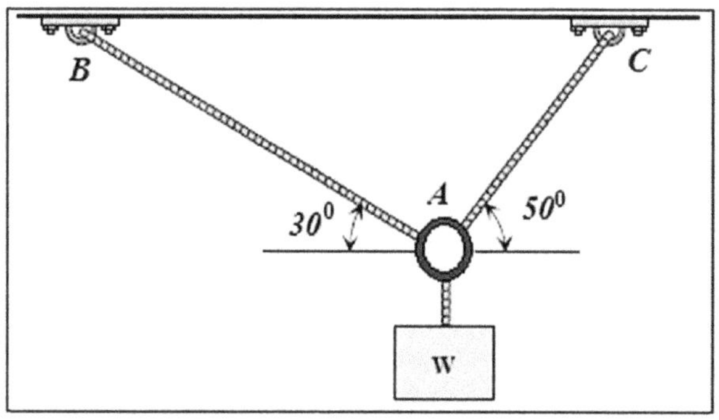

Figure 18.

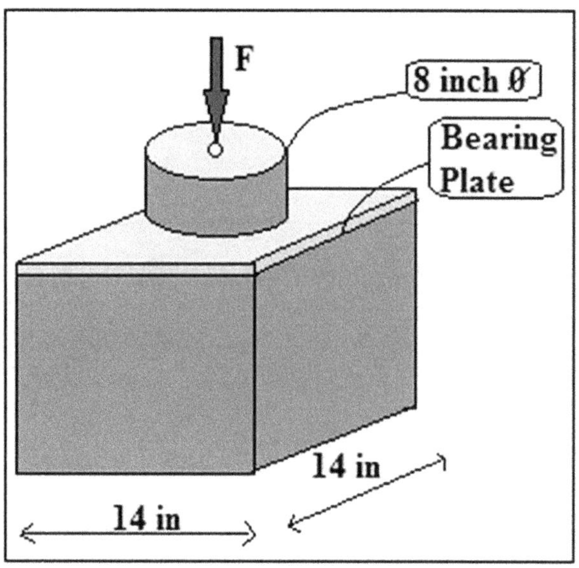

4. An aluminum bar is tightly sandwiched between a steel rod and a bronze rod, as shown in Figure. 16. Axial loads are applied at the positions indicated. Find the maximum value of F that will not exceed stress in the steel of 120 MPa, in aluminum of 100 MPa, or in bronze of 110 MPa.
5. Determine the load W that can be supported by two wires shown in Figure 17. The stress in either wire is not to exceed 241 MPa. The cross-sectional areas of wires AB and AC are 0.5 in^2 and 0.6 in^2, respectively.
6. A 35 cm square steel bearing plate lies between a 20 cm diameter wooden post and a concrete footing, as shown in Figure 18. Determine the maximum value of the load/Force F if the stress in wood is limited to 1600 psi and that in concrete to 600 psi.

REFERENCES

Ashby, M., Shercliff, H., & Cebon, D. (2007). *Materials: Engineering, science, processing and design*. Amsterdam: Elsevier.

Balaji, P. S., Leblouba, M., Rahman, M. E., & Ho, L. H. (2016). Static lateral stiffness of wire rope isolators. *Mechanics Based Design of Structures and Machines*, *44*(4), 462–475. doi:10.1080/15397734.2015.1116996

Balaji, P. S., Moussa, L., Rahman, M. E., & Ho, L. H. (2016). An analytical study on the static vertical stiffness of wire rope isolators. *Journal of Mechanical Science and Technology*, *30*(1), 287–295. doi:10.100712206-015-1232-5

Balaji, P. S., Moussa, L., Rahman, M. E., & Vuia, L. T. (2015). Experimental investigation on the hysteresis behavior of the wire rope isolators. *Journal of Mechanical Science and Technology*, *29*(4), 1527–1536. doi:10.100712206-015-0325-5

Balaji, P. S., & Yadava, V. (2013). Three dimensional thermal finite element simulation of electro-discharge diamond surface grinding. *Simulation Modelling Practice and Theory*, *35*, 97–117. doi:10.1016/j.simpat.2013.03.007

Boresi, A. P., & Chong, K. P. (2000). *Elasticity in engineering mechanics* (2nd ed.). Hoboken, NJ: Wiley.

Boyer, H. F. (1987). *Atlas of stress-strain curves*. Metals Park, OH: ASM International.

Courtney, T. H. (1990). *Mechanical behavior of materials*. New York, NY: McGraw-Hill.

Jenkins, C., & Khanna, S. (2005). *Mechanics of materials*. Amsterdam: Elsevier.

Karthik Selva Kumar, K., & Kumaraswamidhas, L. A. (2015). Experimental investigation on flow-induced vibration excitation in an elastically mounted circular cylinder in arrays. *Fluid Dynamics Research*, *47*(1), 015508. doi:10.1088/0169-5983/47/1/015508

Karthik Selva Kumar, K., & Kumaraswamidhas, L. A. (2015). Experimental investigation on flow-induced vibration excitation in an elastically mounted square cylinder. *Journal of Vibroengineering*, *17*(1), 468–477.

Karthik Selva Kumar, K., & Kumaraswamidhas, L. A. (2018). Investigation on stability of an elastically mounted circular tube under cross flow in inline square arrangement. *Iranian Journal of Science and Technology, Transactions of Mechanical Engineering,* 1-13.

Kumaraswamidhas, L. A., & Karthik Selva Kumar, K. (2016). Experimental investigation on stability of an elastically mounted circular tube under cross flow in normal triangular arrangement. *Journal of Vibroengineering*, *18*(3), 1824–1838. doi:10.21595/jve.2016.16708

Leblouba, M., Altoubat, S., Ekhlasur Rahman, M., & Palani Selvaraj, B. (2015). Elliptical leaf spring shock and vibration mounts with enhanced damping and energy dissipation capabilities using lead spring. *Shock and Vibration*, *12*. doi:10.1155/2015/482063

Ugural, A. C. (2010). *Stresses in beams, plates, and shells* (3rd ed.). Boca Raton, FL: CRC Press.

APPENDIX: UNITS

The authors usually adopt the International System of Units (SI units).

- The SI units of force are the Newton (N).

Stresses and pressures in SI units are in newton per square meter (N/m^2), which is given the special name of Pascal (Pa). Millions of Pascal (Mega Pascal's, MPa) are generally appropriate for the authors' use (i.e., 1 MPa = 1 MN/m2 = 1 N/mm^2).

Chapter 2
Introduction to Stress-Strain Relationship and Its Measurement Techniques

Karthik Selva Kumar Karuppasamy
https://orcid.org/0000-0001-5431-2542
Indian Institute of Technology, Guwahati, India

Balaji P. S.
https://orcid.org/0000-0002-6364-4466
National Institute of Technology, Rourkela, India

Niranjan Sahoo
Indian Institute of Technology, Guwahati, India

ABSTRACT

In this chapter, the basic concepts of strain and the subject of components deformation and associated strain is proposed to be briefly discussed. The strain is a pure geometric quantity so there will be no limitation on the material of the component is required. The significant aspects related to strain and stress-strain relationship briefly discussed in this proposed chapter were as follows: displacement and strain, principal strains, compatibility equations, relationship between stress and strain, two and three dimensional state of stress with respect to strain, stress-strain relationship curve, and overview of various experimental techniques employed for the measurement of stress and strain respectively.

INTRODUCTION

The components that researchers employ in various engineering applications are exposed to numerous loading conditions, which induces stress and component failures. In the same way, the structural components placed in the flow path of fluids with high currents leads to distortion because of the high stress and strain concentration on the material surfaces, which further leads to the catastrophe of the engineering

DOI: 10.4018/978-1-7998-1690-4.ch002

components (Karthik & Dhas, 2015, 2016, 2018). The failures from stress can be caused by excessive vibrations (Balaji et al., 2015; Leblouba et al., 2015; Balaji et al., 2016a; 2016b). Moreover, thermal stresses can be caused from the temperature distribution on the structural component (Balaji & Yadava, 2013). In the previous chapter, the authors discussed that the states of stress developed at a particular point or location in a body due to the force or load acting on the surface of the structure. To estimate the nature of stress-strain on the materials, it is important to know the graphical measure of mechanical properties of various materials. All the design engineers, university students, and academicians, who are all study or analyze the mechanics of materials, have to understand the relationship between stress and strain. This chapter mostly discusses stress and strain and their relationship. The movement of the arbitrary point due to the applied forces is a vector quantity known as displacement. In addition, every single point in a structure undergoes displacement due to applied forces that have a unique signature on displacement vectors. Each displacement component is matched with respect to the Cartesian coordinates in x, y, and z directions. Researchers can represent the motions of the structure by considering the rotational movement and translational movements as a single component. Subsequently, the authors considered the movement of the points in the structures to be relative to one another. In this chapter, the authors present a detailed overview of the numerous experimental techniques employed to measure the stress and strain.

Engineering Stress

Engineering stress is the ratio between the load applied to the original cross-sectional area of the material. Also, researchers consider engineering stress as nominal stress.

True Stress

True stress is the ratio between the load applied to the actual cross-sectional area of the deformed material.

True Strain

Researchers refer to true strain as the natural log of the quotient of deformed length after loading over the original length before loading.

Engineering Strain

Researchers refer to engineering strain as the quantity of material deformation per unit length. This is also referred as the nominal strain.

Lateral Strain

In continuum mechanics, researchers consider the mechanical behavior of the modeled materials to be a single mass instead of distinct masses. Due to this consideration, the distortion of material in the longitudinal direction causes a transverse strain, also called a lateral strain. Lateral strain is the ratio between the variation in lateral dimension to the original lateral dimension.

Shear Strain

Shear strain is the ratio of the variation in distortion to its actual dimension (length) at right angles to the axes of the subjected component under shear load.

TWO DIMENSIONAL ASPECTS OF STRAIN

The two-dimensional aspects of the strain are more or less identical to the understanding of strain in the one-dimensional aspects, in which the authors defined the deformation of material by visualizing the material considered to be the group of line elements in lesser amount (Boresi & Chong, 2000). When a test specimen starts to deform, resulting in the development of stretching on the line elements, alternatively, it starts to diminish to revolve around in the space relative to one another. Thus, researchers comprehend the displacement of the line elements with respect to the indication of strain. The strain at a particular point is simply the outcome of the stretching of the line element, due to the contraction and the rotation of the whole elements originating from that particular position. Figure 1 illustrates the occurrence of strain in the z- and y-axis, respectively.

THREE DIMENSIONAL ASPECTS OF STRAIN

The bending of the beam can be represented in three-dimensional aspects, which the authors further brief in detailed in the following segment. Broadly, the authors characterize strains into three normal strains along xx yy zz as ε_{xx}, ε_{yy}, ε_{xz} and three shear strains along xy yz zx as ε_{xy}, ε_{yz}, ε_{zx}. The authors consider the strains to be present in any materials when they are subjected to certain loading conditions. The ε_{zz}

Figure 1. Occurrence of strain in the y-axis and z-axis

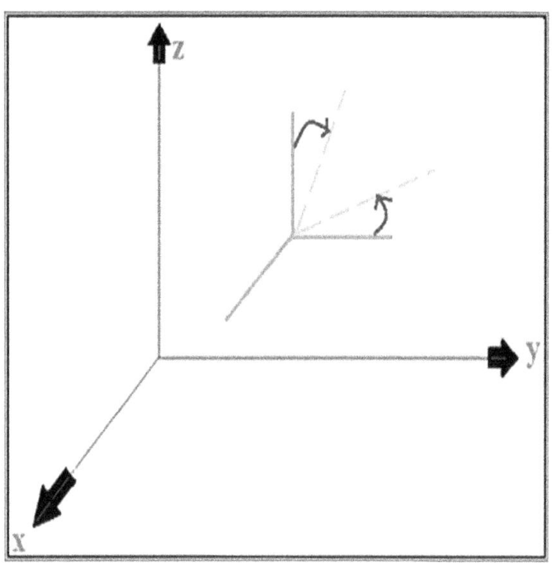

Figure 2. Occurrence of stress and strain in the x-axis, y-axis, and z-axis

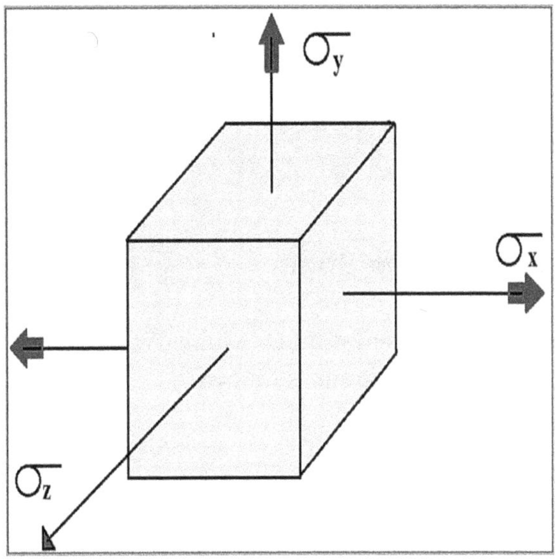

strain represents the change in subjected specimen length along the z axis with respect to the line element at initial stage. The y z ε strain represents the half of the change at the beginning, whereas the two-line elements are observed to be aligned perpendicular with respect to the y and z axes and further to the zx ε strain. The authors illustrate this scenario in Figure 2.

Compatibility Equation

The strain field must satisfy the following equations of compatibility:

$$\frac{\partial^2 \gamma_{xy}}{\partial x \partial y} = \frac{\partial^2 \varepsilon_{xx}}{\partial y^2} + \frac{\partial^2 \varepsilon_{yy}}{\partial x^2} \tag{1}$$

$$\frac{\partial^2 \gamma_{yz}}{\partial y \partial z} = \frac{\partial^2 \varepsilon_{yy}}{\partial z^2} + \frac{\partial^2 \varepsilon_{zz}}{\partial y^2} \tag{2}$$

$$\frac{\partial^2 \gamma_{zx}}{\partial z \partial x} = \frac{\partial^2 \varepsilon_{xx}}{\partial x^2} + \frac{\partial^2 \varepsilon_{xx}}{\partial z^2} \tag{3}$$

$$2\frac{\partial^2 \varepsilon_{xx}}{\partial y \partial z} = +\frac{\partial}{\partial x}\left(-\frac{\partial \gamma_{yz}}{\partial x} + \frac{\partial \gamma_{zx}}{\partial y} + \frac{\partial \gamma_{xy}}{\partial z}\right) \tag{4}$$

$$2\frac{\partial^2 \varepsilon_{yy}}{\partial z \partial x} = +\frac{\partial}{\partial y}\left(\frac{\partial \gamma_{yz}}{\partial x} - \frac{\partial \gamma_{zx}}{\partial y} + \frac{\partial \gamma_{xy}}{\partial z}\right) \tag{5}$$

$$2\frac{\partial^2 \varepsilon_{zz}}{\partial z \partial x} = +\frac{\partial}{\partial z}\left(\frac{\partial \gamma_{yz}}{\partial x} + \frac{\partial \gamma_{zx}}{\partial y} - \frac{\partial \gamma_{xy}}{\partial z}\right) \tag{6}$$

Displacement Calculated from the Strain

When one considers a circular rod with a radius r that is subjected to load in torsion, the outcome of the applied load leads to the development of a strain as follows:

$$\gamma_{yz} = rx \quad \gamma_{zx} = -ry \quad \varepsilon_{xx} = \varepsilon_{yy} = \varepsilon_{zz} = \gamma_{xy} = 0 \tag{7}$$

In general, researchers consider a compatibility check to be the initial condition to solve the issues in the field of displacement:

Condition I - The subjected body of the specimen must be connected simply.
Condition II - The equation represents the relationship of strain must satisfy the compatibility equations.

Dilation of Volume

When examining a line element with rectangular nodes in the deformed body, the authors observe the edges to be oriented along the principal axis. Further, there is a change in the length at all sides of the element. Conversely, the authors have identified that the particular element is not to undergo any deformation because of the absence of shearing strain at the face of the element. Researchers call the volume dilation the ratio between the change in the initial volume of the subjected element with respect to the initial volume.

STRESS-STRAIN RELATIONSHIP CURVE

The tensile test is one of the significant tests to determine the mechanical behavior of any material in which one side of the specimen is fixed in a loading clamp and the other is exposed to a controlled pull or to the displacement (Boyer, 1987). In regard to the displacement, researchers obtain the electronic reading of the loading through a series of transducers attached throughout the specimen. On the other hand, researchers consider the modern testing machines with the servo controller the load as a controlled variable instead of the displacement, and in such cases, the displacement is a function of the applied load. The measurement of stress and strain in engineering aspects are denoted as σ_e and ε_e respectively, as shown in Figure 3. Researchers monitor the stress and strain through the load and deflection over the original cross-sectional area of the specimen A_0 and L_0 as follows:

$$\sigma_e = \frac{P}{A_0} \tag{8}$$

$$\varepsilon_e = \frac{\delta}{L_0} \tag{9}$$

Generally, researchers prefer the engineering stress-strain plot over the true stress-strain plot due to the convenience in practical understanding. Further, the stress- strain plot of any material, if not specially mentioned, it is generally engineering stress-strain plot.

IMPORTANT TERMINOLOGY

A brief discussion on the important terminology with respect to the stress-strain relationship curve (Ashby et al., 2007) for a test specimen depicted in the Figure 4 were as follows.

Proportional Limit

Proportional limit is a position on the stress-strain curve up to which the stress is directly proportional to the strain.

Elastic Limit

Elastic limit is a point on the stress-strain relationship curve until which the material will return to its original shape when unloaded.

Figure 3. Engineering stress-strain relationship curve for an annealed polycrystalline copper

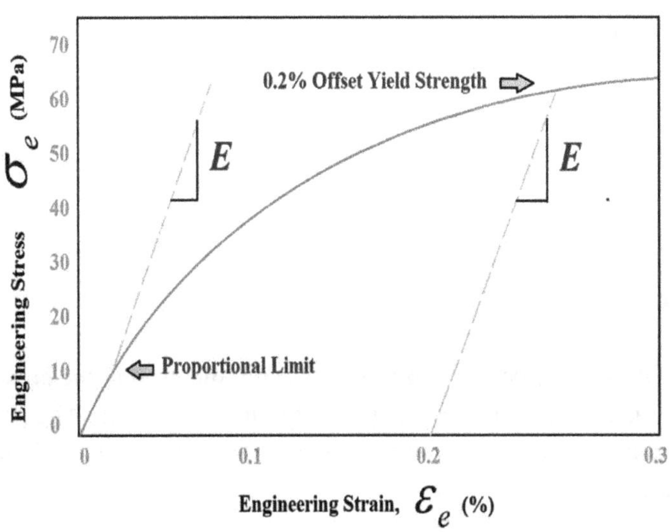

Yield Point

Yield point is the position on the stress-strain relationship curve at which there is a substantial yielding of the material that occurs without any considerable increment of load.

Ultimate Strength

Ultimate strength is the peak ordinate or the maximum stress that a material can withstand when subjected to applied load.

Rupture Strenvgth

Rupture strength is the stress that the material sustains at the instant of breaking or rupture.

Modulus of Elasticity

The slope of the linear portion at the stress-strain curve is the modulus of elasticity. Researchers may also classify the modulus of elasticity as the stiffness of the material's ability to repel the distortion along the ranged lines.

Tangent Modulus

The slope observed above the proportional limit in the stress-strain curve is as the tangent modulus, and it varies with respect to changes in strain.

Shear Modulus

The slope observed at the initial portion of the linear line in the shear stress-shear strain curve is the shear modulus.

Percentage of Elongation

The percentage of elongation is found when the fracture occurs due to the act of tension causes a strain in the material, which is further stated as a percentage = ((gage length at the final – gage length at the initial)/gage length at the initial) x 100. Researchers also consider the percentage of elongation to be a measurement of ductility.

Percentage Reduction in Area

When the fracture of materials happens, researchers measure the drop in the cross-sectional area at the tensile specimen, which is considered to be the percent reduction in area. The percentage reduction in area is given by: ((actual area - area measured after fracture)/ actual area) x 100. It is also considered to be a measurement of ductility.

Isotropic

Researchers consider the most commonly used structural materials to be an isotropic material. Further, isotropic possess equal elastic characteristics in all directions.

Anisotropic

Researchers consider the materials that are not self-determining with the direction to be anisotropic materials. Additionally, researchers consider the fiber-reinforced composite materials to be a suitable example of anisotropic materials.

Homogeneous

The homogeneous materials may or may not have anisotropic properties; Researchers consider it to be have a similar percentage of composition at each part of the material.

CLASSIFICATIONS OF MATERIALS

The materials are classified into two comprehensive groups, namely: ductile materials and brittle materials.

Figure 4. Stress-strain relationship curve

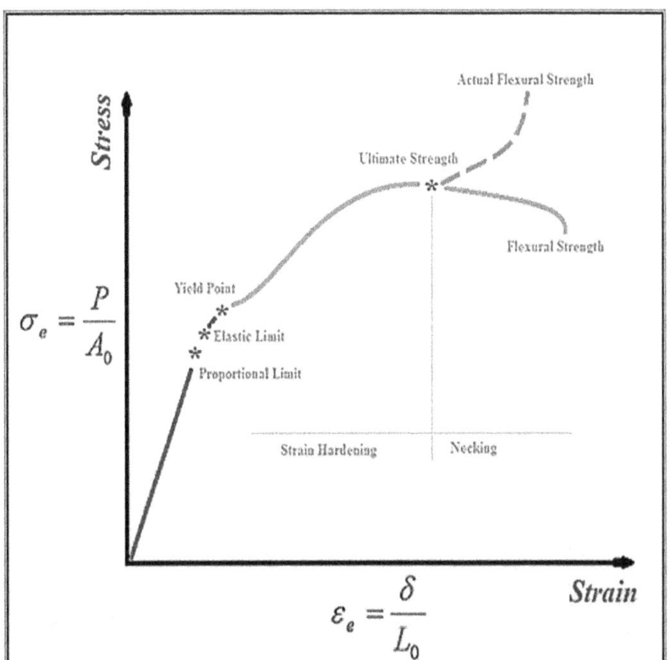

Ductile Material

Ductile materials are materials that have an ability to undergo plastic strains (at standard temperature) before failure.

1 - Represents the Ultimate strength, 2 – Represents the Yield strength (yield point), 3 – Represents the Rupture, 4– Represents the Strain hardening region, 5– Represents the Necking region, A– Represents the Apparent stress and B– Represents the Actual stress

A significant benefit of employing the ductile materials is that it shows the detectable distortions in the materials in advance to the excess of applied loads other than the standard loads. Subsequently, the ductile materials show that the materials have the ability to captivate enormous quantities of energy in advance of failure. Some of the most common ductile materials used in the engineering applications include mild steel, aluminum and its alloys, copper, magnesium, nickel, brass, and bronze. Researchers categorize ductile materials, like structural steel and numerous alloys of many metals, by their capability to yield before fracture at standard temperature conditions. Basically, steel with low carbon content reveals an undeviating stress-strain relationship capable of a distinct yield point, which is clearly illustrated in Figure 5. The linear part of the stress-strain curve is the elastic region, and the slope represents the modulus of elasticity, otherwise known as the Young's Modulus. One can observe the initiation of the plastic flow to happen at the upper yield point and endure at the lower one. At the location of the lower yield point, researchers observe that the uninterrupted distortion is distributed heterogeneously all along the material sample. The distortion band, which is developed at the upper yield point, will proliferate throughout the length of the gauge at the point of lower yield. The group inhabits the entire gauge at the luder's strain. Beyond this point, the initiation of the work hardening on the test components can be observed.

Figure 5. Schematic representation of the stress-strain curve for a steel specimen

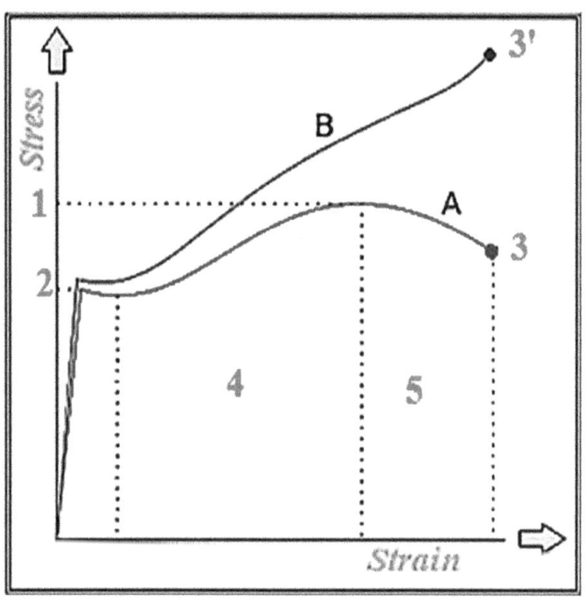

The presence of the yield point is connected with the restraint of disturbances in the structure. One can understand this principle with a suitable example, such as the interaction of the solid solution with dislodgments, which resulted in prevention of the dislocation of the components from moving around by acting as a pin. For that reason, an enormous amount of stress is required to start the movement. On the assumption that the escape of the displacement from the pinning, stress is required to continue until it reaches its anticipated conditions. Subsequently, researchers observe the yield point on the stress-strain curve, to be characteristically decreased to some extent because dislocations evade from the Cottrell atmospheres. Consecutively in continuation of the deformation, researchers observe the stress to increase with respect to the strain hardening until it reaches the ultimate tensile stress. Up until this point, the cross-sectional area of the subjected material is reduced consistently due to the Poisson contractions. Further, it starts to have necking and leads to fracture.

The presence of necking in ductile materials represents the instability in the geometry of the system. Due to the natural inhomogeneity of the subjected material, it is a common phenomenon to observe some sections with minor insertions or permeability within them or on the surface of the subjected materials. Where researchers observe the strain to be concentrated in a gradual mode, this leads to a nearby lesser area other than remaining sections. From this observation, researchers identify that a strain that is found to be smaller than the ultimate tensile strain causes a sudden upsurge in the rate of the work-hardening rate on that particular region, which will be bigger than the area reduction rate. This phenomenon creates a region that is tougher to withstand further deformation compared with others. As a result of that the instability of materials in that region, (i.e., the materials were found to have capacities to deteriorate the inhomogeneity in advance attain ultimate strain).

Figure 6. Comparison of stress-strain relationship curve for brittle materials with respect to the ductile materials

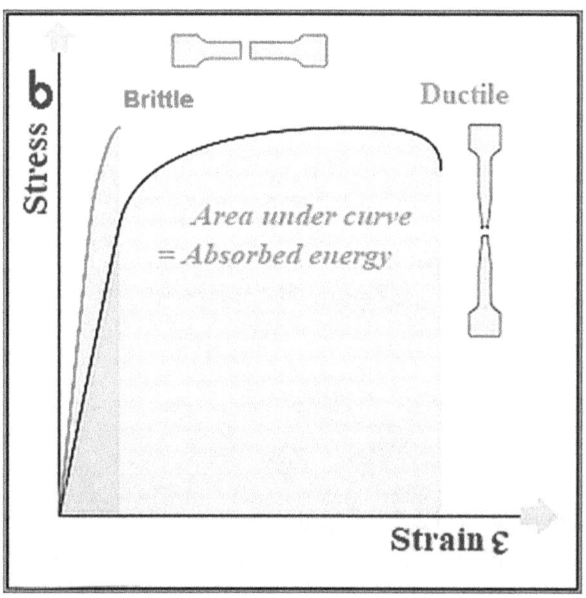

Conversely, as the strain turned out to be higher, the rate of work hardening gradually starts to decrease to facilitate the observation that, for now, the region with the minor area is identified to be weaker than the other region. For this reason, the authors will focus on the reduction in area in this section, and the neck turns out to be more and more noticeable up until the fracture happens. Subsequently, in the formation of the neck in the materials, researchers observe the further concentration of plastic deformation in the neck region despite the fact the residue of the material goes through the elastic contraction on account of the reduction in tensile force. The Ramberg-Osgood equation can be employed to approximate the stress-strain curve for the ductile materials. It is a conventional equation that requires only the general parameters, such as ultimate strength of the material, yield strength of the material, percentage of elongation of the material, and elastic modulus of the material. The authors illustrate a distinctive stress-strain relationship curve in Figure 6.

Brittle Material

Brittle materials are materials that display very small inelastic distortions. The authors deliberated the materials that distort due to tension at a comparatively lesser amount of strain to be brittle. Brittle materials consist of concrete, cast iron, rocks, plaster, and glasses. Sometimes, observation shows that the fracture may have resulted before attain the yielding. Brittle materials such as concrete or carbon fiber do not have distinct yield points and are lacking in strain hardening. Hence, the ultimate strength and breaking strength can be considered to be equal for brittle materials. The failure of brittle materials was basically due to the tensile/normal stresses, and splitting occurs along a surface at right angles to the load applied. Ductile materials frequently undergo deformation on planes that are parallel to the maximum shear stress, which occurs at 45°. The failure in ductile material is distinct, and it is of a cup and cone types with the edges of cup and cone inclined at nearly 45° to the axis of the respective specimen. Researchers observe the distinctive stress-strain relationship curve for the brittle material be linear in nature.

Failure Due to Tension in Materials

Ductile Material: Shear Failure

Causes were listed as follows

- The distortion occurs in the test specimen at an angle of 45°.
- The cup and cone-shaped fracture are commonly presented.

Brittle Material

Causes were listed as follows

- The distortion occurs in the test specimen at an angle of 90°.
- Researchers identify the principal stress to be the main cause of the fracture.

Elements that Influence the Stress-Strain Relationship Curve

The stress-strain relationship curves for countless materials may show extensive discrepancies on account of various basic configurations. Owing to the extrinsic aspects, different tensile tests carried out on the identical materials can be observed to be have different outcomes, predominantly due to the temperature of the subjected specimens and the rate of speed of the applied load (Courtney, 1990). Researchers observe the boundary concerning intrinsic and extrinsic features to be not rigid. Numerous features can have an impact on the stress-strain relationship curve by regulating Young's modulus through toughening or strengthening, on the assumption that they adjust the structure and configurations. So often, researchers do not consider the time to be a significant aspect that influences the relationship curve, but in an extensive nature when one considers higher strain and stress rates, the time plays a significant role in the stress-strain relationships curve (Ugural, 2010). {\displaystyle \mathrm {\sigma _{t}} =K({\dot {\epsilon }}_{T})^{m}}One more principal feature influencing the relationship is the temperature. Temperature controls the initiation of displacements and dispersions. As the temperature rises, researchers observe that the characteristics of the brittle materials transform into ductile materials.

Hooke's Law

When a material that is elastic in nature displays a linear relationship between stress and strain, the condition of the material is known to be linearly elastic. In such case, stress is directly proportional to the strain.

Poisson's Ratio

Poisson's ratio is the ratio of lateral or transverse strain to longitudinal strain. For most materials, the Poisson's ratio ranges from 0.25 to 0.35.

Strain Energy

During strain, researchers observe the subjected test material to show an ability to absorb the energy in the elastic region. Further, researchers consider the extreme quantity of energy-absorbing capability of the test specimen in the elastic region as the proof resilience.

Sign Convention for the Strain

Researchers consider tensile strain to be positive. Researchers consider compressive strain to be negative.

STRESS AND STRAIN MEASUREMENT TECHNIQUES

Stress and strain measurement techniques are different experimental and numerical techniques that are employed to understand the nature of stress and strain in any material. To understand that the authors discuss important techniques on the measurement of stress and strain on the structural components were as follows.

Tensile Test

The tensile test is the most basic technique to understand the behavior of material subjected to tension in the uniaxial direction up until failure of the material. The test further leads to the selection of suitable material for selected loading conditions without compensating the quality of the materials. Researchers can measure certain properties of the materials directly through the tensile stress, which are as follows: ultimate tensile strength, extreme elongation, and contraction in the cross-sectional area of the subjected material. According the tensile test technique, researchers can easily calculate certain additional characteristics, such as the yield strength, Poisson's ratios, strain hardening and the Young's modulus of the tested material.

Strain Gages

Researchers can measure deformation of the physical part experimentally by locating the thin flat resistor to the surface of the particular position from which the strain developed in the particular directions. Subsequently, researchers can calculate the stress state established in a particular position with the help of the strain measured on the surface in different directions. When a test material is subjected to distortion, researchers observe that the length of the stretched element starts to increases, and subsequently, the cross-section of the subjected material is observed to decrease, which leads to an upsurge in its electrical resistance. Researchers consider the so-called change in electrical resistance to be the extent of its motion mechanically. Thus, researchers acknowledge that strain gage is a suitable device to measure the strain with respect to the change in the electrical resistance. To measure the shock or vibration of a related instrumentation, researchers observe the employment of resistance strain gage to be an effective method. Globally, to measure stress, various analysts identify the strain gage to be an effective technique during experimentation. In close observation, the vibration is always accompanied by the strain, which leads to the usefulness of strain gage in broader areas of engineering, such as in the measurement of vibration and shock. Further, the strain gage has the capability to understand the time history other than the magnitude of either shock or vibration with a higher frequency.

Classification of Strain-Gage

Researchers have categorized the strain gages by numerous means. The first one depends on the types of application (i.e., static or dynamic strain measurements). The static types of strain gauges are more often prepared with an alloy comprised of copper-nickel alloy, which is observed have the least change of resistance with respect to the change in the temperature. The dynamic types of strain gages are sometimes prepared with an alloy comprised of iron, nickel, and chrome, respectively, which is responsible for a superior gage factor compared. Researchers observe the dynamic gages to expose a less sensitivity with respect to the change in the temperature.

Selection of Strain-Gages

Regarding the measurement of shock, a transient will be possibly employed to the structure, which is already under examination. Researchers observe the shock to have a very small time duration, and subsequently, it is understandable that the compensation of temperature is not required due to the fact that during the impact, the temperature doesn't require any time to change. Thus, the authors typically applied dynamic gages for shock measurement. Regarding the measurement of vibration, the selection of gages mostly depends on the type of application and information that researchers are required to measure. The authors employed the dynamic type of strain gages on the condition that they only anticipated required vibration frequency and cyclic stress magnitude.

Photo-Elastic Technique

The photo-elastic technique is dependent on the certain circumstances, in which some materials develops birefringence with respect to the application of load. Consequently, the amount of the refractive indices observed to be develop uniquely at every point is solidly associated with the state of stress at a particular point. Simultaneously, researchers can easily calculate the stresses in any structure by employing a prototype model of the respective structures with some photo-elastic materials.

Steps Involved in Stress Measurement Using the Photo-Elastic Technique

Researchers consider the birefringence to be a significant parameter for the measurement of stress in the photo-elastic technique. During the observation, by passing light on test materials, researchers identify subjected stress to reveal characteristic refraction with two indices from the test materials. Researchers most commonly observe in the photo-elastic materials, especially the optical crystals. When a load is applied to certain materials, it starts to reveal the birefringence phenomenon, through which researchers can determine the magnitude of stress on each different location on the material surface. Apart from that, by employing the polariscope, that the maximum shear stress with respect to its orientation can also be identified. Generally, researchers consider circular and plane polariscope setup to be employed under basic categories.

Neutron Diffraction Technique

To determine the subsurface strain in a part of the test material, researchers observe the Neutron diffraction technique to be a suitable method, as illustrated in Figure 7.

Dynamic Mechanical Analysis (DMA)

Dynamic mechanical analysis (DMA) is a method employed to understand and distinguish the viscoelastic materials, predominantly polymers. Researchers observe the viscoelastic property of a polymer material and understand by the dynamic mechanical analysis, which basically entails measuring the resulting displacement/strain by applying a sinusoidal force on the material. When the authors consider a perfectly elastic solid, they anticipate the resulting strains and the stresses to be perfectly in phase. Subsequently, for a purely viscous fluid, is the authors anticipate that there will be a 90-degree phase lag of strain with respect to stress that will be obtained.

Figure 7. Stress in plastic protractor causes birefringence

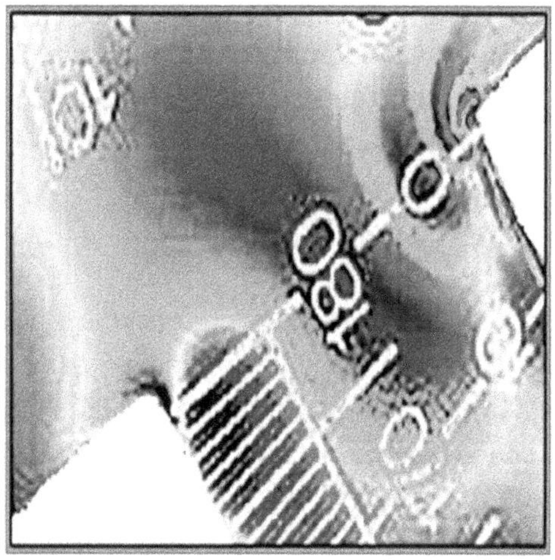

Brittle Coating Technique

The brittle coating technique is one of the effective experimental techniques to understand the nature of stress-strain on a material specimen subjected to different loading conditions. Normally, in this method, researchers coat the subjected test specimen by spraying all over the surface and subsequently cooling it down to reach brittleness. When the test specimen is subjected to the loading, the applied coating over the surface starts to crack, as its alleged onset sensitivity on strain or strain is surpassed. A distinctive pattern crack propagates on the subjected specimen. The propagation of cracking on the surface represents the enhanced strain to facilitate the instantaneous signal that reveals the stress concentrations. Further, the cracks also represent the maximum strain directions at these particular locations; meanwhile, the authors observe to be aligned with maximum principal tensile strain at right angles. Researchers observe this technique to be more efficient in determining the exact location for accurate measurement through strain gages in the respective directions.

Thermal Imaging Technique

Researchers can measure thermal stress by employing the thermal imaging technique, which has the capability of identifying the wavelengths away from the normal eyes of human beings. By employing this technique, researchers observe that thermal imaging is capable of identifying the stress on a test specimen in a darker environment similar to the quality of testing in the daylight. Some of the basic principles of the thermal imaging technique are as follows. First, the infrared waves emitted by the test specimen are concentrated with the help of a distinctive lens. The focused light from the test specimen is further perused by an element to detect the infrared waves. Consequently, it creates a thermogram image, which is considered to be the temperature pattern. Further, the created thermogram is converted into electric impulses. Then, the converted electric impulses are forwarded to a signal processor, which consists of a complicated distinctive chip that is employed to decipher the obtained information into quantitative data. Finally, the obtained quantitative data are displayed in multiple color patterns with respect to the intensity of the infrared (IR) emission.

Relationship between Temperature and Stress in a Test Specimen

The microscopic observation of metals reveals the nature of the crystalline structure. Further, the authors observe the crystals to develop grains, which are identified as having a distinctive crystal with respect to the orientation. The dividing area of two grains is considered to be the grain boundary, and these are known to be the faults in the crystalline structure. Further, when a test specimen is subjected to a tensile load that gradually increases, the defects in the crystalline structures become dislocated and triggered thermally, which further leads to further increases in the temperature at the dislocated area.

Types of Thermal Imaging Devices

There are two types of thermal imaging techniques that the authors employed, which are as follows:

- Cooled Thermal Imaging: Researchers employ this technique to obtain the thermal imaging in the long-wave infrared band and the mid-wave infrared band of spectrum.
- Uncooled Thermal Imaging: It is employed to obtain the thermal imaging in the long-wave infrared band.

CONCLUSION

In this chapter, the authors briefly discussed strain and techniques to identify the strain at particular point. Further, the chapter has included an overview of the various stress and strain measurement techniques for the real-time application. In the subsequent chapters, the authors compiled detailed studies on the stress and strain measurement techniques in an orderly manner so that the reader can easily understand the nature of stress and strain acting in any engineering component and can more effectively predict and examine the stress and strain by employing suitable techniques.

REFERENCES

Ashby, M., Shercliff, H., & Cebon, D. (2007). *Materials: Engineering, science, processing and design.* Amsterdam: Elsevier.

Balaji, P. S., Leblouba, M., Rahman, M. E., & Ho, L. H. (2016a). Static lateral stiffness of wire rope isolators. *Mechanics Based Design of Structures and Machines*, *44*(4), 462–475. doi:10.1080/15397734.2015.1116996

Balaji, P. S., Moussa, L., Rahman, M. E., & Ho, L. H. (2016b). An analytical study on the static vertical stiffness of wire rope isolators. *Journal of Mechanical Science and Technology*, *30*(1), 287–295. doi:10.100712206-015-1232-5

Balaji, P. S., Moussa, L., Rahman, M. E., & Vuia, L. T. (2015). Experimental investigation on the hysteresis behavior of the wire rope isolators. *Journal of Mechanical Science and Technology*, *29*(4), 1527–1536. doi:10.100712206-015-0325-5

Balaji, P. S., & Yadava, V. (2013). Three dimensional thermal finite element simulation of electro-discharge diamond surface grinding. *Simulation Modelling Practice and Theory*, *35*, 97–117. doi:10.1016/j.simpat.2013.03.007

Boresi, A. P., & Chong, K. P. (2000). *Elasticity in engineering mechanics* (2nd ed.). Hoboken, NJ: Wiley.

Boyer, H. F. (1987). *Atlas of stress-strain curves*. Metals Park, OH: ASM International.

Courtney, T. H. (1990). *Mechanical behavior of materials*. New York, NY: McGraw-Hill.

Jenkins, C., & Khanna, S. (2005). *Mechanics of materials*. Amsterdam: Elsevier.

Karthik Selva Kumar, K., & Kumaraswamidhas, L. A. (2015). Experimental investigation on flow-induced vibration excitation in an elastically mounted circular cylinder in arrays. *Fluid Dynamics Research*, *47*(1), 015508. doi:10.1088/0169-5983/47/1/015508

Karthik Selva Kumar, K., & Kumaraswamidhas, L. A. (2015). Experimental investigation on flow-induced vibration excitation in an elastically mounted square cylinder. *Journal of Vibroengineering*, *17*(1), 468–477.

Karthik Selva Kumar, K., & Kumaraswamidhas, L. A. (2018). Investigation on stability of an elastically mounted circular tube under cross flow in inline square arrangement. *Iranian Journal of Science and Technology: Transactions of Mechanical Engineering*, 1-13.

Kumaraswamidhas, L. A., & Karthik Selva Kumar, K. (2016). Experimental investigation on stability of an elastically mounted circular tube under cross flow in normal triangular arrangement. *Journal of Vibroengineering*, *18*(3), 1824–1838. doi:10.21595/jve.2016.16708

Leblouba, M., Altoubat, S., Ekhlasur Rahman, M., & Palani Selvaraj, B. (2015). Elliptical leaf spring shock and vibration mounts with enhanced damping and energy dissipation capabilities using lead spring. *Shock and Vibration*, *12*. doi:10.1155/2015/482063

Ugural, A. C. (2010). *Stresses in beams, plates, and shells* (3rd ed.). Boca Raton, FL: CRC Press.

Chapter 3
An Overview of Stress and Strain Measurement Techniques

Anil Kumar Rout
Indian Institute of Technology, Guwahati, India

Niranjan Sahoo
Indian Institute of Technology, Guwahati, India

Vinayak Kulkarni
Indian Institute of Technology, Guwahati, India

ABSTRACT

Stress and strain are mechanical behaviour of materials, subjected to mechanical or thermal loading. The detrimental effect of such loading is the ultimate failure of materials due to generation of high stress and strain. Therefore, measurement and prediction of stress and strain values help in proper design and maintenance of engineering equipment and structures. The present contents elaborate and summarize different methods adopted by researchers for mechanical stress and strain measurements. The content is focused to provide an overview regarding the measurement techniques adopted for strain measurement. The analysis holds information regarding working principle of different strain measuring technique along with a brief description about the history of strain measurement. Special attention has also been devoted for explanation of thermal stress and strain measurement techniques. The modern non-contact techniques have evolved as a potential tool for such measurements even at higher temperature conditions.

INTRODUCTION

Stress is the quantification of internal forces that neighboring particles exert on each other in a continuous material. Figure 1 shows the tensile loading on a bar. By definition, stress(σ) is defined as the intensity of force at a point, i.e.,

DOI: 10.4018/978-1-7998-1690-4.ch003

Figure 1. Stresses in bar

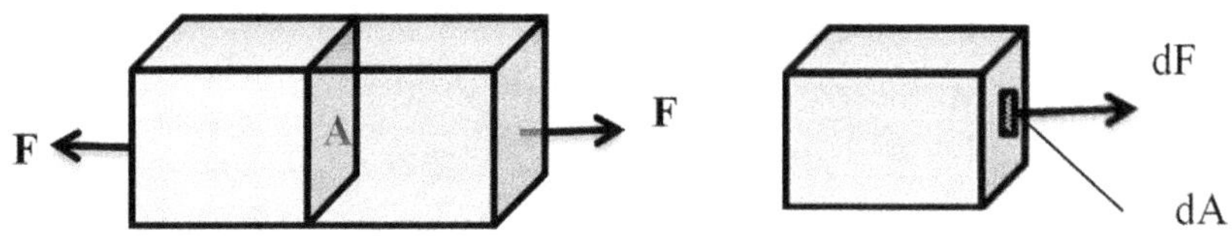

$$\sigma = \frac{dF}{dA} \text{ with } dA \to 0 \tag{1}$$

According to the nature of external loading, the force may be compressive or tensile. The normal component of the force (perpendicular to the cross-section area) is the normal force, and the tangential component of the force is called tangential force. The corresponding stresses are called tensile/compressive stress and shear stress, respectively. However, in real life, combined stresses are seen to be applied many times. Stresses are also generated through thermal effects.

Most of the existing materials in nature expand when subjected to a rise in temperature and contract when cooled. Figure 2 shows the thermal stress induction subjected to no constraint, and Figure 3 shows the thermal stress induction subjected to constraints with rigidly fixed ends.This expansion and contraction bear a proportionality relationship with the change in temperature for a wide range of temperature values. That proportionality constant is expressed in terms of coefficient of linear thermal expansion (alpha) defined as the change in length a bar of unit length experienced when its temperature is changed by 1°.When a homogeneous body is subjected to a non-uniform rise in temperature, different elements of the body tend to expand by different amounts depending upon their local rise in temperature. In a case of restricted free movement, for the body to remain continuous, each element has to expand in a similar manner, which is against the proportionality increment of length as per local temperature. Therefore, the elements of the body exert upon each other a restraining action resulting in continuous unique elongation at every point. For continuity of displacement, the system of strains produced by this

Figure 2. External constraints: No constraint, No thermal stress

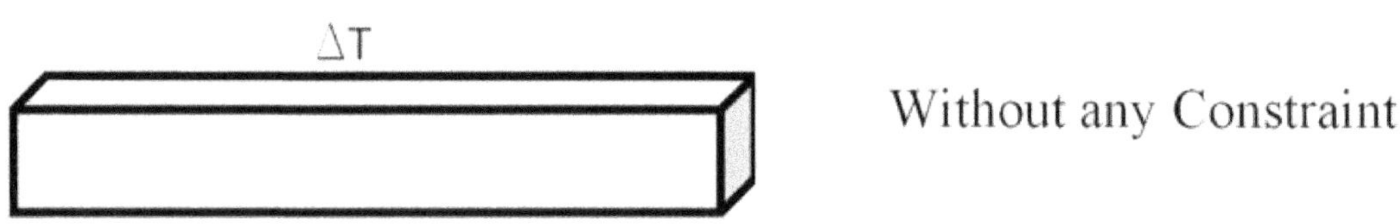

Figure 3 External constraints (Thermal stress-induced with a change in temperature)

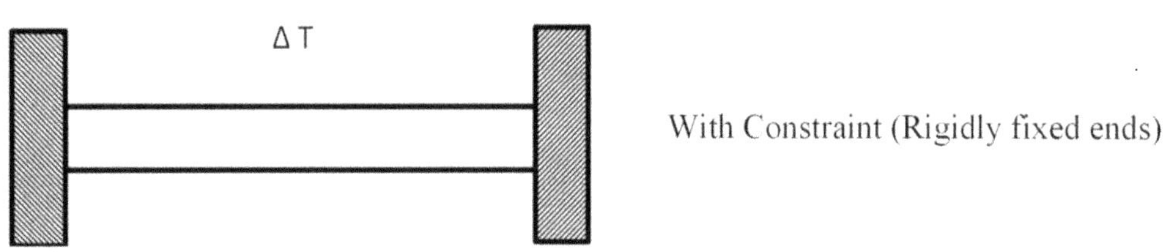

process cancels out some part of free expansion. This system of strains must be accompanied by a corresponding system of self –equilibrating stresses called thermal stresses. If the body is allowed for free expansion, there will be no generation of thermal stress due to change in temperature.This type of stress is often observed in case of composite bodies made of dissimilar materials. For composite structures with non-uniform heating, the thermal stress depends on both temperature difference and geometry of the structure. Thermal strain in the case of uniform structures may result due to uneven temperature change, and uniform heating thermal strain may arise in the case of non-homogenous structures. Literally, thermal stresses are equivalent to mechanical stresses resulting from forces caused by expansion or contraction of a constrained object.

Constraints causing thermal stresses can be external or internal. External constraints prevent the expansion and contraction of the system resulted due to a change in temperature, as shown in Figure 3. Similarly, internal constraints are the restrictions or restraints present within the material due to differential expansion and contractions at various locations. For example, if a heating source is kept inside a pipe capable of conducting heat and made free from any external constraint, then the inner materials of the pipe will tend to expand due to a rise in temperature against a no change in the outer layer of the pipe. This process will generate thermal stress. "Thermal stress" is a misnomer; they are really stress due to thermal effects. Stresses are always mechanical. Stresses in bonded joints are quite helpful for structural designs (Deheeger et al., 2009).

The strain is the quantification of deformation, indicating the displacement between constituent particles of the body compared to the original length. Like stress, the strain can also be defined as normal strain and shear strain depending on the direction of stretch or compression. If the length of the material increases, the normal strain is called tensile strain, and if length decreases, it is called compressive strain. However, engineering strain or Cauchy strain is defined as the ratio of change in dimension (deformation) to the original dimension of the material. The normal strain is positive if the material fibers are expanded and negative if compressed.

$$\varepsilon = \frac{\Delta l}{l} = \frac{l_f - l_0}{l_0} \tag{2}$$

where ε is engineering strain, l_o is the original length, and l_f is the final length of the fiber. The strains are often expressed in ppm or microstrains; however, shear strains are expressed in radians.

All materials exhibit a change in dimension subjected to change in temperature. When a sample is heated, the atoms present in the sample vibrate, causing them to move apart and the opposite effect when it is cooled. The strain (ε_t) is related to change in temperature (ΔT) through the relation given in equation 3,Where, (α) is a Thermal expansion coefficient.

$$\varepsilon_t = \alpha \Delta T \tag{3}$$

The strain in a material is defined by the difference in bulk thermal expansion coefficients. However, the total strain of material in a general sense is the sum of mechanical strain and thermal strain. The actual total strain, which is measured, is the physical deformation of the part. The thermal strain is the direct result of temperature difference, whereas a mechanical strain is the one that is directly related to stress.

If a conducting bar (e.g., steel bar) is heated without any constraint, it has a tendency to expand in all directions. As there is no constraint, the stress (thermal stress) developed is zero, but the strain has a non-zero value. The total strain developed here is equal to the strain due to the thermal effect, leaving mechanical strain as zero. However, if the bar is constrained (say both the ends) and heated, the bar cannot expand along the length. The thermal strain is the same as was in previous case, but now the total strain is zero due to zero physical deformation. Thus, the mechanical strain, in this case, is the same in magnitude and opposite direction to the thermal strain. Stresses will arise in this case due to mechanical strain, which is termed as thermal stress.

Strain Measurement

The magnitude of stress in material under loading cannot be measured directly. It must be calculated from a measurable parameter through some suitable correlations. Therefore, most of the time, strain value is measured and is used in conjunction with other material properties for the prediction of stress. Strain in a given loading condition can be measured through various techniques, e.g., mechanical, optical, acoustical, pneumatic, electrical, digital imaging, etc. Different sensors and measurement techniques have their own advantages and disadvantages. Some of the methods are summarized in Table 1. Most of the time, along with experiments, analytical methods and numerical methods also yield appreciable results for stress and strain modeling.

Table 1. Different strain measurement methods

Methods of strain measurement	**Sub Categories**	
Strain gauges	Mechanical strain gauges Optical strain gauges Electric strain gauges Interferometric strain gauges Acoustical strain gauges	Pneumatic strain gauges Scratch strain gauges Strain rosettes Semiconductor strain gauges Thin-film strain gauges
Photoelasticity	Two-dimensional photoelasticity Three-dimensional photoelasticity	Scattered light photoelasticity
Coatings	Brittle coatings	Photoelastic coatings
Grid method	Circular grid imprint	
Optical Method	Moire's fringe method	
Photo hole elasticity		
Photo orthotropic elasticity		
Piezo-electric sensors	PVDF	PZT
Digital image correlation		
Similarity and model laws method		
Analogous method		
Numerical methods		

STRAIN GAUGE

Strain gauge (often pronounced as strain gage) is a device (sensor) to measure the strain of an object. Initial gauges were used to measure the strain values which were working by mechanical means, popularly called as mechanical strain gauges.

The strain gauge concept was originally proposed separately by two eminent contemporary scientists at the almost same historical time. One among them was Prof. Arthur C. Ruge of Massachusetts Institute of Technology (MIT) and Edward E. Simmons from the California Institute of Technology (Caltech). As a research assistant in Caltech, Prof. Simmons had studied the stress-strain behavior of metals under shock load in 1936. To measure the induced force due to impact on the specimen, a dynamometer with fine wire fabricated from constantan was used. It was a part of research project, which was started in 1936 however; the research works related to it were not published until 1938. If the starting date of the project is taken into granted then it can be said that it was Simmons who invented the principles of strain gauge. In 1940, this work was registered in the United States patent office.

On the other hand, in 1938, prof. Arthur C. Ruge found some concept regarding strain gauge while working in the field of engineering seismology with the help of his assistant, J. Hanns Maier. His work mostly focused on the influence of earthquakes on mechanical structures. The testing was carried out on a small scale model of an elevated tank, mounted on a vibration table. Due to low stress, they failed to record the strain value during the experiment using mechanical and optical methods available during that time. On a fine morning, while doing experiments, he attached a thin wire from potentiometer to the water tank and recorded some nice reproducible results. For ease of measurement and handling, Ruge cemented the measuring wire to a carrier paper, which was further stiffened by cementing the paper to two Plexiglas end pieces with a brass spacer bar, and it was finally patented in 1944. Strain gauges are associated with some popular terms, as mentioned below.

Gauge Length

As per the definition of the strain, the magnitude of change in length is the measurable quantity, mostly used in mechanical methods. That means the measurement of strain is actually the measurement of displacement between two points, which are at some known distance apart. That distance is called Gauge length, and this length plays an important role while comparing various strain measurement techniques. The average value of strain over the gauge length is considered for measurement.

The importance of selecting a suitable gauge length can be cleared with an example. Let a tensile specimen is loaded from both the ends as mentioned in Figure-4; the part length of the specimen can be measured before and after loading. The difference between them will provide the total deformation of the object. The value divided by the original length will provide an average value of strain for the

Figure 4. A tensile specimen

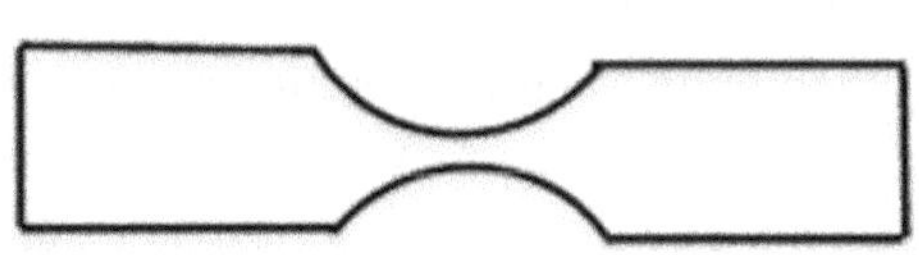

entire object. For the present case, the gauge length will be the total length of the specimen. Now, if one analyses the structure, different parts of the specimen will be loaded differently as the specimen is not symmetrical throughout its length. The same can be more clearly observed if the measurements are done at different parts along the length. The strain in the region with a small width will be locally higher than the average measured value, and hence, the stresses will also be highest in those areas. Therefore, while actual tensile testing, the narrow part will yield before it attains the averaged strain value due to localized stress concentration, which puts a question mark on the averaged strain value.

In ideal conditions, the strain measuring instrument to have the length value as small as possible to capture the localized strain. Apart from size, few other properties of strain measuring devices, which are important for the efficient strain measurement includes:

- Negligible mass so that there should not be any deformation due to self-weight
- Ease of attachment to the specimen under test
- No relative movement between the specimen and strain gauge
- Should not interfere with material properties of the specimen
- High sensitivity
- Ability to measure static, transient and dynamic strain
- Low cost, reliable and easily available

Gauge Factor

The change in resistance per unit strain value is defined as the Gauge factor. The Gauge factor is the measure of the sensitivity of the strain gauge, i.e., .the larger gauge factor leads to a higher change in resistance and hence, a better resolution. Metal wire and foil gauges bear a gauge factor between 2 and 4, which does not permit them to measure small strain values. In contrast, semiconductor gauges vary between 35 and 200; therefore can measure micro strains.

$$\text{Gauge Factor} = \frac{\frac{\Delta R}{R}}{\frac{\Delta L}{L}} = \frac{\frac{\Delta R}{R}}{\varepsilon} \tag{4}$$

The working principle of different strain gauge techniques is explained below.

Mechanical Strain Gauges

A mechanical method of strain gauges mostly measures a change in length before and after the loading of the sample. These are preferably used in the early days for strain measurement (Stern, 1979). Mechanical strain gauges are popularly known as Extensometers used to measure static or gradually varying loads. Two knife edges are clamped through a clamping spring to the test specimen. The knife edges are placed at a distance of known length (gauge length). When the specimen is loaded with tensile force, the knife-edge undergoes displacement. The small displacement is amplified to a reading scale by using a system of levers and linkages and measured on a calibrated scale. Some popular examples are Berry strain gauge, Huggenbeger extensometer, Johansson extensometer (Motra et al. 2014). A minimum gauge length of 0.5 inches and nearly 10 microstrains can be obtained from a purely mechanical strain gauge. Though the process looks simple, these gauges are bulky, provide low resolution and difficult to use.

Electrical Strain Gauges

Electrical strain gauges operated through a notable change in electrical characteristics, which is a measurable parameter corresponding to the change of strain in the sample. These electrical properties may be changed in resistance, change in capacitance, change in inductance, etc.

Resistance Strain Gauges

The most important electrical characteristics are electrical resistance. Resistance strain gauges operate on the principle of development of electrical resistance of the strain gauge due to displacement or development of strain in the sample (Higson, 1964). This is due to the material property of the conductor, i.e., when a conductor is stretched, its length increases, and to keep the volume constant, its diameter decreases. This change in geometrical properties results in the development of strain following the formula in equation 5. The change in resistance per unit strain value is defined as Gauge factor. The Gauge factor is the measure of sensitivity of the strain gauge (Keil, 2017).

$$R = \rho \frac{L}{A} \tag{5}$$

The total change in resistance of the strain gauge material is the sum of resistance due to change in length, resistance due to change in the area, and resistance change due to the piezo-resistive effect.

Capacitive Strain Gauges

The capacitive strain gauges based on the principle of change of capacitance subjected to deformation. The sensor mainly consists of two flexible substrates based on polyimide films and ultra-flexible epoxy resin, which adheres to the two substrates (Spillman Jr, & Weissman, 1998). In some cases, interdigital capacitors are used to obtain large initial capacitance.

Semiconductor Strain Gauges

Typical semiconductor gauges made up of semiconductor materials such as silicon and germanium. After doping with a suitable dopant, the materials can be made in the form of a p-type or n-type semiconductor, which is the basic construction material for semiconductor gauges. When the strain is applied to the semiconductor element, an appreciable change in the resistance value is observed which can be measured. The resistance of p-type strain gauge increases with the applied tensile strain and decreases for n-type gauges. Typically semiconductor wafers or filaments (2mm-10mm length and 0.05mm thickness) are bonded on any suitable substrate, which is insulating in nature. The electrical connections are mostly performed by gold leaves (Hannah, & Reed, 1992). The electrodes are prepared through vapor deposition techniques. Most of the advantages of semiconductor gauges include very high gauge factor, low hysteresis, useful in measurement of very small strains, possess high-frequency response, large fatigue life, and can be manufactured in very small size. Similarly it encounters some demerits, also like it is highly sensitive to temperature, nonlinearity, difficult to mount and degradation of performance due to presence of moisture. For semiconductor gauges, the gauge factor is expressed as ‘

$$\text{Gauge Factor} = \frac{\Delta R}{R_0 \varepsilon} = 1 + 2v + \Pi_L Y \tag{6}$$

where v is Poisson's ratio, Y is Young's modulus of the semiconductor material, Π_L is longitudinal piezo-resistive coefficient, ε is strain ΔR is a strain-induced change in resistance, R_0 is zero strain gauge resistance value. The first two terms of equation 6 represent the change in resistance due to change in dimensions, whereas; the last two terms indicate the change in resistivity with strain.

Acoustical Strain Gauges

The acoustical strain gauges mainly operate on the principle of change in the frequency of the wire. In this case, the variation in the length of wire stretched in between two gauge points is measured, which alters the natural frequency of the wire (Jerrett, 1945). The measuring parameter is the frequency of vibration of the wire when it is plucked by means of an electromagnetic impulse (Sebastian, & Stubbs, 2001). This type of gauge is highly stable, and readings can be done over a period of years without any change of zero drift.

Pneumatic Strain Gauges

The principle of operation of a pneumatic gauge mainly depends upon the relative discharge of air between a fixed orifice and a variable orifice. These gauges are suitable for both static and dynamic strain measurements. The gauge length up to 1mm and magnification up to 1 lakh is possible with these gauges.

Thin-Film Strain Gauges

Thin-film strain sensors are a more advanced form of strain gauge. It is mostly constructed by single deposition (in thin films) or multiple depositions (in thick films) of conductive films. A ceramic layer is deposited on the stressed metal surface and then depositing the strain gauge onto this insulation layer(Lei & Will, 1998). The materials are bonded either through vacuum deposition technique or sputtering method. As the thin film gauge is bonded to the specimen at the molecular level, the installation is very stable, and the resistance values experience less drift(Arshak et al.1994). Additionally, the stressed force detector can be a metallic diaphragm or beam with a deposited layer of ceramic insulation.

Brittle Coatings

The method of brittle coating is one of the oldest methods for strain measurement. Initially, it was used for measurement of yield strength of the specimen through naturally formed coatings on the members, e.g., oxides on heated surfaces, mill scales on hot rolled steel, etc. When the base metal yield under load, these coatings failed by cracking or flaking. Therefore, this method was helping to find out the region of plastically deformed zones in a component. The accuracy of the techniques depends on the properties of the coating. First artificially prepared brittle coating was a mixture of shellac and alcohol, which was used by Suenwald and Wieland in 1925 for finding out regions of plastic strain which was showing deformation at a large strain value. The coating material properties were further improved, which could help the coating materials to fail at a very smaller strain value. The application of these coating materials in the field of engineering was pioneered by A.J. Durelli and his coworkers.

The working procedure mainly focuses on spraying a suitable coating material, which is brittle in nature on the entire test specimen or areas understudy to form a thin coating. After drying, the specified load is applied to the specimen. The focus is mainly on identifying zones of stress concentration, which can help in further studies. In general, the coating cracks at a right angle to the direction of maximum tensile strain. So, the cracks generated due to loading can be analyzed to calculate magnitude as well as the direction of surface strains. The coating can be calibrated to obtain quantitative strain measurements (Craig & Peyton, 1965).After calibration, coatings can be calibrated to obtain quantitative strain measurement. Like other methods brittle coating also has its own advantages. Some important advantages are; its gauge length approaches towards zero; it can give an overall picture of strain distribution and especially highlight the regions of stress concentration; it can be applied to any mechanical parts of the structure irrespective of its shape. The major use of brittle coatings is to quickly identify the high-stress points in a design, to obtain principal stress directions for the subsequent placement of electric resistance strain gauges. Commercially available brittle coatings are mainly used for the following purposes: 1. Identifying small areas with high stresses. 2. Identifying principal stress direction. 3. Measuring the magnitudes of tensile and compressive stress values under static dynamic and impact loading conditions. 5. Indicating localized yielding due to plastic deformation. There are mainly two principal types of brittle coatings available. One category is a series of strain-sensitive coatings prepared from resins by dissolving them in solvents so to be sprayed on the parts under study. Plasticizers are added in varying amounts during the formulating process to produce coatings with differing failure characteristics. After drying, the coatings are designed to crack at strain levels on the order of 500 to 700 microstrain. These coatings are very sensitive to temperature changes. Another series of coatings are ceramic coatings, which are insensitive to temperature changes up to 300 °C and are mostly used for coating on steel and similar materials. The ceramic powder is suspended in a carrier to give it a liquid form. To form a continuous brittle coating, the coated part is required to be glazed by firing approximately 530°C. This technique is quite useful in strain measurement in real-time applications like bridges, civil structures, etc. which does not require any laboratory involvement

Grid Method

The grid method of displacement and hence, strain measurement is an easy way of strain measurement, which involves low cost through manual calculation whereas the cost increases by implementing automatic calculation. Basically, the grid patterns are used in metal forming processes owing to its simplicity and low cost (Badulescu et al.2009).A grid pattern (regular image of some repeating shapes) is imprinted on the surface of the material (sheet metal). After deformation, the dimension of the deformed grids is measured. The change of shape due to deformation as compared to the original shape helps in acquiring the strain measurement. As an example the circular grids after deformation normally becomes elliptical

Figure 5. Grid Method (a) A screen printed grid (b) grid formed by a chemical etching method

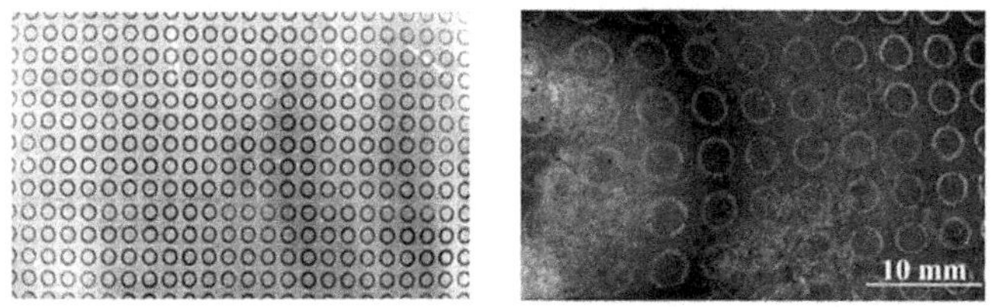

provide one maximum and one minimum radius in two different directions (Yildiz & Yilmaz, 2017).The maximum and minimum diameters of the grids, in many cases provide information regarding critical combinations of major and minor strains. The measurement of such grid structures can be accomplished manually at a minimal cost with the help of Mylar tapes, microscopes, and rulers, but the process is time-consuming and sometimes yields low resolution(Badulescu et al. 2009). Therefore, for a small study region, this method is appreciated. For accuracy and covering large areas with maximum points coverage, automated methods of grid size measurements are preferred(Goldrein et al. 1995).

The marking of grids on the structure can be achieved through different techniques. However, the primary criteria should be frictional resistance, i.e., the imprinted grid should resist frictional effects during the testing process so that the grids can be visible after the test. Apart from this, a grid's accuracy, resolution, cost, contrast, etc. also depend on selection method. Some of the methods are punch marks, scribed shapes (circles, squares), serigraphy (screen printing), electrochemical etching, photochemical etching, laser marking, etc. Each process has its own advantages in terms of cost, material-specific application, color contrast, etc. Punch marks are simply mechanical processes where getting a uniform shape is a challenging task whereas, screen printing methods can impart uniform grids, but the selection of ink, which will stick till the completion of experiment is a crucial task(Grediac et al. 2016). Similarly, in electrochemical etching, photochemical etching and laser marking some materials are lost from the specimen providing markings on them however, getting an appropriate contrast is a challenge (Figure 5). The modified grid pattern is adopted for some specific applications. The modified patterns include implementation of high-density grids around 1000-1200 lines/mm and the deformed patterns are observed through high-end optical microscopes (Zhao et al. 1999).

With the advancement of measurement techniques, the automated grid measurement methods have been implemented in recent times, which include image processing through correlation techniques like Fourier methods, De-convolution technique, averaging image technique, strain energy calculation, etc. (Sirkis & Lim, 1991).Unlike conventional methods of measurement, these statistical methods can measure the grids in a larger surface area and can acquire larger points, which ultimately increase the accuracy of measurement(Morimoto et al. 1988).

Piezo Strain Sensors

There are some materials which develop charge when subjected to stress or pressure. The piezo sensors work following the same principle of operation. The principle of action can be piezoelectric, piezoresistive, inductive, capacitive, and resistive. Each category has its strain sensitivity limit and the threshold value.

Piezoelectricity is defined in terms of the accumulated electric charge in some specific materials due to mechanical stress, and the sensors are categorized under piezo-electric strain sensors. The effect was discovered by Pierre Curie in 1880. However, its actual use started after around 1950 by industry people. The piezo sensors have been successfully applied in many industrial applications such as aerospace, nuclear, automotive, and also in the medical field. Piezoelectric elements are used successfully to monitor combustion process while developing internal combustion engines(Sirohi & Chopra, 2000). The sensors are directly fitted in the cylinder, or the spark plug was attached with a built-in miniature piezoelectric sensor. Any material showing the piezoelectric property can be used for sensor while two types of piezoelectric materials are most popularly used by people for strain measurement are piezoceramics (e.g. PZT Sensors) and polymer piezoelectric film (PVDF Sensors).

PZT Sensors

Lead Zirconate Titanate, commonly called PZT, is a ceramic perovskite material that depicts remarkable piezoelectric effects (Nanda et al. 2017a). Lead zirconate Titanates are mostly composed of solid solutions of lead zirconate and lead titanate combined with other elements to obtain specific properties. This was developed by three prominent physicists; Yutaka Takagi, Gen Shirane, and Etsuro Sawaguchi in Tokyo Institute of Technology around 1952. This is the most commonly used type of piezoceramics. These ceramics are fabricated by heating the mixture containing a proportional amount of lead, zirconium, and titanium oxide powders to around 800-1000 °C. The resulting product after reaction leads to form perovskite PZT powder, which is mixed with a binder sintered to obtain desired shape. There ours some property changes in the material during cooling process (paraelectric and ferroelectric phase transitions), which alters the cubic cell structure to tetragonal structure. Due to this structural change, the unit cell structure becomes elongated in one direction along with a permanent dipole moment orientation along a single axis (long axis). The unpoled ceramic contains many randomly oriented domains and hence, no net polarization. With the application of the electric field, most of the unit cells align themselves parallel to the applied field called poling which imparts polarization characteristics to the ceramic. The material in such a state is able to exhibit both the direct and converse piezoelectric effects.

Being ceramic in nature, PZT sensors has most of the characteristics of an ideal ceramics like high elastic module, brittleness, low tensile strength. In terms of structural isotropy, this material itself is mechanically isotropic, and through the poling process, it is also assumed to be isotropic in the plane perpendicular to poling direction. While dealing with working procedures in application to sensors, it develops a voltage difference across two of its faces while stressed (Nanda et al. 2017b). A similar type of change in voltage is observed corresponding to a temperature change across its two faces, which can help it to use as a thermal sensor like thermocouples.

Commercially, the PZT is not used in its pure form; rather, it is preferred after doping it with either donor or acceptor dopants, which create cations or anion vacancies facilitating domain wall motion in the material. Acceptor dopants make the PZT hard and donor dopants soft. The hard and softness of the PZTs also affect the piezoelectric constants, which are proportional to the polarization or to the electric field generated per stress. In general, soft PZT has higher piezoelectric constant but larger losses in the material due to internal friction. Similarly, in hard PZT, domain wall motion is pinned by the impurities; hence, lower the losses in the material at the expense of reduced piezoelectric constants.

PVDF Sensors

Polyvinylidene fluoride or polyvinylidene difluoride (PVDF) is a fluoropolymer produced through the polymerization of repeating monomers of vinylidene difluoride ($-CH_2 - CF_2-$) in the monomer, with respect to the carbon atom, the hydrogen atom is positively charged and fluorine atom is negatively charged which imparts a positive and negative end and hence, a natural dipole moment.

It is a highly non-reactive and thermoplastic in nature. PVDF is plastic with special characteristics for which its dominating applications include places with a requirement of the highest purity, resistance to solvents, acids, and hydrocarbons. PVDF film is manufactured by solidification of the film from a molten phase, which is then stretched in a particular direction and finally pooled. While in the liquid phase, in a given volume of liquid, the dipoles cancel each other, making zero net dipole moment. after solidification it is still in the non-piezoelectric alpha phase, which needs to be stretched or annealed for getting piezo properties. While stretching the film to one direction the polymer chains are mostly aligned in the stretching direction and in this phase, it is in the piezoelectric beta phase. After poling, the film provides a permanent dipole moment to the film, which ultimately gets the characteristics of piezoelectric material.

The young's modulus of PZT material is of the same order as in aluminum, whereas PVDF is of the order of $1/12^{th}$ that of aluminum. Hence, PVDF can be suitable for sensing applications as it will have very little influence on the cost structure in terms of stiffness. The PFDF resins are very resistive to heat. PVDF was tested by keeping it under 150 °C for 10 years, and no thermal or oxidative breakdown occurred.

Optical Methods for Strain Measurement

The principle of optical measurement uses the ray and wave nature of light to create some measurable entity. However, small optical arrangements like light beams and mirrors can be used to improve the measurement quality of mechanical strain gauges. Some common examples are Marten's optical gauge and Tuckerman optical gauges. However, optical strain gauges mostly refer to strain gauges involving fiber optics principles or fiber grating sensors(Roosbroeck et al. 2009). Unlike another type of strain, gauges optical strain gauges do not need electricity rather, it is based on light, which propagated through a fiber. Therefore, these sensors are passive in nature and immune to other electric and electromagnetic magnetic interference, which is its superior quality than other electrical strain gauges at certain applications.

The method of interference of lights is sometimes used as a tool for measurement of strain. The interference is a phenomenon where two light rays superimpose, resulting in a series of constructive and destructive fringes in the form of alternative light and dark patches at a particular interval. Moire's fringe method is one of the most popular methods which use the interference fringe pattern for measurement of strain(Aldrich et al. 2017). This method measures material deformation through interference patterns with the help of a spatial mismatch of two optical rulings. Among them, one ruling is obtained from the integral to the structural material, and another one is a separate analyzing grid. If both the grids (structural and analyzer) are originally having equal spatial density and alignment between them is proper there is no observance of optical interference between the two grids and hence, no fringe. When the material deforms, due to the change in orientation of the structural grids, interference fringes are developed, which bear a relationship with the deformation.

The holographic interference method is another utilization of optical principle for the strain measurement (Polyaev et al.1992). Holograms are photographs of three-dimensional impressions on the surface of light waves(Thalmann & Dändliker,1987). Two optical phenomena govern the principle of holography; interference and diffraction of light. The interference of light sources helps in generation of a pattern of standing wave which helps in easy imaging(Sharpe,1968). Laser lights with constant wavelength serve the purpose for interference of light, which produces crests and troughs(Füzessy,1977). Two basic waves are engaged for the generation of interference patterns; one of them is the wave that bounces off the object we are making the hologram of is called an object wave. The object wave is interfered with a reference wave to produce a standing wave pattern of interference, which is photographed and called as hologram. However, for three dimensions Bragg's diffraction effect is used in which the interference waves allowed through thick emulsion(Moyer & Gillespie,1972). On deformation, the fringe pattern changes appreciably which can be imaged and analyzed.

Like other optical methods, Photoelasticity is one of the promising techniques used for stress and strain measurement, which is most efficient for structures with complicated geometry, difficult loading conditions even both(Doyle et al. 1989). The name photoelasticity may be analyzed as a grouping of two words photo and elasticity. Photo implies the use of light rays and optical techniques while elasticity refers to the study of stress and deformation in elastic bodies. The stress and strain information in the extended regions of any component can be easily predicted through this method of analysis. Highly stressed areas and peak stress values at the surface and at the interior points can be quantitatively estimated. When the structure under study made from photoelastic material is stressed, and a ray of light is entered along one of the directions of the principal stress, the light gets divided into two wave components each with its plane of vibration (plane of polarization) parallel to one of the remaining two principal planes (planes having shear stress zero). The light travels along these two paths with different velocities, which depend upon the magnitudes of the remaining two principal stresses in the material. The waves travelling with different velocities merge with a new phase relationship or relative retardation which is the difference between the no of wave cycles experienced by the two rays travelling inside the body. This phenomenon is called double refraction. The two waves are brought together in the photoelastic polariscope and brought to optical interference, where due to destructive and constructive interference, the photoelastic patterns of light and dark bands appear. Therefore, through photoelastic directions the stresses can be derived. Similarly, Shearography is another optical technology used for non-destructive testing and strain measurement(Hung, 1982).

Digital Image Correlation

The concept of using a correlation technique to measure a set of data has been applied for a long time. However, its application to digital images was made in 1975. In the traditional methods of strain measurement, the generation of strain maps is too costly and not even practical. While dealing with measurements with the involvement of principles of optics and lights, Digital Image Correlation (DIC) has evolved as an important non-contact technique for the measurement of displacement and strain. Through this technique, the contour map of strains can be easily obtained for the entire specimen under study. This technique is mostly defined as an image identification technique(Depuydt et al. 2017). The working principle of DIC mostly focuses on comparison of digital photographs of the test piece at different stages of deformation. Through tracking of blocks of pixels, the system can be able to measure surface displacement and generate 2D, 3D deformation vectors and strain maps. For an effective measurement, the pixel blocks need to be random and unique with a range of pixels as well as intensity. DIC mostly uses image correlation technique by analyzing the movement of marked points on the specimen. It requires neither any special light nor any special surface preparation. Through efficient execution of algorithm integrated with software interface, detailed sub-pixel resolutions can be obtained which helps in high-resolution measurements. The DIC is a state of art technique which can be implemented for measurement of strain values accurately (Hensley et al. 2017). Due to fast data acquisition, this technique can be suitably implemented for measurement of strain in both elastic and plastic range. All these techniques can lead to extract surface deformation value in Parts per million scales from a commercially available digital photography. In recent times, this technique is mostly used for experimental stress-strain measurement due to its distinct advantages over other techniques. DIC can be implemented on almost all materials with a large inspection area without any need for pre-treatment like other optical methods. DIC is widely used in many areas of science and engineering. The application areas have broadened due to the involvement of computer technology and developments in high-resolution cameras. For static as well as dynamic applications especially in deformation analysis, DIC has proven to be a flexible and useful tool. DIC is simple to use and cost-effective compared to other optical techniques.

Comment on Thermal Stress and Strain Measurement

Ideally, the manufacturers prefer to avoid any change in the measurable quantity due to a change in temperature. In most of the strain measurements, effect of temperature is considered having an extra influence on the measured quantity. It is practically difficult to design strain gauges with zero temperature influence (Cha & Kim, 2014). Therefore, a compensation factor is generally adopted for such measurements (Hertl et al.2006). Many times, the manufacturer supplies a compensation chart for nominal gauge factor and tolerance with each gauge(Bertodo,1959). However, there are many industrial and practical applications where stresses developed due to thermal loading cannot be compensated rather; thermal stresses play an important role in determining safety and durability of the material (Vijayasankar et al.2016). Common examples for such cases are thermal stresses developed in railway tracks; thermal cracks developed in the concretes due to development of thermal stresses (Kim et al.2002). Sometimes, the thermal stresses can be predicted in railroad rails with the help of ultrasonic SH waves(MacLauchlan & Alers, 1987). Apart from these, there are many industrial requirements like measurement of thermal stresses developed in welded structures, porous combustion materials, etc(Dour et al. 2004). Modeling of strain measurement during welding process is also a highlighting area for making proper weld joints (Chen et al. 2015).

Different methods are adopted for thermal stress and strain measurement depending on the location, possibility, and cost (Kasahara et al. 2003). For example, the hole drilling method is mostly used in continuously welded rails (Zhu et al.2015). A hole drilling method is basically a semi-destructive method which measures residual stress close to the specimen surface along the in-plane directions (Huang et al.2013). The method of stress measurement through-hole drilling involves three basic steps; (i) drilling of a small hole at the specimen surface to relieve the localized existing stress(Zhu & di Scalea, 2017). (ii) Measurement of deformation caused by stress relaxation with the pre-installed strain-gauge rosette or optical measurement system.(iii) Calculation of residual stress using the deformation field with the help of calibration coefficients (Bjøntegaard, & Sellevold, 2004). In case of concretes structure studies sometimes load cells are used for measuring restrained forces(Haines & Wright, 1969).

For measurement of stress during high-temperature loading conditions optical imaging techniques (e.g. holography) are mostly adopted. Through this imaging technique a pattern of distribution of thermal deformations over the specimen is obtained (Seo et al.2011). Through analysis of this image and using correlations stresses developed due to thermal effect can be estimated. However, in many cases numerical simulations are providing quite accurate predictions (Jiang et al.2013). With the advancement of computing facilities many reported works involve computational simulations for different stresses using finite element method.

CONCLUSION

A brief description of different stress and strain measurement techniques has been explained in this chapter. The contents have been designed in a way to provide the reader with an overview of the strain measurement techniques adopted from early age till today, along with a little insight into the history of strain measurement. All the methods are described in brief without going into detailed mathematical explanations. Special attention has been given for strain developed due to thermal effects and its measurement technique.

REFERENCES

Aldrich, D. R., Ayranci, C., & Nobes, D. S. (2017). OSM-Classic: An optical imaging technique for accurately determining strain. *SoftwareX*, *6*, 225–230. doi:10.1016/j.softx.2017.08.007

Arshak, K. I., Ansari, F., Collins, D., & Perrem, R. (1994). Characterisation of a thin-film/thick-film strain gauge sensor on stainless steel. *Materials Science and Engineering B*, *26*(1), 13–17. doi:10.1016/0921-5107(94)90180-5

Badulescu, C., Grédiac, M., & Mathias, J. D. (2009). Investigation of the grid method for accurate in-plane strain measurement. *Measurement Science and Technology, 20*(9), 1-17.

Badulescu, C., Grédiac, M., Mathias, J. D., & Roux, D. (2009). A procedure for accurate one-dimensional strain measurement using the grid method. *Experimental Mechanics*, *49*(6), 841–854. doi:10.100711340-008-9203-8

Bertodo, R. (1959). Development of high-temperature strain gauges. *Proceedings - Institution of Mechanical Engineers*, *173*(1), 605–622. doi:10.1243/PIME_PROC_1959_173_052_02

Bjøntegaard, Ø., & Sellevold, E. J. (2004). The temperature-stress testing machine (TSTM): Capabilities and limitations. In *First International Rilem Symposium on Advances in Concrete through Science and Engineering. E&FN Spon.*

Cha, S. L., & Kim, J. K. (2014). Application of Thermal Stress Device for Measuring Thermal Stresses. *The 2014 World Congress on Advances in Civil, Environmental, and Materials Research (ACEM'14)*, 268-275.

Chen, J., Yu, X., Miller, R. G., & Feng, Z. (2015). In situ strain and temperature measurement and modelling during arc welding. *Science and Technology of Welding and Joining*, *20*(3), 181–188. doi:10.1179/1362171814Y.0000000270

Craig, R. G., & Peyton, F. A. (1965). Measurement of stresses in fixed-bridge restorations using a brittle coating technique. *Journal of Dental Research*, *44*(4), 756–762. doi:10.1177/00220345650440042201 PMID:14321410

Deheeger, A., Badulescu, C., Mathias, J. D., & Grédiac, M. (2009). Experimental study of thermal stresses in a bonded joint. *Journal of Physics: Conference Series*, *181*(1), 012041. doi:10.1088/1742-6596/181/1/012041

Depuydt, D., Hendrickx, K., Biesmans, W., Ivens, J., & Van Vuure, A. W. (2017). Digital image correlation as a strain measurement technique for fibre tensile tests. *Composites. Part A, Applied Science and Manufacturing*, *99*, 76–83. doi:10.1016/j.compositesa.2017.03.035

Dour, G., Medjedoub, F., Leroux, S., Diaconu, G., & Rezai-Aria, F. (2004). Normalized thermal stresses analysis to design a thermal fatigue experiment. *Journal of Thermal Stresses*, *28*(1), 1–16. doi:10.1080/014957390523651

Doyle, J. F., Phillips, J. W., & Post, D. (1989). *Manual on Experimental Stress Analysis*. Society for Experimental Mechanics.

Füzessy, Z. (1977). Methods of holographic interferometry for industrial measurements. *Periodica Polytechnica Mechanical Engineering*, *21*(3-4), 257–263.

Goldrein, H. T., Palmer, S. J. P., & Huntley, J. M. (1995). Automated fine grid technique for measurement of large-strain deformation maps. *Optics and Lasers in Engineering*, *23*(5), 305–318. doi:10.1016/0143-8166(95)00036-N

Grediac, M., Sur, F., & Blaysat, B. (2016). The Grid Method for In-plane Displacement and Strain Measurement: A Review and Analysis. *Strain*, *52*(3), 205–243. doi:10.1111tr.12182

Haines, D. J., & Wright, G. P. (1969). An experimental method of determining thermal strains. *Experimental Mechanics*, *9*(7), 327–331. doi:10.1007/BF02325139

Hannah, R. L., & Reed, S. E. (Eds.). (1992). *Strain gauge users' handbook*. Springer Science & Business Media.

Hensley, S., Christensen, M., Small, S., Archer, D., Lakes, E., & Rogge, R. (2017). Digital image correlation techniques for strain measurement in a variety of biomechanical test models. *Acta of Bioengineering and Biomechanics*, *19*(3). PMID:29205227

Hertl, M., Fayolle, R., Weidmann, D., & Lecomte, J. C. (2006). Thermal Stress Failures: A New Experimental Approach For Prediction and Prevention. THERMINIC 2006, 169-174.

Higson, G. (1964). Recent advances in strain gauges. *Journal of Scientific Instruments*, *41*(7), 405–414. doi:10.1088/0950-7671/41/7/301

Huang, X., Liu, Z., & Xie, H. (2013). Recent progress in residual stress measurement techniques. *Guti Lixue Xuebao*, *26*(6), 570–583.

Hung, Y. Y. (1982). Shearography: A new optical method for strain measurement and nondestructive testing. *Optical Engineering (Redondo Beach, Calif.)*, *21*(3), 391–395. doi:10.1117/12.7972920

Jerrett, R. S. (1945). The acoustic strain gauge. *Journal of Scientific Instruments*, *22*(2), 29–34. doi:10.1088/0950-7671/22/2/303

Jiang, T., Ryu, S. K., Zhao, Q., Im, J., Huang, R., & Ho, P. S. (2013). Measurement and analysis of thermal stresses in 3D integrated structures containing through-silicon-vias. *Microelectronics and Reliability*, *53*(1), 53–62. doi:10.1016/j.microrel.2012.05.008

Kasahara, N., Jinbo, M., & Hosogai, H. (2003). *Mitigation method of thermal transient stress by thermalhydraulic-structure total analysis*. Academic Press.

Keil, S. (2017). *Technology and practical use of strain gauges: with particular consideration of stress analysis using strain gauges*. John Wiley & Sons. doi:10.1002/9783433606667

Kim, J. H. J., Jeon, S. E., & Kim, J. K. (2002). Development of new device for measuring thermal stresses. *Cement and Concrete Research*, *32*(10), 1645–1651. doi:10.1016/S0008-8846(02)00842-6

Lei, J. F., & Will, H. A. (1998). Thin-film thermocouples and strain-gauge technologies for engine applications. *Sensors and Actuators. A, Physical*, *65*(2-3), 187–193. doi:10.1016/S0924-4247(97)01683-X

MacLauchlan, D. T., & Alers, G. A. (1987). Measurement of thermal stress in railroad rails using ultrasonic SH waves. In *Review of Progress in Quantitative Nondestructive Evaluation*. Boston, MA: Springer. doi:10.1007/978-1-4613-1893-4_175

Morimoto, Y., Seguchi, Y., & Inatani, T. (1988). Strain Analysis by Grid Method and its Automization Using Image Processing. *Nondestructive Testing Communications*, *4*(2-3), 74. doi:10.1080/02780898808962125

Motra, H. B., Hildebrand, J., & Dimmig-Osburg, A. (2014). Assessment of strain measurement techniques to characterise mechanical properties of structural steel. *Engineering Science and Technology, an International Journal, 17*(4), 260-269.

Moyer, R. G., & Gillespie, G. E. (1972). *Displacement measurement by holographic interferometry* (No. AECL--4129). Academic Press.

Nanda, S. R., Agarwal, S., Kulkarni, V., & Sahoo, N. (2017). Shock Tube as an Impulsive Application Device. *International Journal of Aerospace Engineering*.

Nanda, S. R., Kulkarni, V., & Sahoo, N. (2017). Apt strain measurement technique for impulsive loading applications. *Measurement Science & Technology*, *28*(3). doi:10.1088/1361-6501/aa577a

Polyaev, V. M., Genbach, A. N., & Genbach, A. A. (1992). An experimental study of thermal stress in porous materials by methods of holography and photoelasticity. *Experimental Thermal and Fluid Science*, *5*(6), 697–702. doi:10.1016/0894-1777(92)90113-J

Roosbroeck, J. V., Vlekken, J., Voet, E., & Voet, M. (2009). A new methodology for fiber optic strain gauge measurements and its characterization. *Proceedings of the SENSOR+ TEST conference Opto,* 59-64.

Sebastian, J. R., & Stubbs, D. A. (2001). Strain measurement using surface acoustic waves. AIP Conference Proceedings, 557(1), 1467-1471. doi:10.1063/1.1373926

Seo, C. H., Shi, Y., Huang, S. W., Kim, K., & O'Donnell, M. (2011). Thermal strain imaging: A review. *Interface Focus*, *1*(4), 649–664. doi:10.1098/rsfs.2011.0010 PMID:22866235

Sharpe, W. N. Jr. (1968). The interferometric strain gauge. *Experimental Mechanics*, *8*(4), 164–170. doi:10.1007/BF02326343

Sirkis, J. S., & Lim, T. J. (1991). Displacement and strain measurement with automated grid methods. *Experimental Mechanics*, *31*(4), 382–388. doi:10.1007/BF02325997

Sirohi, J., & Chopra, I. (2000). Fundamental understanding of piezoelectric strain sensors. *Journal of Intelligent Material Systems and Structures*, *11*(4), 246–257. doi:10.1106/8BFB-GC8P-XQ47-YCQ0

Spillman, W. B., Jr., & Weissman, E. M. (1998). *U.S. Patent No. 5,747,698*. Washington, DC: U.S. Patent and Trademark Office.

Stern, F. B. (1979). Brittle coatings. *Experimental Mechanics*, *19*(6), 221–224. doi:10.1007/BF02324986

Thalmann, R., & Dändliker, R. (1987). Strain measurement by heterodyne holographic interferometry. *Applied Optics*, *26*(10), 1964–1971. doi:10.1364/AO.26.001964 PMID:20454429

Vijayasankar, A., Garg, H., & Karar, V. (2016). A Novel Experimental Approach for Thermal Stress Measurement in Elastomerically Mounted Optics of Head-Up Display System. *Journal of Display Technology*, *12*(8), 859–868. doi:10.1109/JDT.2016.2540720

Yildiz, R. A., & Yilmaz, S. (2017). The verification of strains obtained by grid measurements using digital image processing for sheet metal formability. *Journal of Strain Analysis for Engineering Design*, *52*(8), 506–514. doi:10.1177/0309324717734669

Zhao, B., Asundi, A. K., & Oh, K. E. (1999). Grid method for strain measurement in electronic packaging using optical, electronic, and atomic force microscope. Advanced Photonic Sensors and Applications, 3897, 260-271. doi:10.1117/12.369315

Zhu, X., & di Scalea, F. L. (2017). Thermal stress measurement in continuous welded rails using the hole-drilling method. *Experimental Mechanics*, *57*(1), 165–178. doi:10.100711340-016-0204-8

Zhu, X., Lanza di Scalea, F., & Fateh, M. (2015). On the study of the in-situ thermal stress measurement using a Hole-Drilling method. In *11th International Workshop on Advanced Materials and Smart Structures Technology*. University of Illinois Urbana-Champaign.

Chapter 4
Introduction and Application of Strain Gauges

Balaji P. S.
https://orcid.org/0000-0002-6364-4466
National Institute of Technology, Rourkela, India

Karthik Selva Kumar Karuppasamy
https://orcid.org/0000-0001-5431-2542
Indian Institute of Technology, Guwahati, India

ABSTRACT

Strain gauge method is one of the essential and fundamental methods in experimental stress techniques that uses the resistance of the material to determine the stress at a point. The strain gauges can be used in a different combination called Rosette to obtain stress in various directions. This chapter intends to cover types of strain gauges, materials, and Rosette arrangements to provide the reader with an overview of the techniques. The chapter will discuss the basic physics behind the resistance measurement and take the reader into insights on how the developments were made to the application of strain gauges as experimental techniques.

INTRODUCTION

Static and dynamic loadings act on any components during the operations. In addition, other minor factors (e.g., loading of wind and vibrations) also impact on components (Karthik & Dhas, 2015a, 2015b 2016, 2018). The stresses from these various factors provide challenges to the mathematical techniques for stress evaluation (Balaji, Leblouba, Rahman, & Ho, 2016; Balaji, Moussa, Rahman, & Ho, 2016; Balaji, Moussa, Rahman, & Vuia, 2015; Balaji & Yadava, 2013; Leblouba, Altoubat, Ekhlasur Rahman, & Palani Selvaraj, 2015). Hence, experimental stress evaluation technique is required in such practical cases. The strain gauge is a reliable technique that is employed for stress evaluations. The concept of strain gauge in experimental stress analysis can be understood by understanding the resistance of the conductors, since strain measurement principles using strain gauges rely on the concept of resistance.

DOI: 10.4018/978-1-7998-1690-4.ch004

This chapter covers the basics of strain measurement and its application. First, the authors discuss the resistance of a conductor. Subsequently, they illustrate strain sensitivity, gauge factors, Wheatstone bridge, and rosettes in order to detail the experimental procedure for strain measurements.

RESISTANCE OF A CONDUCTOR

The ability of the material to resist the flow of electric current is measured as resistance (R). The unit of resistance is given in ohms (Ω), to credit the German scientist Georg Simon Ohm (1784-1854) who studied the relation between voltage, current, and resistance (Bhattacharya, 2011). The resistance of a conductor is given below:

$$R = \frac{\rho L}{A} \tag{1}$$

where L is the length of the conductor, and A is the cross-sectional area (Figure 1) and the specific resistance of the materials. The specific resistance is material-dependent and the values for general materials are available in the literature.

STRAIN SENSITIVITY OF THE WIRE

The strain sensitivity of a material refers to the change in the resistance of the material for a change in the strain. The strain sensitivity of the wire can be evaluated by differentiating Equation 1 as follows:

$$\left(\frac{dR}{R}\right) = \left(\frac{d\rho}{\rho}\right) + \left(\frac{dL}{L}\right) - \left(\frac{dA}{A}\right) \tag{2}$$

In Equation 2, the last term dA/A can be written in terms of dL/L by considering Poisson's ratio. For a wire of diameter D, area A is given as follows:

Figure 1. Wire conductor

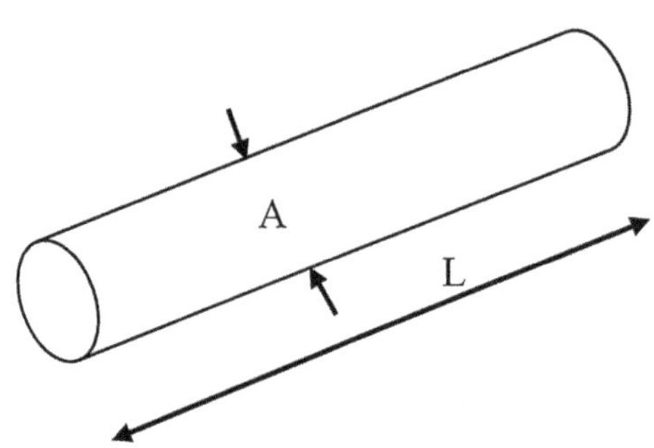

$$A = \frac{\pi}{4} D^2 \tag{3}$$

The result of the differentiation of Equation 3 is:

$$\left(\frac{dA}{A}\right) = 2\left(\frac{dD}{D}\right) \tag{4}$$

Further, by the Poisson's ratio (υ) of the material:

$$\left(\frac{dD}{D}\right) = -\upsilon\left(\frac{dL}{L}\right) \tag{5}$$

By substituting Equation 5 into Equation 4, the result is:

$$\left(\frac{dA}{A}\right) = -2\upsilon\left(\frac{dL}{L}\right) \tag{6}$$

Now, Equation 1 can be rewritten as follows:

$$\left(\frac{dR}{R}\right) = \left(\frac{d\rho}{\rho}\right) + \left(\frac{dL}{L}\right) + 2\upsilon\left(\frac{dL}{L}\right) \tag{7}$$

The strain sensitivity (S_A) of a material is the fraction change in resistance to the fraction change in the length of the material, and it is given by the following equation:

$$S_A = \frac{\left(\frac{dR}{R}\right)}{\left(\frac{dL}{L}\right)} = 1 + 2\upsilon + \frac{\left(\frac{d\rho}{\rho}\right)}{\left(\frac{dL}{L}\right)} \tag{8}$$

The strain sensitivity (S_A) of a material plays a key role in the measurement of strain and it is required to understand that the quantity that is measured by the strain gauge is in the scale of μ (microns), hence the measurement procedure has to be conducted carefully. Indeed, any negligence in the procedure can easily result in error in the strain evaluated. The strain sensitivity of the strain gauge should be given by the manufacturer. When working with a strain gauge, it is necessary to understand the gauge construction. The following section details the required knowledge to appreciate the strain gauge design.

GAUGE CONSTRUCTION

The working of the strain gauge is based on the measurement of resistance change in the wire materials as a result of the applied load. Thus, it is imperative to understand how much resistance is required to use wire materials in the application of the strain gauge. As per the working knowledge in industries, the minimum amount of necessary resistance to have a scientific measure of the strains is 100 Ω. Further, if, for example, the wire is 0.025mm thick and has 1000 Ω per meter, then a wire of 100 mm is necessary to obtain the minimum amount of the resistance (i.e., 100 Ω). However, from the point of view of the mechanics of materials, strain at a point is the focus of the authors' study. Therefore, the wire 100 mm long needs to be innovatively configured to measure the strain at a point (Ramesh, 2018). The authors used a loop configuration n (Figure 2).

Figure 2 shows the parts of the strain gauge, grid length is the operative length of a strain gauge that is useful in strain measurement, and the strain gauge measures the strain in the gird or gauge length direction. The thin wire with the required resistance is looped to form a strain gauge. In current practice, standard resistance of 120 Ω and 350 Ω are used for general applications, while high resistance of 500 Ω, 1000 Ω, and 3000 Ω are used in special-purpose applications (Ramesh, 2018). The end loop and the solder tabs of the strain gauge are made thick, so their resistances can be reduced and their influence in the strain measurement can be minimized. Further, the thick region of strain gauges has less resistance, due to the increased area, and the effect due to transverse strain can be minimized in the longitudinal strain measurements. The strain gauge is a very thin structure, thus care needs to be taken in the usage, and proper handling can minimize the damage. A damaged strain gauge may malfunction, which, in turn, can result in an error in the data from the material. The strain gauge is constructed by bonding a thin electric resistance wire or photographically etched metallic resistance foil on an electrically insulated base, using appropriate bonding materials. The strain gauge is also covered with a thin film to avoid any atmospheric effects on the strain measurements.

Further, a proper and accurate measurement of strain at a point needs a speck of a material with high resistance; however, it is practically difficult to have such speck of the materials. Therefore, the coil of wire has to be foiled in order to comply with the practical constraints. Hence, the gauge length of the strain gauge is a critical factor in strain measurement. The gauge length that is generally used in the applications ranges from 0.2 mm to 100 mm, and the range required depends on the applications. A general-purpose strain gauge has gauge length of 3 mm. However, the gauge length needs to be identified based on the application, which can require a smaller or larger gauge length. The gauge for concrete

Figure 2. Strain gauge and its parts (SHOWA MEASURING INSTRUMENTS CO., 2018)

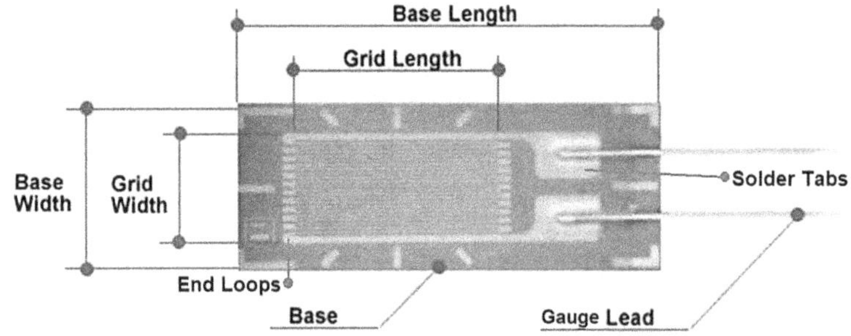

materials is usually longer than 3 mm, while a gauge of 0.25 mm is used in very special applications. The gauge length strongly influences the measure of the strain, so it is important to have an understanding of errors due to it.

ERROR DUE TO GAUGE LENGTH

Prior knowledge of stress distribution is essential and can assist in the proper estimation of stress at a point. In order to know the stress distribution, methods such as optical or brittle coating technique can be followed. When the stress distribution is linearly varying (Figure 3), such as in the case of loading on a cantilever beam, then the strain can be estimated with high accuracy and the strain shown by the strain gauge is the average strain along the gauge length. However, when the stress distribution is high, as in the case near the crack tip, then the stress distribution can be highly nonlinear. In such application, the strain gauge length needs to be small to capture the required details, and the larger strain gauge length can give lower values of a strain than the real value. Hence, importantly, the gauge length is one of the predominant factors in the accuracy of the results.

Figure 3. Effect of gauge length in the strain measurement

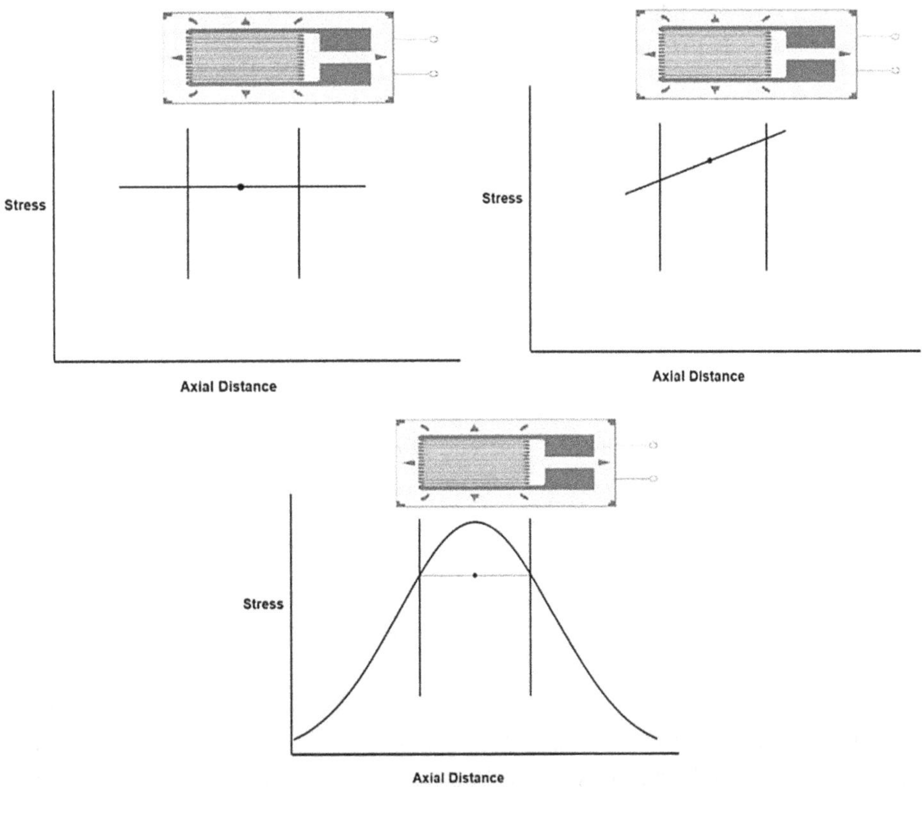

STRAIN SENSITIVITY OF A STRAIN GAUGE

High resistance for strain measurement can be obtained in a wire of considerable length, hence such wire needs to be foiled in order to measure the strain at a point along the gauge length direction. When the wire is foiled, additional factors come into play on strain measurement. As a result, it is necessary to understand strain sensitivity of the strain gauge. The sensitivity of the strain can be given as the sum of strain sensitivity along axial (S_a), transverse (S_t), and shear (S_s) directions:

$$\frac{dR}{R} = S_a\varepsilon_a + S_t\varepsilon_t + S_s\gamma_{at} \tag{9}$$

where ε_a, ε_t, and γ_{at} are the axial, transverse, and shear strain in the material, respectively. The wire of the foil is very thin, thus the shear effects can be neglected and the above equation becomes as follows:

$$\frac{dR}{R} = S_a\varepsilon_a + S_t\varepsilon_t \tag{10}$$

The strain of the materials is very small. Besides, the user should understand and appreciate the manufacturer's procedures, in order to have an effective measurement of strain. The proper understanding of the working of the strain gauge is crucial to fully benefit from it. In this chapter, S_A is the strain sensitivity of wire, while S_a is the strain sensitivity of the wire when foiled as a strain gauge.

Transverse Sensitivity Factor and Gauge Factor

The ratio of transverse sensitivity to axial sensitivity is the transverse sensitivity factor, which is denoted by K_t. From Equation 10, it is possible to write the following:

$$\frac{dR}{R} = S_a(\varepsilon_a + K_t\varepsilon_t) \tag{11}$$

where $K_t = \frac{S_t}{S_a}$, which is called transverse sensitivity factor.

The transverse sensitivity factor is kept as small as possible, in order to effectively measure the strain along the gauge length. However, K_t also plays a critical role: If K_t can be made equal to the Poisson's ratio of the base material, then the gauge can be called as stress gauge. In this case, the electrical signals in the gauge are proportional to the stress, hence the gauge can be calibrated to measure the stress directly in the base material.

However, ideally, K_t is desired to be zero or, practically, it is made the minimum. K_t is generally given by the manufacturer as a part of the strain gauge specification. The strain gauge being the foiled wire, the strain sensitivity along axial, transverse, and shear has to be considered for the strain measurement;

however, such complex calculation can be difficult to the user. Further, it is challenging to identify and use sensitivity for each direction. Consequently, for simplicity, the strain gauge is calibrated using the so called gauge factor S_g, which is given by the following equation:

$$\frac{dR}{R} = S_g \varepsilon_a \tag{12}$$

The gauge factor is experimentally obtained for a specific kind of strain gauge material and type for a uniaxial stress field. This test incorporates a common factor for all directions and calibrates the electrical signal to the strain on the material.

Experimental Test for Gauge Factor

The manufacturer performs the cantilever test to determine the strain gauge factor S_g (Figure 4). The results of the experiments are compared with the analytical solutions to calibrate the electrical signals to the strain in the material. The key factor to understand in the uniaxial stress is that the stress may be uniaxial, but the resulting strain is biaxial. As a result, Poisson's ratio also plays a critical role in the calibration of the strain gauge. The calibration experiment is performed on a standard material whose Poisson's ratio is 0.285.

The transverse strain can be written in terms of axial strain using Poisson's ratio (υ_o), as follows:

$$\varepsilon_t = -\upsilon_o \varepsilon_a \tag{13}$$

Using Equations 11 and 12, the relation between gauge factor and axial strain sensitivity is expressed as follows:

$$S_g \varepsilon = S_a(\varepsilon_a + K_t \varepsilon_t) \tag{14}$$

Upon writing the transverse strain in terms of axial strain, the result is:

$$S_g = S_a(1 - K_t \upsilon_o) \tag{15}$$

Figure 4. Cantilever setup for the strain gauge calibration

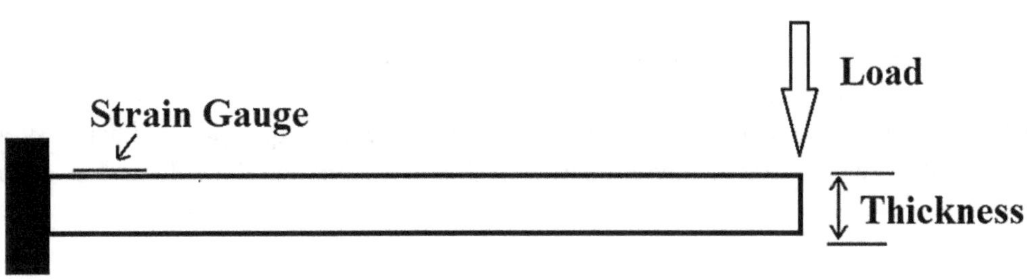

The above equation allows to understand that, for materials whose K_t is zero, the gauge factor is the axial strain sensitivity of the material. Further, Equation 15 is valid for the material whose Poisson's ratio is 0.285, since it is calibrated with standard material. Actually, the specimen material can be different from the calibrated material, and, since the strain is very small, the small changes in the operating conditions can cause an error in the strain measurement. Thus, in order to have a proper measurement of strain, it is necessary to calibrate S_g using the same material as specimen material. Importantly, S_g is an experimental quantity resulting from the experiment test. The mismatch in S_g between the calibrate and real-time application can affect accuracy. The manufacturer does the calibration with different materials and provides the average of S_g, in order to have effective accuracy in the strain measurement. Further, the mismatch of S_g can be eliminated when K_t of a gauge is zero, but, practically, K_t is some value and kept as minimum as possible. K_t is kept small by adding more wire material in the end loop and soldering tab of the strain gauge.

MEASUREMENT OF STRAIN

The equation of the gauge factor (Equation 12) highlights that the fractional change in the resistance is what was required for the strain measurement, and it is not the absolute value of the resistance. It would be useful to understand how much resistance change occurs in the strain measurement, thus the following example may help. If the initial resistance of the gauge is 120 Ω, and the gauge factor is +2 and requires to measure 1με, then Equation 12 allows to estimate the following:

$$dR = S_g \times \varepsilon \times R = 2 \times 0.000001 \times 120 \; = 0.000240\;\Omega = 240 \mu\Omega \tag{16}$$

It can be seen that it is required to measure 240 μΩ of resistance to evaluate the 1με strain. In order to measure such a small fraction of strain, a very sensitive measuring instrument with microohm sensitivity is required. Further, in order to measure the change in resistance, the absolute resistance of the strain gauge needs to be measured accurately. As the available materials have resistance with tolerance in microohm, it is difficult to measure the small change in the resistance. The effective method used in strain gauge is the Wheatstone bridge, and most of the instrumentation uses this bridge circuit to measure the strain (OMEGA, 2018).

Wheatstone Bridge Circuit

The unknown electrical resistance in the circuit can be measured with high sensitivity using a Wheatstone bridge (Stout, 1960) (Figure 5). A major advantage of this circuit is its ability to measure resistance with high accuracy. This bridge circuit was invented by Samuel Hunter Christie in 1833 and subsequently improved by Sir Charles Wheatstone in 1843. It is popularly known as Wheatstone bridge. This circuit is very used in many strain gauge applications. Besides, this instrumentation can be connected to a computer through the multichannel system to handle and store large quantities of data. By applying a voltage V at the points AC, the voltage E between BD is measured and the voltage E is given by the following equation:

$$E = V_{BD} = V_{AB} - V_{BD} \tag{17}$$

where the voltage *AB* and *AD* is given as follows:

$$V_{AB} = \frac{R_1}{R_1 + R_2} V \tag{18}$$

$$V_{AD} = \frac{R_4}{R_3 + R_4} V \tag{19}$$

Finally, the voltage *E* is given as follows:

$$E = V_{BD} = V_{AB} - V_{BD} = \frac{R_1 R_3 - R_2 R_4}{(R_1 + R_2)(R_3 + R_4)} V \tag{20}$$

The bridge circuit is initially balanced (i.e., $E = 0$); this is achieved when $R_1R_3=R_2R_4$. However, when the balance of the bridge is disturbed, then the output voltage E for an incremental resistance change is given as follows:

$$\Delta E = V \frac{R_1 R_2}{(R_1 + R_2)^2} \left(\frac{\Delta R_1}{R_1} - \frac{\Delta R_2}{R_2} + \frac{\Delta R_3}{R_3} - \frac{\Delta R_4}{R_4} \right) \tag{21}$$

Figure 5. Wheatstone bridge circuit

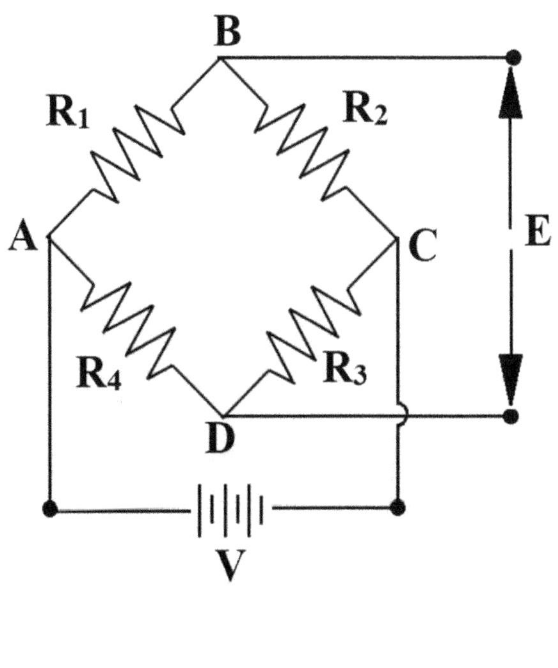

Strain Measurement Using Wheatstone Bridge

Strain gauges can be connected to one or more arms of the Wheatstone bridge circuit, and, by connecting known resistances on the other arms of Wheatstone bridge, the voltage ΔE can be measured. From the voltage ΔE, a corresponding change is resistance, thus further strain can be evaluated. A computer can be connected to the measurement instrumentation to store a large quantity of data and to process the data to get the strain as output. The measurement of strain can be carried out in the following two ways:

1. Measure ΔE directly and obtain strain from the measured ΔE. This technique is useful to measure strain at both static and dynamic conditions. The main requirement is that the bridge has to be balanced initially, before making the strain measurement.
2. Upon loading the specimen, after initially balancing the bridge, the strain is induced in the specimen. Due to the strain, the bridge balance is disturbed. Hence, the resistive balance of the bridge can be adjusted to make the bridge balance again (i.e., ΔE =0). From the value of the resistive balance required for balancing the bridge, the strain can be evaluated. This technique is applicable mainly for static strain measurements. In addition, this method can provide high accuracy, even though the process is very slow.

The measuring ΔE directly is a very convenient method and it also applies to both static and dynamic strain measurements.

BRIDGE SENSITIVITY

The wire is foiled to form a strain gauge, which is connected to a bridge to measure the electrical signal, so the corresponding strain is estimated. In every step, sensitivity comes into play, and it is necessary to understand sensitivity when each form is taken by the wire to measure the strain. The sensitivity of the Wheatstone bridge also needs to be determined, as it is used to convert the resistance change into an output voltage signal. The bridge sensitivity is a function of magnitude of bridge voltage (*V*), gauge factor (*Sg*), bridge factor (*n*), and ratio of resistance (*m*), which is *R2/R1*. As a result, the output voltage from the Wheatstone bridge can be rewritten by incorporating the factors as follows (Bakshi & Bakshi, 2008):

$$\Delta E = V\frac{R_1R_2}{(R_1+R_2)^2}n\left(\frac{\Delta R}{R}\right) \tag{22}$$

where the term $\Delta R/R$ represents the resistance ratio of Equation 12 and can be written in terms of *Sg*; also, introducing *m* (=R2/R1), Equation 22 is rewritten as follows:

$$\Delta E = V\frac{m}{(1+\mathrm{m})^2}nS_g\varepsilon \tag{23}$$

The current in the bridge is very small and in the order of 0.1 *A*; the voltage is in the order of 3-5 volts and *Sg* for most of the strain gauge is in the order of 2. Hence, the above equation allows to observe that the magnitude of the output signal can be increased by suitably controlling *m* and *n*, which is the bridge factor that can be 1, 2, and 4 for the quarter, half, and full-bridge mode, respectively. It is important to determine *m,* as it can be controlled to achieve the maximum signal to measure the strain. Figure 6 highlights that the ratio *m*/(1+*m*)2 is maximum when the $m = 1$, hence, by achieving such condition, the output signal can be enhanced to ease the strain measurement. Therefore, for a given bridge having constant *V*, *Sg,* and *n,* sensitivity can be increased by controlling the resistance ratio (*m*).

BRIDGE FACTOR

Apart from the resistance ratio, another factor which can be used to increase the output voltage is the bridge factor. Three configurations of the bridge are currently used in the applications, namely quarter, half, and full-bridge circuit. Each bridge circuit is discussed in the subsequent subsections.

Quarter Bridge Configuration

Figure 7 shows the quarter bridge configuration; where the resistance arms are replaced with the strain gauge. This configuration can be used to measure both bending strain and axial strain in the specimen. The strain gauge has to be placed in a way that the gauge length is along the longitudinal axis of the specimen for the strain measurement. Further, the bridge factor for this configuration is $n = 1$. In the examples in Figure 7, the strain gauge is configured to measure $+\varepsilon$ (i.e., tensile strain or elongation due to the applied load). This configuration suffers a major disadvantage, that is the temperature effect can affect the accuracy. As a result, this configuration can be less preferred for the strain measurement in ap-

Figure 6. Influence of resistance ratio in the output voltage

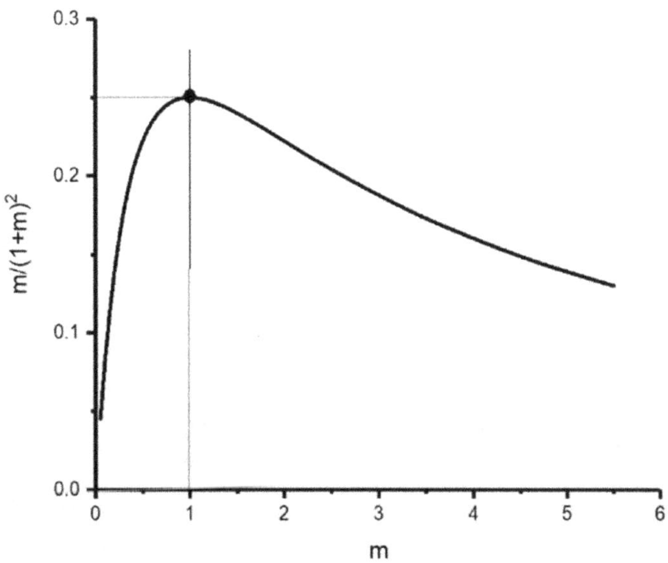

plications where the temperature is significant. However, in order to compensate the temperature effects in the quarter bridge, a dummy strain gauge (R_2) can be connected to the bridge; the dummy bridge is not bonded to the specimen (Figure 8). Both the active (R_1) and dummy (R_2) strain gauge undergo the same temperature, and its effect can be neutralised in the resistance ratio.

Half-Bridge Configuration

In the half-bridge configuration, two active strain gauges are used to measure the strain (Figure 9), and the bridge factor is $n = 2$. Figure 9 allows to observe that the two strain gauges R_1 and R_2 are connected to the top and the bottom of the specimen, respectively. For a given loading direction, R_1 can read the tensile strain and R_2 can read the compressive strain, and they are connected to the adjacent arms in the bridge circuit. However, if they are connected to the opposite side as R_1 and R_3, then the bridge may not read any output, though there will be strain in the loaded specimen. This is due to the configuration of the bridge; the induced change in resistance cancels them in the bridge (Equation 21) and may read zero output voltage. Hence, knowledge of mechanics of material is required in the application of strain gauge to measure strain in the material. Generally, two strain gauges are used only when the signal is low and amplification is needed. Otherwise, for the application of cantilever for bending strain, a quarter bridge can satisfy the need. Further, the half-bridge configuration has better temperature advantage than the quarter bridge.

Figure 7. Wheatstone bridge circuit for quarter bridge configuration

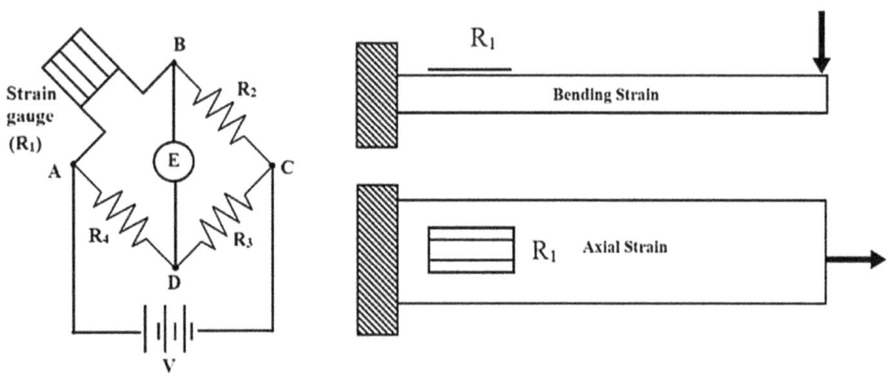

Figure 8. Quarter bridge configuration II

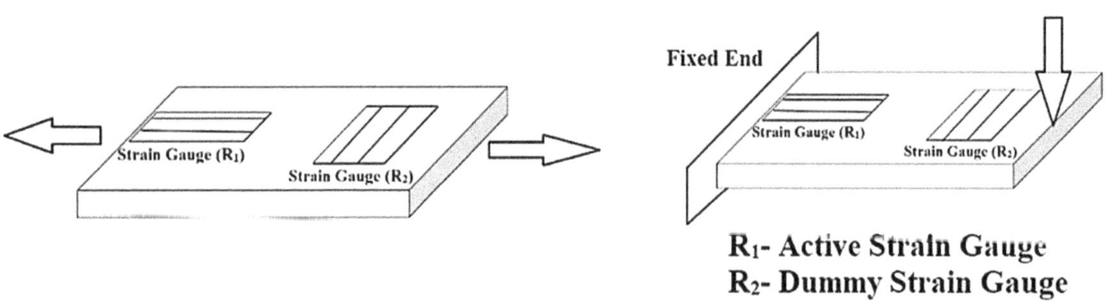

The half bridge can also be used in configuration II, which can measure both axial strain and bending strain, depending on the input load (Figure 10). The R_2 resistance is provided to compensate for Poisson's effect in the strain measurement. This configuration is similar to the quarter bridge configuration II; however, in this the strain gauge is active unlike in the quarter bridge, in which one is active and the other is dummy. Moreover, the dummy strain gauge of quarter configuration II is not bonded to the specimen, while in the half-bridge configuration II both the strain gauges are bonded. For the axial loading, R_2 can also be used to measure Poisson's ratio of the specimen.

Full Bridge Configuration

In the full-bridge configuration (Figure 11), the strain gauges are connected to all four arms of the Wheatstone bridge. The bridge factor for full-bridge configuration is n=4. This configuration is advantageous because the temperature effect is minimal, when compared with other configurations. The concept of connecting the strain gauges in the bridge should follow the principle the authors discussed in earlier sections. The strain gauge measuring the tensile and compressive strain for the given bending load needs to be connected to the adjacent arms, rather than to the opposite arms of the bridge circuit. Knowledge of mechanics of materials is mandatory to connect the strain gauges to the bridges. The full-bridge configuration can also be extended, as the half-bridge configuration II, to measure bending strain, to minimize the temperature effect, and to measure Poisson's ratio.

Figure 9. Wheatstone bridge circuit for half-bridge configuration

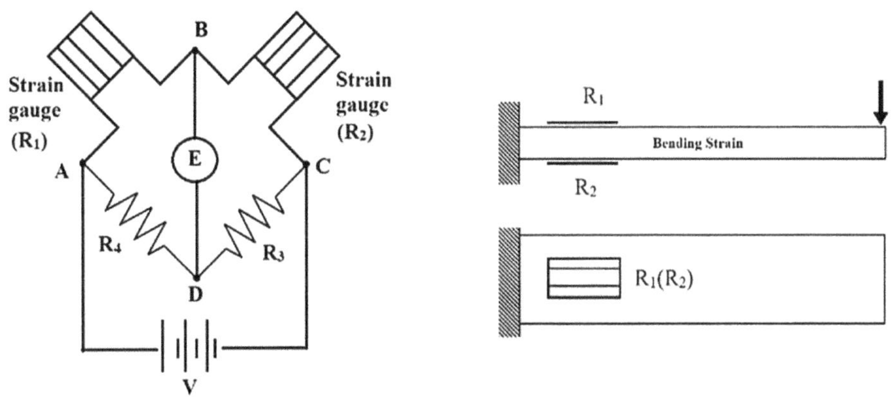

Figure 10. Half-bridge configuration II

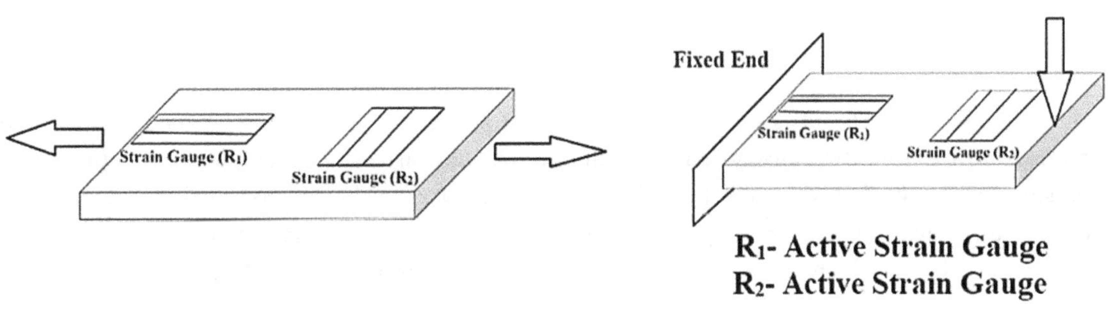

Figure 11. Wheatstone bridge circuit for full bridge configuration

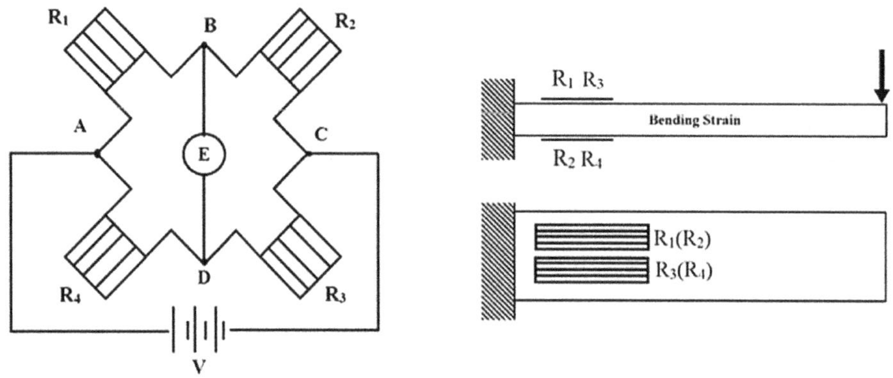

STRAIN GAUGE ACCURACY

The accuracy of the strain measurement depends on several factors. As the authors discussed earlier, it is necessary to be very careful from the beginning and till the measurement, since the strain measurement is very small, hence any small error can be significant in that microscale. The strain gauges are produced with resistance accurate to ±0.3% and the gauge factor accurate to ±0.5%. Further, it is important to bear in mind that strain measurement is also a function of the installation procedure, state of strain being measured, and environment condition. In recent times, the strain gauge technology has reached a development to measure even 0.5 με with high accuracy.

Strain Gauge Linearity, Hysteresis, and Zero Drift

The response of the strain gauge is desired to be linear, and to allow loading and unloading paths to be the same. However, due to factors such as temperature effects, the strain gauge may exhibit hysteresis. This means that the loading and unloading paths are different; further, upon fully unloading the strain gauge, reading will not return to zero and will have a nonzero value called zero drift (Figure 12). In the strain gauge, due to the input voltage V, current flows into the wires of gauge and can induce heat loss called I^2R loss. When the strain gauge is mounted on material with low thermal conductivity, the heat developed in the strain gauge, due to the applied current, is not efficiently dissipated and accumulated in the strain gauge. Hence, such thermal accumulation can cause hysteresis and zero drift. Therefore, the selection of material, adhesive, and backing material has to be conducted carefully, in order to avoid such errors. In this regard, calculations can have a provision to compensate for zero drift.

Strain Rosette

A strain gauge is capable to measure to the strain along its gauge length. However, in many applications, it is required to measure the plane strain at a point on the free surface. In such conditions, three strain components need to be determined. The strain at a point on a free surface is obtained by measuring the strain along three directions. For example, three strain gauges A, B, and C are bonded to the specimen at an angle, and θ_a, θ_b, and θ_c are angles of each strain gauge with respect to the reference x-y axis (Figure

Figure 12. Response of strain gauge

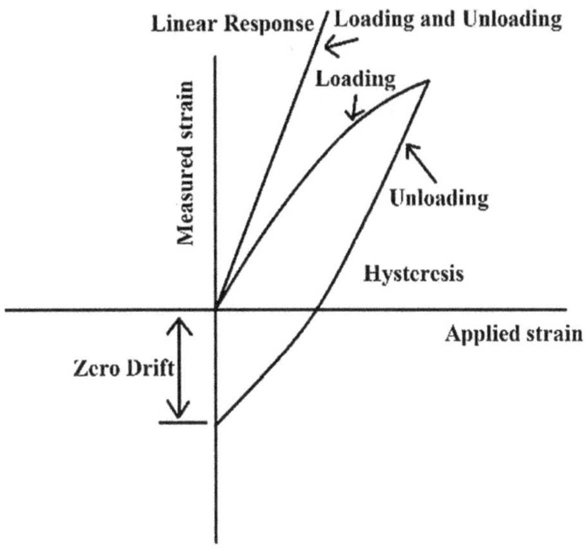

13). In this case, the strain measured by the each strain gauge is ε_a, ε_b, and ε_c, and can be related to the normal and shear strain using the strain transformation law as per Equation 24:

$$\varepsilon_i = \frac{\varepsilon_x + \varepsilon_y}{2} + \frac{\varepsilon_x - \varepsilon_y}{2}\cos 2\theta + \gamma_{xy}\sin 2\theta \tag{24}$$

where ε_i is the strain along any axis *I*, ε_x and ε_y are the normal strains along the reference x and y axis respectively, γ_{xy} is the shear strain, and θ is the angle of the axis ' with respect to the reference x-y axis. For the strain gauge arrangement in Figure13, Equation 24 can be written as follows:

$$\varepsilon_a = \frac{\varepsilon_x + \varepsilon_y}{2} + \frac{\varepsilon_x - \varepsilon_y}{2}\cos 2\theta_a + \gamma_{xy}\sin 2\theta_a \tag{25}$$

$$\varepsilon_b = \frac{\varepsilon_x + \varepsilon_y}{2} + \frac{\varepsilon_x - \varepsilon_y}{2}\cos 2\theta_b + \gamma_{xy}\sin 2\theta_b \tag{26}$$

$$\varepsilon_c = \frac{\varepsilon_x + \varepsilon_y}{2} + \frac{\varepsilon_x - \varepsilon_y}{2}\cos 2\theta_c + \gamma_{xy}\sin 2\theta_c \tag{27}$$

From Equations 25–27, the normal strain along x and y and shear strain can be computed. The transformation law for the strain needs to be well understood, in order to effectively use the strain gauge rosette. The two mainly used strain rosettes are 45° strain gauge rosette and 60° strain gauge rosette. However, the combination of strain gauges can be connected at any angle to find the strain at a point

Figure 13. General strain rosette

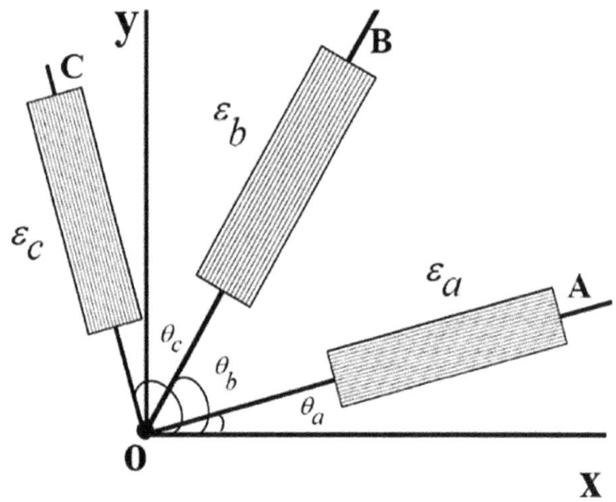

on the free surface, and, accordingly, the transformation law can be applied using the aligned angles of the strain gauges.

Three Element 450 Rectangular Rosette

In this arrangement, two strain gauges are aligned along reference x and y axis and a third strain gauge is kept at 45° to the x-axis (Figure 14). For this configuration, the angle θ_a=0°, θ_b=45°, and θ_c=0°. Hence, the transformation law can be applied by Equation 24, and solved for the normal strains and shear strain (i.e., three equations and three unknown variables). The result is:

$$\varepsilon_x = \varepsilon_a \tag{28}$$

$$\varepsilon_y = \varepsilon_b \tag{29}$$

$$\gamma_{xy} = \varepsilon_b - \frac{\varepsilon_a + \varepsilon_c}{2} \tag{30}$$

60° Delta Strain Gauge Configuration

In this configuration, the three strain gauges are placed as Figure15 shows. The angles are θ_a=30°, θ_b=90°, and θ_c=150°. Upon applying the strain transformation law and solving for normal and shear strain, the result is:

$$\varepsilon_x = \frac{2}{3}\left(\varepsilon_a - \frac{1}{2}\varepsilon_b + \varepsilon_c\right) \tag{31}$$

Figure 14. Rectangular strain gauge rosette

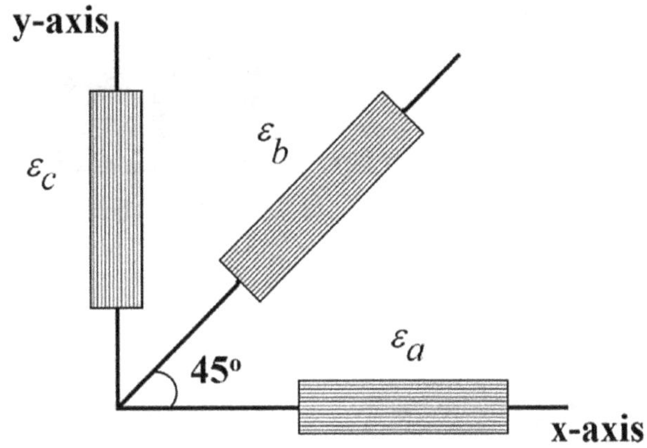

$$\varepsilon_y = \varepsilon_b \tag{32}$$

$$\gamma_{xy} = \frac{1}{\sqrt{3}}\left(\varepsilon_a - \varepsilon_c\right) \tag{33}$$

Prealigned Strain Gauges

The development in strain gauge has reached a level in which prealigned strain gauges are available commercially. This prealignment eases the user in obtaining the strain measurement in the applications. Figure16 shows some of the prealigned strain gauges. The strain measurement can be done by applying the strain transformation law using each strain gauge reading and angle of inclination.

Figure 15. Delta strain rosette

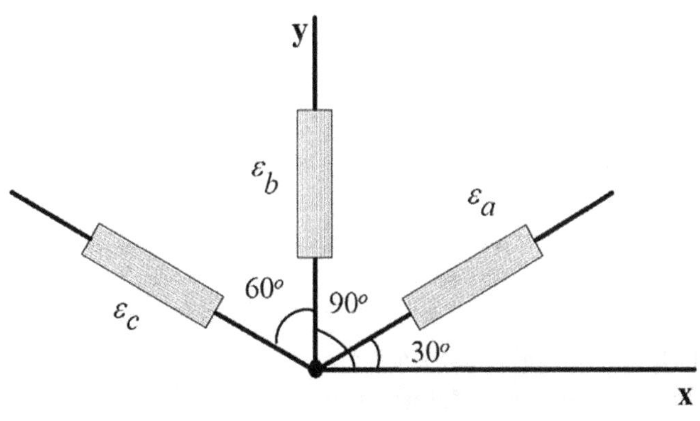

Figure 16. Prealigned strain gauges available commercially

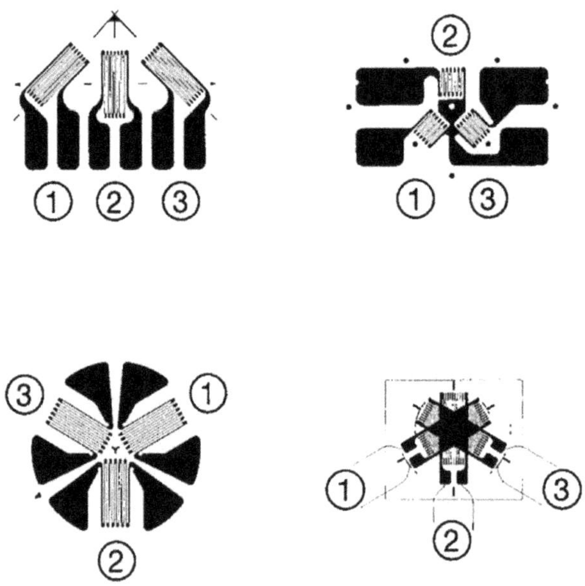

STRAIN GAUGE MATERIALS

The strain gauge materials directly influence the strain gauge sensitivity, hence the knowledge on the strain gauge material is imperative. The materials of the strain gauge backing and adhesive are involved in the proper estimation of strain at a point. The materials used are listed below, with their properties and application area.

Wire Materials

Constantan

Constantan is one of the oldest gauge materials that is widely used, due to its high electric resistivity (0.49 μΩ.m) to achieve a proper resistance for a small gauge length. The strain due to thermal effect is relatively low in the temperature range of 30 to 193ºC (-20 to 380ºF), thus it can be said that it has self-temperature compensation. The strain gauge from this material also has constant sensitivity across wide ranges of strain. The annealed constantan can be used in the plastic region, where strain is greater than 5%. The major drawback comes from its continuous drift in the resistance, when the temperature is raised to about 65°C (150°F), which can affect the strain measurement over a long period of time. Key application areas of constantan alloy wire are:

1. It is suitable mainly for static and quasi-static strains.
2. It is also suitable for plastic or larger deformation.
3. It has limitations in high-temperature applications
4. It is widely used and the least expensive of other gauge materials.

Isoelastic Alloy

Isoelastic alloy has the capability to measure dynamic strain and is used mainly in vibration and impact applications. This material also possesses higher sensitivity, which is 3.6, when compared with 2.1 of constantan; hence, it has improved signal to noise ratio. Further, the key advantage of isoelastic alloy over constantan is its high resistance, which is 350 Ω compared with 120 Ω of constantan. The fatigue property is also better than other strain gauge materials. However, a major drawback of this material is the impact of temperature on its resistance. Indeed, as its resistance is very sensitive to changes in temperature, isoelastic alloy lacks self-temperature compensations, unlike constantan or karma. Secondly, its strain sensitivity reduces from 3.1 to 2.5, when the strain exceeds 7500 μ. Key application areas of isoelastic alloy wire are:

1. It is suitable to achieve high signal to noise ratio.
2. It is suitable for dynamic strain measurement.
3. It is also suitable to measure strains under fatigue or cyclic loading.
4. It needs to be operated in a temperature-controlled room (e.g., air-conditioned laboratory).
5. It is not suitable for strain measurement for a long period of time in temperature fluctuation condition.

Karma

Karma possesses overall properties similar to constantan. Its main advantage is its effective self-temperature compensation property from -73 to 260ºC (-100 to 500ºF). It also has higher cyclic strain resistance than constantan. However, this material is difficult to solder. Key application areas of karma alloy wire:

1. It can be used in low temperature (as low as -269°C/-452°F) or in applications where temperature varies with time.
2. It is difficult to solder.

Backing Materials

A strain gauge is very thin and fragile, hence it is difficult to handle. Backing materials need to be used in order to enable proper handling and damage-free usage of strain gauge. Backing materials are generally dielectric in nature to provide good electrical insulation between the strain gauge wires and the specimen. Polyimide is a widely used backing material and the default standard. Polyimide is used for thin gauges, preferably for 0.025 mm or thinner, and when no high temperature or extreme operating condition occurs. This backing material is suitable for static strain applications. Epoxy is also used as backing material. A major advantage is that the installation and handling of epoxy is well understood, thus the error due to the backing can be reduced. However, it is brittle in nature and requires skilled workmanship for its installation. Glass-reinforced epoxy, which has moderate temperature conditions (~400°C/750°F), is also being developed for fatigue analysis applications.

Adhesives

Adhesives are used to secure a strain gauge to the specimen. This plays a key role in the measurement, as the adhesive should transfer the strain from the specimen to the strain gauge wire materials. The following adhesives are used in strain gauge applications.

Cyanoacrylate Cement

Cyanoacrylate cement is used when it is required to soon bond the strain gauge material to the specimen. This adhesive can take approximately 10 minutes to secure the strain gauge; it requires gentle pressure for the first 2 minutes to facilitate proper adhesion. However, this adhesive is not suitable for long-period applications and is preferred for a wide range of applications.

Epoxy

Epoxy is used in the applications that require higher bond strength and when higher strains are needed to measure which is at the failure of the specimen. Epoxy needs a clamping pressure of 350 kPa for thin bond lines, while clamping pressure of 35 to 145 kPa is applied during the curing process. It needs sufficient time to cure and a temperature of about 120°C for several hours to complete polymerization.

Cellulose Nitrate Cement

Cellulose nitrate cement is used for paper-backed gauges and can be cured by blowing warm air. It is used when the environment is dry, that is when no water or higher moisture content is present. Ceramic cement is preferred in high-temperature, up to 980°C, applications.

CONCLUSION

This chapter discussed the basics of strain gauges and the principle of operations. This chapters also provided the information on the strain gauge construction, errors, strain rosette and the strain gauge materials. This chapter is intended to introduce the basic of strain gauge to the reader and the discussions on the performance of the strain gauge is given in the subsequent chapters. This chapters also shows that the strain gauge is effective in the strain measurement and they are one of the useful and essential gauge in the experimental stress analysis techniques.

REFERENCES

Bakshi, U. A., & Bakshi, A. V. (2008). *Electrical measurements and measuring instruments*. Pune: Technical Publications.

Balaji, P. S., Leblouba, M., Rahman, M. E., & Ho, L. H. (2016). Static lateral stiffness of wire rope isolators. *Mechanics Based Design of Structures and Machines*, *44*(4), 462–475. doi:10.1080/15397734.2015.1116996

Balaji, P. S., Moussa, L., Rahman, M. E., & Ho, L. H. (2016). An analytical study on the static vertical stiffness of wire rope isolators. *Journal of Mechanical Science and Technology*, *30*(1), 287–295. doi:10.100712206-015-1232-5

Balaji, P. S., Moussa, L., Rahman, M. E., & Vuia, L. T. (2015). Experimental investigation on the hysteresis behavior of the wire rope isolators. *Journal of Mechanical Science and Technology*, *29*(4), 1527–1536. doi:10.100712206-015-0325-5

Balaji, P. S., & Yadava, V. (2013). Three dimensional thermal finite element simulation of electro-discharge diamond surface grinding. *Simulation Modelling Practice and Theory*, *35*, 97–117. doi:10.1016/j.simpat.2013.03.007

Bhattacharya, S. K. (2011). *Basic electrical and electronics engineering*. New Delhi: Pearson Education India.

Karthik Selva Kumar, K., & Kumaraswamidhas, L. A. (2015a). Experimental investigation on flow-induced vibration excitation in an elastically mounted circular cylinder in arrays. *Fluid Dynamics Research*, *47*(1), 015508. doi:10.1088/0169-5983/47/1/015508

Karthik Selva Kumar, K., & Kumaraswamidhas, L. A. (2015b). Experimental investigation on flow-induced vibration excitation in an elastically mounted square cylinder. *Journal of Vibroengineering*, *17*(1), 468–477.

Karthik Selva Kumar, K., & Kumaraswamidhas, L. A. (2018). Investigation on stability of an elastically mounted circular tube under cross flow in inline square arrangement. *Iranian Journal of Science and Technology, Transactions of Mechanical Engineering,* 1-13.

Kumaraswamidhas, L. A., & Karthik Selva Kumar, K. (2016). Experimental investigation on stability of an elastically mounted circular tube under cross flow in normal triangular arrangement. *Journal of Vibroengineering*, *18*(3), 1824–1838. doi:10.21595/jve.2016.16708

Leblouba, M., Altoubat, S., Ekhlasur Rahman, M., & Palani Selvaraj, B. (2015). Elliptical leaf spring shock and vibration mounts with enhanced damping and energy dissipation capabilities using lead spring. *Shock and Vibration*, *12*. doi:10.1155/2015/482063

Omega. (2018). *Practical Strain Gage Measurements*. Retrieved May 24, 2019, from https://www.omega.co.uk/techref/pdf/StrainGage_Measurement.pdf

Ramesh, K. (2018). *Experimental stress analysis*. Retrieved June 20, 2019, from https://nptel.ac.in/courses/112106068/

Showa Measuring Instruments. (2018). *Strain Gauges*. Retrieved April 14, 2019, from http://www.showa-sokki.co.jp/English/index_e.html

Stout, M. B. (1960). *Basic electrical measurements*. Englewood Cliffs, NJ: Prentice-Hall.

Chapter 5
Performance of Strain Gauge in Strain Measurement and Brittle Coating Technique

Balaji P. S.
https://orcid.org/0000-0002-6364-4466
National Institute of Technology, Rourkela, India

Karthik Selva Kumar Karuppasamy
https://orcid.org/0000-0001-5431-2542
Indian Institute of Technology, Guwahati, India

Bhargav K. V. J.
National Institute of Technology, Rourkela, India

Srajan Dalela
National Institute of Technology, Rourkela, India

ABSTRACT

The strain gauge system consists of a metallic foil supported in a carrier and bonded to the specimen by a suitable adhesive. Previous chapters discussed the construction, configuration, and the material of the strain gauge. The strain gauge has advantages over the other methods. A strain gauge can give directly the strain value as output. However, in optical methods, it is required to interpret the results. It is also required to be aware that the strain gauge technology is majorly used, and it can also be easily wrongly used. Hence, it is required to obtain the proper knowledge of the strain gauge to get the full benefit of the technology. This chapter covers the majorly on the performance of the strain gauge, its temperature effects, and strain selection. Further, this chapter also covers the brittle coating technique that is used to decide the position of the strain gauge in the applications.

DOI: 10.4018/978-1-7998-1690-4.ch005

INTRODUCTION

Engineering structures are subjected to various loading conditions that results in the development of stress and strain. In the same way, the structures which are in the path of flowing fluid is more susceptible to distortion as a result of the high fluid current, which further causes the concentration of stress and strain on the surface of component (Karthik & Dhas, 2015, 2016, 2018). The stress evaluation is essential for the mechanical components to validate the design and the vibration and thermal effects can also cause stress in the materials and possess challeges to the mathematical methods (Balaji & Yadava, 2013; Balaji et al., 2015; Leblouba et al., 2015; Balaji et al., 2016a; 2016b;). Hence the strain gauge is considered more relialbe techniuque in many of the practical applications. The strain gauge system consists of a metallic foil supported in a carrier and bonded to the specimen by a suitable adhesive. Previous chapters discussed the construction, configuration, and material of the strain gauge. The strain gauge has advantages over the other methods that a strain gauge can give directly the strain value as output however in optical methods, and it is required to interpret the results. It is also required to be aware that the strain gauge technology is majorly used, and it can also be easily wrongly used. Hence it is required to obtain the proper knowledge of the strain gauge to get the full benefit of the technology (Karl., 2000). This chapter covers the majorly on the performance of the strain gauge, its temperature effects, and strain selection, and special gauges for various applications. Further this chapter also covers the brittle coating technique which is used to decide the position of the strain gauge in the applications.

Temperature Compensation in Strain Gauge

In the strain gauge instrumentation, it is required to exhibit ample care right from the selection of the strain gauge until the measurement of strain to obtain the results that can be useful (Hannah, 1992). Any error in the process can easily result in an error in strain value from the strain gauge, and they may not represent the real specimen strain condition. One of the main factors that affect strain reading is the temperature effect. The temperature effect has to be compensated in the strain reading or by bridge circuit connection methodology. This section covers the techniques to handle the temperature effects properly. The strain gauge resistance change due to temperature is given as,

$$\left(\frac{\Delta R}{R}\right)_{\Delta T} = S_g\left(\alpha_s - \alpha_g\right)\Delta T + S_T \Delta T \tag{1}$$

where S_T is gauge sensitivity to temperature, α_s is the thermal coefficient of expansion of specimen material and α_t is the thermal coefficient of expansion of gauge material. Eq.(1) shows how the change in resistance is influenced by temperature. The first term is the differential thermal expansion coefficient between gauge material and specimen. The second term incorporates the change in resistance from the temperature change. There are two methods in which the temperature effect is compensated. Firstly, by adjusting the gauge parameters, so the terms in Eq.(1) cancels each other, and Eq.(1) becomes zero. For this, certain design changes in material or gauge parameters are made in the strain gauge so that the temperature is compensated. The compensation using this method is called Self Temperature Compensation. The strain gauges are generally tailored made for an application so that the temperature change is adjusted by itself without manual involvement. Secondly is by cancelling the influence of the temperature by means of signal conditioner. Wheatstone bridge circuit can be configured suitably so that the overall temperature change is nullified from the resistance ratio.

Self-Temperature Compensated (STC) Gauges

The Self-Temperature Compensated (STC) gauges have the capability to adjust for the temperature change within a range of temperatures. The properties α_s and S_T plays a key role and they are depending on the impurities in the strain gauge alloy and degree of cold work in the manufacturing. In general applications, the impurities are desired to be minimized or completely avoided however in the strain gauge application, and impurities can be controlled to influence the temperature effect. Further from the amount of cold working on the strain gauge material in the production process, α_s and S_T can be controlled suitably. Hence during the metallurgical process, the α_s and S_T can be estimated for the batch of strain gauge and can categorize them suitable for a specific base material accordingly.

Further, the strain gauge material such as Karma and Constantan has self-temperature compensation and suitable for a variety of base material as STC gauge. However, Isoelastic is very sensitive to the temperature, and they lack the ability of an STC gauge. Table 1 gives the thermal coefficients and recommended the STC number for strain gauge applications. These numbers indicate the thermal coefficients of the base material. A strain gauge made of constantan or karma is given with an STC number in industries, and for specific base material, a strain gauge with corresponding STC can be used. For example, for Aluminium alloy, STC recommended is 13. Hence, a strain gauge with STC 13 can be used to avoid temperature effects in the strain reading. The strain gauges are designed and fabricated according to a particular base material to compensate for the temperature effects. Moreover, it is required to observe that the STC number varies from one base material to another and hence STC is base material-specific, and corresponding strain gauge has to be selected. The STC number for many other materials is available in VPG Sensor.

STC gauges are effective only for a range of temperatures, and they do not remove the temperature effects completely however they minimize the temperature effects to provide strain with improved accuracy. The user should be aware that in strain gauge application, it is not possible to blindly use any type of strain gauge for any base material. However in such act, the strainmeter would give a strain reading, but the readings would be of no practical use. Hence the user should be aware of the base material, and appropriate strain gauge must be selected in order to measure the strain with reasonable accuracy. In

Table 1. STC number for various base materials (VPG Sensor, 2019)

Material Description	Temperature Coefficients		Recommended STC number
	Per °F	Per °C	
Aluminium Alloy (2024-T4, 7075-T6)	12.9	23.2	13
Beryllium Copper (Cu 75, Be 25)	9.3	16.7	09
Cast Iron (Cu 90, Sn 10)	10.2	18.4	09
Glass, Soda, Lime, Silica	5.1	9.2	05
Steel, Stainless (Austenitic 304)	9.6	17.3	09
Titanium Alloy (6Al-4V)	4.9	8.8	05

case of absence of STC gauges, then it is also possible to use the normal strain gauges and compensate the temperature effects manual. The thermocouple can be used to measure the temperature in the system then manually temperature effects can be compensated. The manufacturer also provides the calibrating curve indicating thermally induced apparent strain for every strain gauge and this curve can be used for manual temperature compensation. Further, STC is also effective only for a certain range of temperature, and for strain measurement outside the temperature, then calibration curves can be used to compensate the temperature manually.

INSTRUMENTATION FOR STRAIN MEASUREMENT

The strain gauge response is evaluated by connecting it to a strainmeter or strain indicator (Dally et al. 1993). The strain indicators show the strain value as output. Various types of strainmeters are commercially available. A typical strain indicator is shown in Figure 1. It is also required to understand the details of the strainmeter, to fully understand the strain measurement procedure. The strainmeter has inbuilt all the circuits needed for the strain measurement. The manufacturer also provides the manual for step-by-step procedure. It may slightly vary from one strainmeter to another. However general steps would be the same. The first step is selecting the Wheatstone network resistance which is 120 Ω or 350 Ω. This also depends on the resistance of the strain gauge. It is also discussed earlier that to get maximum output, and it is required to have same resistance in all the four arms of the Wheatstone Bridge circuit. Hence this is a critical step in the procedure. Second is the selection of gauge factor of the strain gauge.

Gauge factor is generally provided by the manufacturer for a batch of strain gauges. Further, the strain gauge factor can also be changed to compensate for the transverse effect. Once all the connections and strain gauge factor is set, then comes the important step, balancing the bridge. It is required to balance the bridge at the beginning of the strain measure. Upon loading, the strain value will be indicated in the LCD or LED display. Most of the commercially available strain meter indicates the strain in micro-strain. The strainmeters are required to calibrated periodically using the shunt resistance so as to maintain the proper working and accurate reading. The commercially available strain indicators have provision for storing a large quantity of data and they can portable with battery operated. However, the basic function in all strain indicators remains same.

TWO-WIRE CIRCUIT

In order to measure the strain, the strain gauge needs to be connected to the Wheatstone bridge circuit. The strain gauge is connected to the bridge circuit through the lead wire. As discussed earlier, to obtain the maximum output signal, the Wheatstone required to have all the four arms be the same resistance value. However when connecting the strain gauge via lead wire to the bridge circuit, the resistance of the lead wire is neglected. This resistance of a lead wire can also affect the strain reading. For most of the application, quarter bridge is sufficient for the strain measurement except for the transducer application which uses the full-bridge configuration. A two-circuit is shown in Figure 2. it can be seen from Figure 2, how the strain gauge is connected to the quarter bridge circuit using the lead wire. The resistance of the lead wire is represented as R_L. When connecting the strain gauge to the Wheatstone bridge, the

Figure 1. Typical strain meter used in the applications

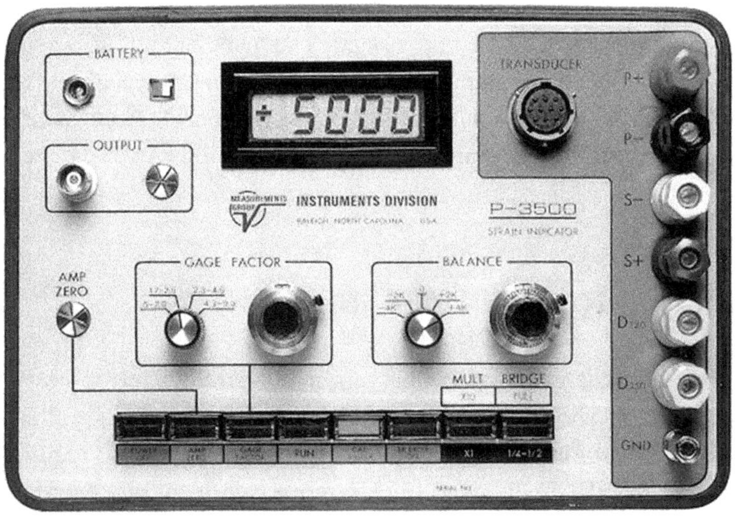

resistance of the arm should be same in order to achieve the initial balancing of the bridge. Further, if the lead wire resistance is small then balance of the bridge is not disturbed, and any addition of lead wire resistance can affect the initial balance and strain reading.

Desensitisation of Gauge Factor

The lead wires used for connecting the strain gauge to the Wheatstone bridge is desired to have zero resistance, but in practice, they have some measurable resistance (R_L). This lead wire resistance is considered as parasitic resistances to the gauge arms of the bridge which effectively reduces or desensitize the gauge factor of the strain gauge. In order to measure a strain at a point, it is ideally desired to have a speck of the material with high resistance, however practical limitation in such high resistance to be in small speck. Generally it is achieved by the wire of material. Hence they have to be foiled to measure the strain. The moment it is foiled it also gets influenced by both axial and transverse strain. With all these

Figure 2. Two circuits for strain gauge applications

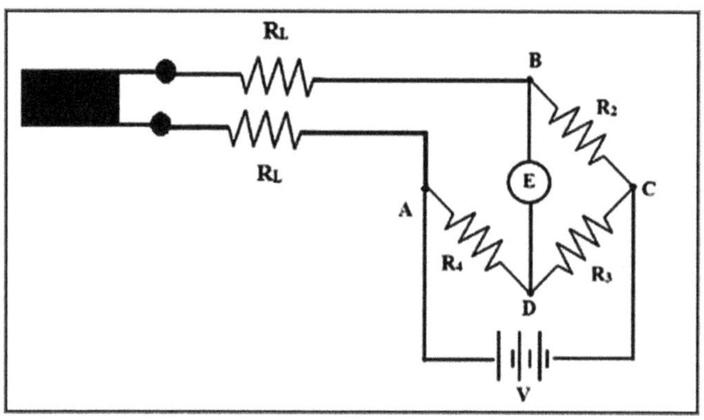

challenges already in place for the strain gauge, now lead wire also adds to these challenges, and this results in the reduced signal output when the specimen is loaded. The lead wires are generally made of copper alloy, which has very less resistance however even such low resistance can influence the strain measurement. The lead wire resistance also adds to the total resistance and hence the bridge balance is affected. When the lead wire resistance is modest, the percentage loss of signal is approximately equal to the lead wire resistance to strain gauge resistance ratio.

The effect of lead wire resistance is understood by measuring the resistance addition and its corresponding strain value. The two-wire circuit adds to the gauge resistance and hence, on the arm of Wheatstone where the strain gauge with resistance (R_G) is connected becomes R_G+2R_L. Lets us consider a 120 Ω strain gauge is used in the specimen, which is located at a distance of 6 m from the strainmeter and connected with a pair of AWG26 (0.4 mm wire diameter) lead wire made up of copper. The total resistance of the specified lead wire at room temperature is 1.7 Ω. The 1.7 Ω of lead wire can be small compared with the 120 Ω of strain gauge, and in general engineering terms it can be assumed negligible. However, in the field of strain, which is in microns, this can be significant. This 1.7 Ω of the lead wire can cause the imbalance of the bridge, and for the strain gauge with the gauge factor 2, the 1.7 Ω is corresponding to 7000με which is very high. Generally material yields around a strain of 2000 με and hence, in this context, 7000 με is very high. The example considers was at a distance of 6 m, which can be when test is conducted inside a laboratory. However, in the real-time evaluation such as on the aircraft body or on top of a tall machine or structure, the lead wire can be of length in terms of several 100 m, and they can significantly affect the reading. This is also evaluated as *loss of sensitivity,* which is the ratio of lead wire resistance to the strain gauge resistance. Further long lead wire can also possess the problems of heating up due to I^2R loss, and the temperature effects can be added up with the errors. Hence in such cases, the initial imbalance of the bridge circuit has to be taken into consideration, and the compensation can be done for the lead wire resistance influence on the strain reading (Ramesh, 2019).

Role of Change in Temperature

The temperature effects on the strain gauge are already discussed in the previous chapter. This section deals with the temperature effect on the lead wire. Generally used lead wire is made up of a copper alloy, which changes its resistance to 22% of their resistance at room temperature for temperature change of 100°F (55°C). Let us consider as a small in change in temperature, which is 10°F (5.5°C) to a lead wire connected to a 120 Ω gauge circuit, can result in an error in strain value of about 156 με. The temperature effect due to the lead wire, loss of sensitivity and gauge factor desensitization are the three main factors from the lead wire that can affect the strain gauge reading. The above factor can be minimized by using high resistance strain gauge such as 350 Ω however; still the lead wire effect cannot be eliminated completely, it can only be minimized. Hence the two-circuit configuration possesses such above discussed difficulties, and this can be overcome by using the three-wire circuit.

THREE-WIRE CIRCUIT

In the three-wire circuit, three lead wires were taken from the strain gauge tab and connected to the Wheatstone bridge circuit (Figure 3). With the addition of one more lead wire in the circuit than the two-wire circuit, the three-wire circuit offers various advantages such as a reduction in loss of sensitiv-

Figure 3. Two circuit for strain gauge applications

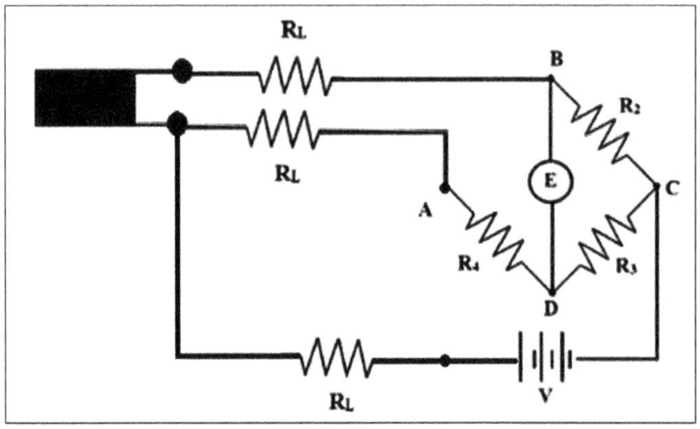

ity, eliminating the bridge imbalance and error due to the temperature effect is minimized. This is a kind of perfect circuit for a single strain gauge in quarter bridge configuration. In this circuit, it can be observed that the lead wires are connected to the adjacent arms R_2 and R_4. Both the lead wires, which are connected to R_2 and R_4, undergo same effect due to the temperature on their resistance, and hence the effects get cancelled in the bridge circuit. This also improves the initial balancing of the bridge. The third lead wire connected to voltage does not play a role in this circuit.

In the three-wire circuit, the first lead wire is in series with the strain gauge, and the second lead wire is in series connection with the arm resistance R_4. If both the lead wires are of the same type, length, and material then their resistance will be equal. Hence such resistance addition will add equally to both sides of the equation in the resistance ratio ($R_1/R_2=R_4/R_3$). Now the bridge is also resistively symmetrical about the horizontal line through the output corners. Even when a strain rosette is used in an application, each strain gauge needs to be connected to a quarter bridge circuit. Hence quarter bridge circuit is a common circuit in the strain gauge application. The three-wire circuit for the quarter bridge circuit is better than the two-wire circuit. Though STC gauges are used for temperature compensation, such STC gauges also need to be connected to bridge through the three-wire circuit to get high accurate strain measurement.

The major advantage of the three-wire circuit is that the bridge remains balanced irrespective of the temperature changes as long the two lead wire connecting the strain gauge and resistance are in the same respective temperatures. The three-wire circuit also reduces the lead wire desensitization by 50% compared to the two-wire circuit. The third wire in the three-wire circuit is for the voltage sensing only, and it is not in series with any of the arm resistance, and hence it does not affect the bridge balance or temperature stability (Ramesh, 2019)

Though a three-wire circuit has serval advantages than the two-wire circuit, for some special application both can also be used combinedly. for example in the rotating shaft slip rings are used to avoid the twisting of the lead wire. These slip rings have a limited number of channels and the three-wire system may have difficulty in connecting with the slip ring due to the increased number of lead wire. Hence in such case, the two-wire circuit can be used to connect the strain gauge to the connecter further three-wire circuit can be between the connector and the measurement instrumentation to minimize the length of the two-wire lead wires.

The three-wire circuit has many benefits compared with two cire circuit. The three-wire circuit maintains initial bridge balance, and temperature effect on the lead wire is compensated. The three-wire circuit also increases the measurement sensitivity, and they are useful in static measurements. However the two circuit also has certain advantages, it can be used to measure purely dynamic strain to measure peak to peak amplitude of a time-varying signal is required and static strain application in which length of the lead wire is kept very small. The two-wire circuit is what user apply by default since the strain gauge has two solder tab. However when the knowledge of the strain gauge is enhanced then it can be understood that the three-wire circuit gives strain reading with better accuracy.

STRAIN GAUGE SELECTION PROCESS

The selection strain gauge plays a key role in the measurement of strain at a point. Hence the stress analyst needs to appropriately select the strain to suit the applications. A single strain gauge can measure the strain alone its gauge length directions and in order to measure the strain at the point on a free surface, it is required to find three components of strain. Each component of strain needs a separate strain gauge to be deployed. In practical application, it is required to find the strain at various points on the structure and each strain gauge needs a channel of the measuring instruments. However, the measuring instruments have limitations in the number of channels. Hence it is challenging to acquire more data with the given limitations in the number of channels (Neubert, 2000).

Due to the limitations in the number of channels and the strain gauge available, it is desirable to identify the stress distribution from others method and apply the strain gauge for the measurement. For example, at some points the stress can be uniaxial, hence at those locations one strain gauge can be deployed further, if the principal stress direction known then the strain gauges can be applied effectively for the measurement. In aerospace applications, photoelastic coating technique is used to identify the critical zone and to estimate the number of strain gauges and their gauge length. Upon identifying the type of gauge required for the measurement, the additional factors that are needs to be considered such as optimizing the performance of the gauge for a specific environment, ease of installation, cost of the gauges and reliability in the strain measurement.

There are challenges involved in the strain measurement using strain gauges. The choice of strain gauges is influenced by many factors and it would be difficult to satisfy all the factors, hence certain factors. In the practical cases the parameters such as strain sensitive alloy, backing material, gauge length, gauge pattern, self-temperature compensation, grind resistance etc, needs to be comprised one for others. Further in addition to the strain gauge parameters, there are operating constraints which need to be taken care of. The operating constraints such as accuracy, stability, temperature, elongation, test duration, ease of instalment and environment can affect the strain gauge selection. There are general guidelines which can be followed for the strain gauge selection. The guideline considers four factors for the selection namely Type of stress distribution, type of strain measurement, characteristics of the anticipated stress field and heat dissipation characteristics of the test materials.

The type of stress defines whether stress is uniaxial or biaxial. Since, based on the type the number of strain gauges can be employed for the strain measurement. The type of strain measurement refers to the static or dynamic strain for a long duration. In these cases, suitable strain gauge needs to be selected and suitability for static or dynamic is usually marked on the strain gauges. The characteristics of the anticipated stress field assist in the selection of strain gauge with proper gauge length. The stress concentration generally needs smaller gauge length. The heat dissipation characteristics of the base material need to be properly understood since for the poor conductor materials, the accumulation of heat can affect the accuracy of the strain measurement. In such case self-temperature compensation strain gauges can be used.

BRITTLE COATING TECHNIQUE

The strain gauge applications to measure the strain requires the knowledge of the point of maximum stress to position the strain gauges. The number of strain gauges that can be used in the experimental work is limited due to variety of reasons hence there was a requirement to reduce the number the strain gauges required and to optimally use the available strain gauges. The basic concept of the brittle coating is by applying a thin brittle coating on the specimen and subjecting it to a loading condition. The failure strain of the brittle coating is designed to be much lesser than the specimen and hence the brittle coating fails before, when the loading is applied. From the place of the first crack origin, the point of maximum stress can be identified. By further allowing the loading on the coated specimen, the pattern of the crack can be obtained which can be used to predict the type of loading and principal stress distribution. The knowledge of the coating strain can also be used to evaluate the stress on the specimen (Hearn, 1971). The brittle coating process is discussed in detail in this section.

Types of Brittle Coating

The brittle coating that is used can be of three types.

Resin-based brittle coating: this is made from 1/3rd of zinc resinate as a base which is dissolved in 2/3rd of the carbon disulphide and a small amount of plasticizer. The plasticizer is used to control the brittleness of the coating, the more plasticizer increases the brittleness of the coating. The dibutyl phthalate is generally used plasticizer in the resin-based coating. The strain sensitivity of the coating is about 0.0003 to 0.0030 and the thickness of about 0.1 mm to 0.15 mm are used on the specimen. This coating can be used up to 60°C.

Ceramic based coating: This coating consists of ceramic particles that are finely grounded and suspended in a suitable solvent. This coating can be sprayed on the specimen. At room temperature these coating gets a chalk-like appearance on the specimen hence this coating needs to be fried at 540°C which makes the particles to coalesce. However, this high temperature can cause detrimental effects on the specimen material such as plastics, aluminium, magnesium etc.

Tens-lac Brittle Lacquer coating: this is the most preferred coating in recent times and it is a highly sensitive, non-flammable and low toxic coating. This coating can be sprayed over the specimen and can measure strain as low as 500 μ cm/cm. This coating may require undercoat if the surface of the test specimen is dark.

Brittle Coating Evaluation Characteristics

In order to select a brittle coating for the application, there are certain parameters that can be used to evaluate the coating. The characteristics such as Strain sensitivity, continuous crack, crazing, the closing of cracks after unloading, crack density and flaking. The minimum strain which is necessary to cause cracking in the coat is defined by the strain sensitivity of the coat. The calibration specimen can be used to measure the strain sensitivity. Further, the first crack that was generated in the coating needs to grow from one boundary to another. This crack pattern can provide the details of the stress distribution and principal stress direction.

Basic Procedure in the Brittle Coating

The basic procedure for the brittle coating is discussed in this section.

Step 1: *Specimen surface preparation*. The test specimen surface should be cleaned for dirt, grease or any other impurity on the surface.

Step 2: *Select the appropriate coating*. It is highly essential to select the suitable for application which is based on the temperature of the condition, strain sensitivity required, humidity, type of specimen materials etc.

Step 3: *Application of the brittle coating*. The coating can be applied to the test specimen using the spraying technique. The spraying is preferred to be carried out at 15 cm from the test surface.

Step 4: *Curing of the Coating*. The coating applied needs to be cured. The curing time depends on the type of coating. Usually manufacture provides the curing time required for the coating.

Step 5: *Loading the coated specimen*. The maximum load that can act on the test specimen needs to evaluated and upon estimated, the load increments to be applied on the specimen needs to decide. In the brittle coating, the main purpose is to identify the point of the first crack to determine the point of maximum stress region in the specimen. Hence the load increment needs to be small and after every increment of the load, the specimen needs to be inspected for the crack formation.

Step 6: *Loading the calibration strip*. In order to calibrate the brittle coating, a calibration specimen also needs to be coated along with the test specimen. The calibration specimen also needs to be dried and a similar procedure which has to be simultaneously followed on the calibration specimen. The calibration can be applied with the known load and the failure strain of the brittle coating can be evaluated and can be used to estimate the stress on the specimen after the crack starts to appear on it.

Step 7: *Inspection of the cracks*. The techniques such as dye etchants or statiflux can be used to improve the visibility in the crack inspection.

Crack Inspection Method

The crack in the brittle coating is very small and the width is in the order of 0.05 mm to 0.075 mm. At this small dimension, the crack is difficult to visible to the naked eye. Hence methods are generally used for the crack inspection.

Focus light method: A focused light is directed on the surface and inspection is carried out on the test specimen. This method is suitable for the small area however for the large area, it can be time-consuming.

Statifulx Method: This is kind of electrified particle inspection methods. In the methods a statiflux penetrant is applied on the brittle coated surface after loading is applied. The statiflux penetrant gets into the crack formed from the loading. Then the statiflux penetrant is cleaned superficially on the surface however the penetrant trapped in the crack remains on the specimen. The statiflux penetrant possesses negative potential. An electrified charge particle is blown over the surface. The positive charge particle is attracted by the penetrant trapped in the crack and hence particles accumulate on the crack region.

Dye etching Method: The visibility of the crack can be increased by applying a suitable red dye over the coated surface. The dye etchant consists of a mixture of turpentine, machine oil and red dye. The dye is applied on the coated specimen and the dye gets penetrated into cracks. After five minutes, the dye can be removed using the emulsifying agent. The crack will be distinctively appearing on the coating. This dye needs to be used after all the loading since the dye can reduce the sensitivity of the brittle coating.

Refrigeration Technique

In the cases where the loading is small and not sufficient to cause the cracking on the brittle, the refrigeration technique can be used to identify the point of maximum stress. In this technique, the coated specimen is loaded as per the requirement, then the loaded specimen is subjected to the rapid temperature drop. The drop in temperature induces the hydrostatic tension on the coating which is superimposed over the existing load. The combined is generally causes the crack on the coating. In order to achieve the rapid temperature, drop either of the two methods are used.

1. Ice water can be flushed over the coating however this method is not very successful
2. Compressed gas can be made to flow over the dry ice and focused on the test coated surface. The cold compressed gas is produced at a rapid temperature drop. This technique is relatively successful over the ice water method.

Calibration Technique

The calibration technique is used to evaluate the failure strain of the coating. The cantilever loading is performed by applying known load on the calibration specimen to evaluate the failure strain of the coating. The calibration specimen is also together coated and made to dry to the same time along with the test specimen. The calibration specimen is tested using the cantilever set up in which one end of the calibration specimen is fixed and the other end is kept free. The load is applied on the free end in small increments to observed for the crack. The care should be taken in not to coat at the ends of the calibration specimen which are fixed end and loading end. The loading time of the calibration specimen is maintained as the test specimen.

Variables Influencing the Coating

There are many factors that influence coating behaviour in this technique. The major factors are listed and discussed.

Number and type of coating: The type of coating has to be appropriately selected for the given application since each coating has its own strain sensitivity. The majorly the coating is expected to fail before the test specimen and the failure strain of the coating is decided by the user.

Effect of spraying unit: the coating is generally sprayed over the test specimen. The spraying needs to be uniform over the surface. Type of coating can vary the spraying method. The poor spraying can cause the trapping of air bubbles and hence care should be taken in the spraying.

Effect of application of coating: the speed of the spraying and distance of the spraying unit from the surface can also affect the coating behaviour. Generally, a distance of 15 cm is maintained to spraying the coating and laterally, the spraying is moved uniformly at a velocity of 1.5 m/s to 2 m/s.

Influence of heat treatment: The coating sprayed on the test specimen is made to dry for a specified time as mentioned by the coating manufacturer. The rapid cooling of the coating can cause cracks on the coating up to drying and hence suitable drying time has to provide for every coating.

Testing condition: the humidity and test temperature can affect coating behaviour. The humidity can cause a delay in drying and the coating when observe humidity can behave differently. The manufacturer prescribes humidity condition and temperature range for every coating and hence care should be taken in performing the coating as per the manufacturer recommendation.

Advantages of the Brittle Coating

This method can provide the point of maximum stress and the direction of the principal stress.

1. The same prototype can be used to test and it doesn't require a model for the test. Hence the results are highly accurate and represent the real behaviour of the test specimen.
2. There is no load simulation problem.
3. The analysis of data is simple and accurate. However the accuracy of the data also depends on the number of variables.

REFERENCES

Balaji, P. S., Leblouba, M., Rahman, M. E., & Ho, L. H. (2016a). Static lateral stiffness of wire rope isolators. *Mechanics Based Design of Structures and Machines*, *44*(4), 462–475. doi:10.1080/15397734.2015.1116996

Balaji, P. S., Moussa, L., Rahman, M. E., & Ho, L. H. (2016b). An analytical study on the static vertical stiffness of wire rope isolators. *Journal of Mechanical Science and Technology*, *30*(1), 287–295. doi:10.100712206-015-1232-5

Balaji, P. S., Moussa, L., Rahman, M. E., & Vuia, L. T. (2015). Experimental investigation on the hysteresis behavior of the wire rope isolators. *Journal of Mechanical Science and Technology*, *29*(4), 1527–1536. doi:10.100712206-015-0325-5

Balaji, P. S., & Yadava, V. (2013). Three dimensional thermal finite element simulation of electro-discharge diamond surface grinding. *Simulation Modelling Practice and Theory*, *35*, 97–117. doi:10.1016/j.simpat.2013.03.007

Bhattacharya, S. K. (2011). *Basic Electrical and Electronics Engineering*. Pearson Education India.

Dally, J. W., & Rilley, W. F. Mc Connell and Kenneth G. (1993). Instrumentation for Engineering measurements (2nd ed.). John Wiley &Son.

Hannah, R. L., & Reed, S. E. (1992). *Strain Gage users Manual*. Society for Experimental Mechanics.

Hearn, E. J. (1971). *Brittle Lacquers for Strain Measurement*. Watford, UK: Merrow Publishing Co.

Holister G. S. (1967). *Experimental Stress Analysis* (C.V.P) 63rd yr. 2, 33-41,

Karl, H. (2000). *An Introduction to Measurements using strain gauges*. Druckerei Drach Press.

Karthik Selva Kumar, K., & Kumaraswamidhas, L. A. (2015). Experimental Investigation on Flow-Induced Vibration Excitation in an Elastically Mounted Circular Cylinder in Arrays. *Fluid Dynamics Research*, *47*(1), 015508. doi:10.1088/0169-5983/47/1/015508

Karthik Selva Kumar, K., & Kumaraswamidhas, L. A. (2015). Experimental investigation on Flow-Induced Vibration Excitation in an Elastically Mounted Square Cylinder. *Journal of Vibroengineering*, *17*(1), 468–477.

Karthik Selva Kumar, K., & Kumaraswamidhas, L.A. (2018). Investigation on Stability of an Elastically Mounted Circular Tube under Cross Flow in Inline Square Arrangement. *Iranian Journal of Science and Technology, Transactions of Mechanical Engineering*, 1-13.

Kumaraswamidhas, L. A., & Karthik Selva Kumar, K. (2016). Experimental investigation on Stability of an Elastically Mounted Circular Tube under Cross Flow in Normal Triangular Arrangement. *Journal of Vibroengineering*, *18*(3), 1824–1838. doi:10.21595/jve.2016.16708

Leblouba, M., Altoubat, S., Ekhlasur Rahman, M., & Palani Selvaraj, B. (2015). Elliptical Leaf Spring Shock and Vibration Mounts with Enhanced Damping and Energy Dissipation Capabilities Using Lead Spring. *Shock and Vibration*, *12*. doi:10.1155/2015/482063

Neubert, H. K. P. (2000). *Strain Guages: Kinds and Uses*. Macmillian.

Ramesh, K. (2018). *Experimental Stress Analysis*. IIT Madras.

Chapter 6
Measurement of Strain Using Strain Gauge and Piezoelectric Sensors

Abhishek Kamal
Indian Institute of Technology, Guwahati, India

Vinayak Kulkarni
Indian Institute of Technology, Guwahati, India

Niranjan Sahoo
Indian Institute of Technology, Guwahati, India

ABSTRACT

Today, measurement of strain plays a crucial role in different areas of research such as manufacturing, aerospace, automotive industry, agriculture, and medical. Many researchers have used different types of strain transducers to measure strain in their relevant research fields. Strain can be measured using mainly two methods (i.e., electrical strain sensors and optical strain sensors). Electrical strain sensors consist basically of strain gauges, piezo film, etc. In electrical strain sensors, the strain gauge is one of the oldest and reliable strain sensors which are available in different types (i.e., wire strain gauge, foil strain gauge, and semiconductor strain gauge). Piezofilm is also playing an important role in the field of strain measurement due to easy availability and less cost.

INTRODUCTION

Measurement of strain plays a crucial role in various areas of research, including manufacturing, aerospace, automotive, agriculture, and medical. Many researchers have used strain transducers to measure strain within their relevant fields. Strain can be measured using the electrical strain sensor and optical strain sensor. Electrical strain sensors are the oldest and most reliable sensors. These include the wire strain gauge, foil strain gauge, and semiconductor strain gauge. Piezo film, which also plays a vital role in the field of strain measurement, is readily available and more affordable. Optical strain sensors, which are

DOI: 10.4018/978-1-7998-1690-4.ch006

based on fiber Bragg grating (FBG) technology, are often used in monitoring fields like civil engineering structures and wind turbines. This chapter discusses strain measurement sensors and their applications. Most researchers have studied the measurement of strain using the sensors discussed in this paragraph.

Regarding hypersonic force measurement, strain gauges and piezo films are commonly used to detect stress wave propagation in a solid bar. Storkmann et al. (1998) designed a six-component strain gauge force balance for three models: (1) pointed cone; (2) Apollo CM capsule; and (3) delta wing configuration ELAC I. A pointed cone and Apollo CM capsule model were tested in the Aachen shock tunnel TH2. The capsule model was also tested in the von Karman Institute Longshot Wind Tunnel facility.

Robinson, Martinez Schramm, and Hannemann (2011) designed a three-component stress wave force balance using the cone model. The stress wave balance, in coordination with the model, was mounted in the high-enthalpy free-piston shock tunnel for experiments.

Wang, Liu, and Jiang (2016), using a finite element method, designed several pulse-type strain gauge balances for three components to optimize characteristics. Force tests were conducted for a large-scale cone with a 10° semivertex angle and a length of 0.75m in the JF12 long-test-duration shock tunnel. The finite element method was used for the analysis of the vibrational characteristics of the model-balance-sting system (MBSS) to ensure a sufficient number of cycles, particularly for the axial force signal during a shock tunnel run. The higher-stiffness strain gauge balance in the test shows excellent performance. The frequency of the MBSS increases due to the stiff construction of the balance. Based on the similar concept of strain measurement, another strain sensors, polymer piezoelectric film, was employed by various researchers in the field of strain measurement.

Duryea and Martin (1968) developed a six-component piezoelectric force balance (slender configuration only) for measuring aerodynamic forces and moments in high-speed flow at Cornell Aeronautical Laboratory, New York. The balance was designed using H-type, E-type, and K-type piezoelectric sensor (PZT) material. It was compared based on feasibleness and high load capabilities.

Min, Yang, Qiu, Zhong, and Pi (2018) designed a new three-component fiber optic force balance based on the micro-electro-mechanical-systems (MEMS) Fabry-Perot (FP) strain sensor and test in a low-density hypersonic wind tunnel. The researchers aimed to achieve a faster response, higher sensitivity, and antielectromagnetic interference as compared to conventional strain gauge balance. It was observed that the fiber optic balance calibration and wind tunnel test had good agreement with the results of traditional strain gauge balance. In addition, the MEMS FP strain sensor was suitable for the application of hypersonic force measurement.

STRAIN

When objects like metallic bars, columns, or beams are subjected to external forces, an internal resistance is generated inside that object. This, in turn, resists external forces. This internal resistance is termed "stress." The dimensions of the object (i.e., length, width) change due to this stress. The ratio of change in dimensions to original dimensions is termed "strain." For example, if a metallic bar with length "L" and Diameter "D" is subjected to an external force, the the amount of change in length is "dL." Therefore, the strain produced in the metallic bar is defined as:

Strain (ϵ) = change in length/original length
= dl/l

Figure 1. Tensile, compressive, and shear strain

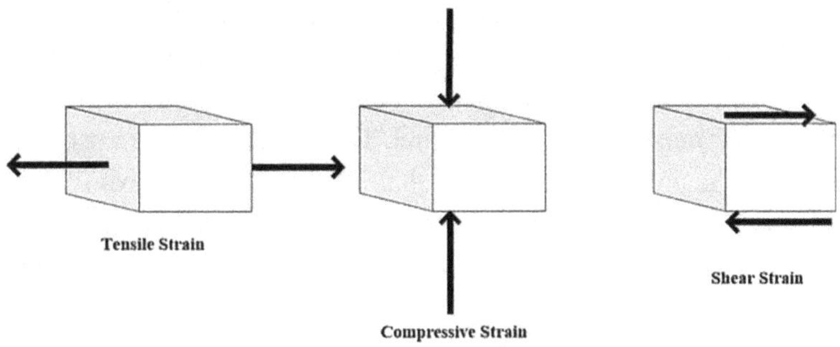

There are two types of strain:

1. **Normal Strain:** This strain develops due to an external force that is applied normally to the surface of the object. It can be tensile, compressive, and volumetric.
2. **Shear Strain:** This strain develops due to an external force that is applied tangentially to the surface of an object. The object undergoes an angular displacement, which measures the shear strain. Strain in the material is shown Figure 1.

MEASUREMENT OF STRAIN

Measurement of strain is necessary to find the level of stress in structural components, forces on the aerodynamic model, etc. The need to measure strain becomes crucial per the area of research. Empirical relations, including Hook's law, are used to evaluate stress. Strain, a dimensionless parameter,can be measured in microns (μm/m or 10^{-6} m/m). There are several types of strain sensors used to measure strain. The following subsections discuss the two broad categories found in strain.

Strain Gauge Sensors

Strain gauges are the most common sensors used to measure strain in materials. At one point, strain gauges were broadly used for experimental stress analysis of objects at a state of loading. Today, strain gauges are exposed to aerodynamics and manufacturing, as well as the medical field. In the field of stress analysis, stress can be calculated directly from the measured strain using material laws like Hook's law in the elastic region (Keil, 2017). However, it cannot be used for elastoplastic deformation. When the load on a material extends beyond the yield point, material laws cannot be applied directly to the calculated stress.

As a mechanical quantity, stress cannot be directly measured. An advantage of strain gauges is the capability to convert strain into an electrical signal. When a material experiences deformation due to external load, the strain measured as a change in resistance produces a strain gauge. This resistance changes the results to a voltage output of a Wheatstone bridge circuit. It is very easy to process the mechanical quantity (or strain in the form of electrical quantity).

Wheatstone bridge circuit is used to acquire a strain signal in terms of electrical signals like a half-bridge, full-bridge, and quarter bridge. Each Wheatstone bridge has an individual capability to acquire the strain signal with less signal-to-noise ratio. It also compensates for undesirable effects like temperature changes. Wheatstone bridge circuit is an excellent technique for compensation. Today, the development of compensation techniques is crucial to the proper signal. These techniques are used to design transducers because it is calibrated into the unit of measured mechanical quantity. Regarding the operating principle of strain gauge, the changes of material surface behavior are transferred to an electrical conductor. This conductor is glued over the surface, changing resistance according to changes in material behavior. These changes in resistance result in precise measuring changes in the strain.

The first strain gauge was developed in the United States in 1940. It was fabricated with a mean-shaped metal wire and glued to a paper carrier. Afterward, many manufacturers designed with different shapes and sizes. Another manufacturer designed a new type of strain gauge called the wrap-around strain gauge. This gauge consisted of a flat auxiliary carrier and resistance wire. The resistance wire was wrapped to the flat auxiliary carrier. The combined unit was glued to the carrier. In a wrap-around strain gauge, the wire's length was not parallel to the measuring direction.

Another development in the field of strain measurement, the foil strain gauge, was introduced in the 1950s. The foil-type strain gauge exposed an area of experimental stress analysis. It has dominated the field over the past several years due to the construction of transducers. This type of strain gauge is easily fabricated and similar to previous gauges. However, it has differences in the grid size and shape. The material of the foil-type strain is reliable for strain measurement. Today, the foil-type strain gauge is the most usable and secure sensor in the field of stress analysis and transducer construction. A five-micrometer thickness of the measuring grid material is supported with a carrier material made of plastic. At 25 micrometers thick, its shape is produced by a photochemical etching process.

Strain gauge measures the mechanical strain in a longitudinal direction of the measuring grid. The strain developed in a material produces a change in strain gauge resistance. This is connected in the Wheatstone bridge circuit. The electrical bridge circuit is imbalanced due to the change in resistance of the strain gauge. This imbalance in the data acquisition system is termed "voltage signal."

Generally, a strain gauge identifies the effective strain developed on the surface of a measuring object at the point of gauge installation. Strain gauge measures a mean strain along the measuring direction using an integral calculation of a strain gradient along the same direction of an inactive length of the measuring grid (see equation 1).

$$\varepsilon_{SG} = \int_{x_1}^{x_2} \frac{\varepsilon_x}{x_2 - x_1} dx \tag{1}$$

where ε_{SG} = average strain of strain gauge at point of installation, x_1 and x_2 = coordinate along the longitudinal direction. This defines the active length.

Material of Strain Gauge Components

Strain gauge is broadly made up of the carrier and measuring grid components. Initially, paper was used as a carrier material because it was easily bonded to a measuring object. A significant disadvantage of a paper carrier is its moisture sensitivity. It was found that paper is not a suitable carrier for measuring a strain gauge.

An ideal carrier material should not be thick. It must easily transfer the strain from measuring object to measuring grid. It should also have the capability to hold a measuring grid and transfer disturbance without a loss. The thickness of the carrier material must be uniform and fixed on a measuring object with proper adhesive. The fixing of the carrier material should be chosen in such a way that there should not be any tear or damage under the static and dynamic loading. The carrier material of strain gauges is varied based on the field of application. For example, the carrier material for experimental stress analysis differs from the transducer.

Similarly, the selection of measuring grid material is crucial. Many scientists carried out experiments to find suitable materials for measuring the grid. In the early 1930s, Czerlinsky found the relationship between change in resistance and mechanical strain in metal. It was found that the Cu-Ni alloy and constantan performed well for measuring purposes. The first metal measuring grid material (or constantan) was used by Ruge and de Forest for the manufacturing of strain gauge.

Measuring grid material should have the following properties for better responses:

1. High sensitivity so that small disturbances in material give a massive change in resistance
2. The capability to obtain a linear relationship between mechanical strain and change in resistance with minimum hysteresis
3. Resistance to measuring grid material should not affect change in temperature
4. Measuring grid material should have high fatigue strength.

Nickel chromium, isoelastic, platinum, and tungsten alloy are generally used as measuring grid materials.

Table 1. Carrier materials and their advantages

Carrier Material	Advantages
Phenolic Resin (Bakelite)	Resistant to moisture High creep resistance Working temperature (-70°C to 200°C) Young modulus 7500MN/mm^2
Acrylic Resin	Young modulus 3500Mn/mm^2 Sustains up to 100°C
Epoxy Resin	Suitable for high-quality transducers High mechanical strength Working temperature (-45°C to +95°C) Young modulus 3500Mn/mm
Epoxy Phenolic Resin	Properties of both Epoxy and Phenolic Works at high elevated temperature (-75°C to +205°C) without glass fiber reinforcement and (-269°C to +290°C) with glass fiber High mechanical strength. High creep resistance
Polyimide	Robust and flexible Suitable for experimental stress analysis Working temperature (-195°C to +175°C) Maximum extensibility up to 20%
Metal Carriers	Suitable for high-temperature applications Very resistant to radioactive radiation Working temperature limits up to 650°C Usually welded to measuring object

Sensitivity of a Strain Gauge

The sensitivity of an instrument is the change in output signal divided by the change in input signal (measuring quantity). In the case of the strain gauge, the input signal is the strain ε developed in a material. Output signal is the change in resistance ΔR/R. These quantities have a proportionality relation to each other. The proportional constant is termed as "gauge factor" of the strain gauge. It is denated by k:

$$k = \frac{\frac{\Delta R}{R}}{\varepsilon} \tag{2}$$

where k is the gauge factor.

In 1856, Lord Kelvin exposed that metal wire changes resistance due to a disturbance in the material (the strain). Afterward, Different researchers used other materials, including Cu-Ni, to measure the strain. Grid material with gauge factor is given in Table 2.

Strain Gauge Behavior on Shock Load

Strain gauge can monitor material's strain due to shock load. Examples include testing of the aerodynamic model in a shock tunnel or high strain rate measurement. The proper estimation of material behavior due to shock waves is complicated in a high-speed application. For example, in a shock tunnel experiment, the shock wave travels with a higher Mach number. This results in a short duration for the interaction of the shock wave with the aerodynamic modelfor small period of time. Strain gauge should have the capability to sense material disturbance within the short duration.

Gauge factor, a precious parameter of the strain gauge, can be used as a proper selection in different areas of an application. The strain gauges, with a lower gage factor (approximately 1-10), are used in longer tests like the bending of material. The strain gauges with higher gauge factors (> 200) are used for high-speed applications like force measurement over an aerodynamic model.

TYPES OF STRAIN GAUGE

Mechanical Strain Gauge

The mechanical strain gauge is a simple, straightforward strain gauge used to measure deformations in a material. It measures the elongation of blades. The value of this elongation scale is magnified 1,000

Table 2. Gauge factors of grid material

Material	Composition	Gauge Factors
Constantan "Eureka"	56% Cu, 44% Ni	Around 2.15
Nichrome	80% Ni, 20 Cr	In between 2.1 and 2.5
Platinum – Tungsten	92Pt, 8W	Around 4.0

times. Huggenberger developed a strain gauge with both a movable and immovable blade. It also has a pointer and scale in which we can observe the value of deformation in the material. The movement of a lever activates as the deformation occurs. However, these strain gauges are massive and tough to use.

Semiconductor Type Strain Gauge

This strain gauge is easily manufactured using semiconductor filaments like silicon or germanium. It is based upon the piezoresistive effects. In 1970, the first semiconductor was developed for the automotive industry using silicon and germanium crystal. The length and thickness of the filament were around 10 mm and 0.05 mm, respectively. The filament of the semiconductors is bound on insulating materials like Teflon. Gold leads connect the semiconductor filament with the electrodes. The process of vapor deposition creates electrodes for the semiconductor type of strain gauges. When the strain is experienced by the semiconductor filament, a massive change in resistance occurs within the strain gauge. This can be measured with the Wheatstone bridge circuit. The change in resistance depends on the material of the semiconductor element. The massive change in resistance gives a high degree of accuracy in strain measurement. Generally, the semiconductor strain gauge has a gauge factor and resistance around 130 and 350ohm respectively. The semiconductor type strain gauge has the following advantages:

1. Higher sensitivity like higher unit resistance
2. Small strains measured in microns due to high gauge factor
3. Few hystereses
4. Very high output due to change in resistance is very high even with slight deformation
5. Works with high-frequency responses (up to 1012 Hz)

Resistance-Type Strain Gauge

These strain gauges are often used for experimental stress analysis. The operating principle of the strain gauge was given by British physicist William Thompson in 1856. According to the principle, the cable resistance (copper or iron) will change accordingly when the wire is compressed or stretched by an external disturbance. The resistance-type strain gauge consists of a thin wire in the form of the grid. This wire is glued between thin sheets of paper. The strain gauge is formally bonded with the surface where the strain is to be measured to provide energy with electrical current. When the deformation occurs over a surface, the gauge follows the deformation in the form of stretching or contracting (depending on type of load acting on a surface). Due to this deformation, the change in resistance is found in the strain gauge. This change in resistance is amplified with a proper amplification factor. Then, it is converted into the strain after calibration. The resistance-type strain gauge has the following advantages:

1. Inexpensive
2. Less mass and highly sensitive
3. Small gauge length and physical size
4. Works in variety of environmental conditions
5. Suitable for measuring rapidly varying strain like a rotating shaft

STRAIN GAUGE INSTALLATION

Strain gauge performance depends on proper installation. An adequately installed strain gauge gives reliable output data. A suitable adhesive is needed for a proper installation between the strain gauge and measuring object material. This material varies per the application. For example, a ceramic bonding agent is used for elevated temperatures and spot welding. Therefore, the quality of installation affects the accuracy of the measurement.

There are important characteristics related to strain gauge, gauge factor, and creep. These important components are present at the measurement point (i.e., strain gauge, adhesive, cleaning agent, solder, etc.). Reliable output requires proper tools and careful installation of the strain gauge at the measurement point. All dust particles at the measurement point of the object must be removed before the installation of the strain gauge. Dust can reduce the effect of the adhesive material, leading to incorrect measurements.

After mounting the strain gauge, the wire is soldered on a strain gauge solder terminal. The soldering wire should not touch the carrier film. During the soldering, both solder terminals of strain gauge cannot touch. After the connection of the strain gauge, the Wheatstone bridge circuit connects the strain gauge to the data acquisition system.

Wheatstone Bridge Circuit

English scientists, Hunter-Christie and Wheatstone, invented the bridge circuit to measure the resistance of metal connecting wires. The main advantage of the bridge circuit is that it can measure resistance with unstable voltage sources. Today, the Wheatstone bridge circuit is widely used in the field of strain measurement. Its structure consists of four arms. Each arm has one resistance, which defines as R_1 to R_4. The resistances are arranged on four sides of the parallelogram. The working principle of the bridge is that if two opposite points of the circuit are excited by an external voltage source, then the voltage on the other two points is dependent on the ratio of resistances of bridge arms. If the ratio of resistances is equal or identical, the output voltage is shown as zero. At that time, the bridge is fully balanced. If there is a change in one resistance, the effect will be shown on the output voltage.

This circuit is useful in strain measurement. One arm of the bridge is replaced by the strain gauge. As discussed, due to change in one resistance, the output voltage will have a disturbance. The disturbance in a circuit is measured by the strain. If the voltage supply from point 1 and point 4 is termed an "input voltage" (V_I), the "output voltage" (V_O) is obtained at point 2 and point 3. The value of V_O becomes zero when all four resistances are identical (i.e., $R_1 = R_2 = R_3 = R_4$). The wiring of bridge circuits should work in a proper manner so that it can be easily identified. All wires should be different colors. When V_O has some value, the circuit is unbalanced for the given input voltage (i.e,. V_I). By applying the Kirchhoff law, the Wheatstone bridge circuit gives the following relationship:

$$\frac{V_1}{V_0} = \frac{R_1}{R_1+R_2} - \frac{R_3}{R_3+R_4} \tag{3}$$

When a disturbance is found in the Wheatstone bridge circuit during strain measurement, the following is a basic equation for measurement with strain gauges:

$$\frac{V_1}{V_0} = \frac{\Delta R_1 - \Delta R_2 + \Delta R_3 - \Delta R_4}{2(2R_0 + \Delta R_1 + \Delta R_2 + \Delta R_3 + \Delta R_4)} \tag{4}$$

where R_0 is a nominal resistance of all strain gauges before the resistance changes occur (i.e., $R_1 = R_2 = R_3 = R_4 = R_0$).

When measuring strain in the Wheatstone bridge circuit, all bridge arms should have strain gauges. However, this is not always required. Therefore, it is possible that one or two indicators of strain will mount on bridge arms and other arms fixed with constant resistance. This combination develops different types of Wheatstone bridge circuits, including half-, full-, and quarter-bridge circuits. These circuits can be used for reliable results in strain measurement.

In high-speed applications, it is found that the half-bridge circuit gives better results. Full-bridge circuits have all arms with a strain gauge. However, in a half-bridge circuit, only two arms of bridge circuits have strain gauges. The other arms have constant resistances. Similarly, in the quarter-bridge circuit, one arm of the bridge circuit is mounted with the strain gauge. The other arms have constant resistances.

Piezoelectric Strain Sensors

Piezoelectric sensors work on direct and converse piezoelectric effects. The piezoelectric effect was observed by the two brothers, Pierre and Jacques Curie, in 1880. They discussed the relationship between mechanical load and electrical polarization. Lippmann predicted the existence of a converse piezoelectric effect based on thermodynamic considerations. Lippmann noted that crystal will deform when an electrical voltage is applied to certain faces of a crystal (Gautschi, G., 2002). The Curie brothers also confirmed this statement.

The first successful experiment was conducted in Japan by Okochi. It measured the cylinder pressure in an internal combustion engine. Quartz was used in the experiment as the piezoelectric material of the pressure sensor. The experiment was carried out at speeds over 3,000 rpm (Okochi, Hashimoto, & Matsui, 1925).

The selection of piezoelectric material is crucial to improved responses. Materials used as piezoelectric sensors should have the following properties:

1. High piezoelectric sensitivity
2. High mechanical strength
3. Linear relationship between mechanical stress and electric polarization
4. High electric insulation resistance at low and elevated temperatures
5. No hysteresis in the material
6. Machinability (easy to manufacture)
7. Low anisotropy of mechanical properties

As per these requirements, it is found that quartz (SiO_2) is the most reliable piezoelectric material for piezoelectric sensors. Today, piezoelectric elements are commonly used in different areas of structural systems like sensors and actuators (Chopra, 1996).

Piezoelectric materials have the characteristic of utilizing the converse piezoelectric effect to excite the structure. They have a direct effect to sense the structural deformation. Piezoelectric material categories include piezoceramics (PZT) and polymer piezo films (PVDF). PZT are used as actuators; PVDF are used as sensing materials. PZT can also be used for sensing and actuation of material like self-sensing actuators (Inman et al., 1992).

Lee, O'Sullivan, and Chiang (1991) investigated the piezoelectric strain rate sensors. They developed the correlation between the piezoelectric gauge reading and conventional foil strain gauge measurement. However, the comparison was only performed at a frequency of 25Hz (Lee et al., 1991). Sirohi and Chopra (2000) carried out a calibration experiment with piezoelectric sensors and foil strain gauge sensors over a beam in the frequency of 5-500Hz. In it, strain measurement from piezoelectric sensors and foil strain gauge are compared in terms of sensitivity and signal to noise ratio (Sirohi and Chopra, 2000). There are mainly two piezoelectric sensors that are widely used in the field of strain measurement, piezoceramic (PZT) and polymer piezofilm (PVDF).

PZT Sensors

PZT is a common piezoceramic sensor for strain measurement. It has a composition of lead zirconate and lead titanate. These solid solutions are doped in other elements to achieve specific properties. PZT ceramics are formed by adding lead, zirconium, and titanium oxide in a proportional amount. This mixture is heated at 1000°C (Sirohi and Chopra,2000). At this temperature, they react to each other and form a PZT powder. Mixing and sintering achieve the desired shape. During the cooling process, the material is in a phase transition state (paraelectric to ferroelectric). As a result of this transition, the cubic cell of material is converted into tetragonal. This unit cell becomes elongated in a particular direction. It also has a dipole moment along its long axis. The unpoled ceramic has randomly oriented domains. This random orientation leads the net polarization become zero.To overcome this problem, a high electric field is applied to align the unit cells parallel to the applied field. The process of aligning the unit cell (poling) brings a net polarization to the ceramic. The ceramics have direct and converse piezoelectric effects. These properties are necessary for better performance, including high elastic modules, brittleness, and low tensile strength.

PVDF Sensors

Polyvinylidene fluoride (PVDF) sensors are polymers with a long chain of repeating monomer (-CH_2-CF_2). In the monomer chain, the hydrogen and fluorine atoms are positively and negatively charged, with respect to the carbon atoms. PVDF sensors are manufactured by solidification from the molten phase. In the liquid phase, the polymer chains are arranged in a random manner for a given volume of liquid. Due to this random arrangement, the net dipole moment does not exist. After solidification, all polymer chains stretch and arrange in the direction of the stretch. The proper arrangement of polymer imparts a dipole moment in the material. Afterward, the material behaves like a piezoelectric material.

CONCLUSION

This chapter consists of a brief description of types of strain measurement techniques used in various applications. The contents provide an overview and history of the use of strain gauge and piezoelectric sensors in the field of strain measurement. The methods described include mathematical expiations.

Special attention has been given to piezoelectric sensors because they play an important role in strain measurement with high accuracy at low cost.

REFERENCES

Chopra, I. (1996). Review of current status of smart structures and integrated systems. *SPIE Smart Structures and Integrated Systems*, *17*, 20–62.

Curie, J., & Curie, P. (1880). Development by pressure of polar electricity in hemihedral crystals with inclined faces. *Bull. soc. min. de France, 3*, 90.

Duryea, G. R., & Martin, J. F. (1968). An improved piezoelectric balance for aerodynamic force. *IEEE Transactions on Aerospace and Electronic Systems*, *3*(3), 351–359. doi:10.1109/TAES.1968.5408988

Gautschi, G. (2002). Piezoelectric sensors. In *Piezoelectric Sensorics* (pp. 73–91). Berlin: Springer. doi:10.1007/978-3-662-04732-3_5

Inman, D. J., Dosch, J. J., & Garcia, E. (1992). A self-sensing piezoelectric actuator for collocated control. *Journal of Intelligent Material Systems and Structures*, *3*(1), 166–185. doi:10.1177/1045389X9200300109

Keil, S. (2017). *Technology and practical use of strain gages: With particular consideration of stress analysis using strain gauges*. John Wiley & Sons. doi:10.1002/9783433606667

Lee, C-K., & O'Sullivan, T., & Chiang. (1991). Piezoelectric strain rate sensor and actuator designs for active vibration control. *Structures, Structural Dynamics, and Materials Conference*, *I32*, 1064.

Min, F., Yang, Y., Qiu, H., Zhong, S., & Pi, X. (2018). Hypersonic aerodynamics measurement with fiber-optic balance based on MEMS Fabry-Perot strain sensor. *Fiber Optic Sensing and Optical Communication*, 10849.

Okochi, M., Hashimoto, S., & Matsui, S. (1925). Tokyo. High-Speed Internal Combustion Engine and Piezoelectric Pressure Indicator. *Inst. Phys. Chem. Research*, *4*, 85–97.

Robinson, M. J., Martinez Schramm, J., & Hannemann, K. (2011). Design and implementation of an internal stress wave force balance in a shock tunnel. *CEAS Space Journal*, *1*(1-4), 45–57. doi:10.100712567-010-0003-5

Sirohi, J., & Chopra, I. (2000). Fundamental understanding of piezoelectric strain sensors. *Journal of Intelligent Material Systems and Structures*, *4*(4), 246–257. doi:10.1106/8BFB-GC8P-XQ47-YCQ0

St-ograve, V. (1998). Force measurements in hypersonic impulse facilities. *AIAA Journal*, *36*(3), 342–348. doi:10.2514/2.402

Wang, Y., Liu, Y., & Jiang, Z. (2016). Design of a pulse-type strain gauge balance for a long-test-duration hypersonic shock tunnel. *Shock Waves*, *6*(6), 835–844. doi:10.100700193-015-0616-x

Chapter 7
Optical Methods in Stress Measurement

Karpagaraj Anbalagan
https://orcid.org/0000-0002-0323-5732
National Institute of Technology, Patna, India

ABSTRACT

Stress will be produced in most of the engineering components related to their manufacturing process or because of their loading condition. For some special cases, both types also combined together and produce stress. Manufacturing processes like casting, welding, machining, and hot forming are creating the stress in the components. This stress will produce due to alteration of the microstructure (size, shape, phase composition, and orientation). Loading conditions are also produced stresses in the engineering components. This stress may be classified into compression, shear, tension, and fatigue. These depend on the load. Measuring the stresses in the components is very important because it can save a lot in terms of money, material, and manpower. A lot of techniques are used in industries to measure the stresses. Based on that, measurement techniques are broadly classified into two category, namely destructive and non-destructive techniques. Each method has its own advantages and limitations too. In this chapter, the optical method of measuring stress is discussed briefly.

INTRODUCTION

Stress measurement techniques are essential in the engineering field, because various processing methods are involved in product manufacturing and life cycle. The final product must be stress-free to serve its function. Practically, it is impossible to produce stress-free products. However, it is possible to measure the amount of stress present in the final product. This measurement may be compared with the base metal or initial conditions to check the condition of the final product. Several stress measurement techniques have been devised over time. New technologies are developed based on the existing principles. Hybrid methods are built by combining with two or more methods together to achieve accurate results. The Optical Method (OM) can measure both in-plane and out-of-plane displacements, depending upon the type of OM. This is a noncontact type of stress measurement technique.

DOI: 10.4018/978-1-7998-1690-4.ch007

The concept of an OM to determine stress is simple. If stress is applied to a specimen or sample, their surface will deform. This deformation will give interference patterns with some changes. These changes are analized to study the amount of stress in a particular component. This method is a quick and time-saving process as compared to other methods (e.g., strain gauge techniques). In this chapter, the author discusses the OM for measuring stress briefly, based on the review of past and present research works. In addition, the author illustrates the developments in this fields and new applications.

BACKGROUND

The basic concept of stress measurement using OMs is the mechanisms of the light interferometer. The monochromatic light waves are passed through the lens and to the objects. Reflected light patterns from the objects are analyzed to predict the results. In order to satisfy the above concepts, accessories such as various light sources and newly updated detecting and analyzing sensors are used. Nevertheless, the concept of measurement remains unchanged.

Walker (2001) discussed various stress measurement methods, including individual as well as combined techniques, to find out the residual stress. This research was aimed tomeasure the stress in the surface level and volumetric level. It was cleared that the selection criteria for measuring stress depends on the situation. It was also informed that formulating the Finite Element Method (FEM) all the necessary data can be collected from various OMs.

Baldi (2014) highlighted that vibration of the body makes OMs to measure stress in the engineering component those were with maximum complexity. The advantages of digital image correlation (DIC) are useful for handling such complex data. It was mentioned that DIC has accurate correlation and the ability to cover 2D as well as 3D. In a single attempt, the full specimen can be measured by using DIC. It was reported that the uncertainty of the component depends on the camera, not on the sample.

Chen et al. (2014) mentioned that strain gauge rosette and hole drilling methods are the two basic principles widely used in strain measurement. They reported that newly developed methods such as Digital Speckle Pattern Interferometry (DSPI) could be affected by the environment. The listed drawback may overcome by using DIC method. The authors concluded that the future focus would be on overcoming this limit and on developing a portable type of measurement system. Lord, Penn, and Whitehead (2008) discussed the remaining obstacle of the conventional methods, stating that alignment and surface preparations are the two significant parameters. They discussed deeply the development in DIC by optimization techniques. They proved that a FEM could be developed based on these results.

Rossini, et al. (2012) described the different methods for measuring residual stresses in components. In contrast to nondestructive residual stresses, they sustained diffraction methods have higher penetration power. The magnetic Barkhausen Noise Analysis (BNA) has penetration power of 100 times more than X-rays. They found that cost is the strong bottleneck of this process. On the other hand, the hole drilling method is cheap, fast, popular, and semidestructive, too. Rossini et al. reported that diffraction method at higher penetration power was good for getting the accurate results. But the contour method was the economical process, as compared to other methods.

Martínez-García et al. (2014) developed a new approach to residual stress analysis using hole drilling. This approach have various steps, first step was applying laser power, to remove quasi nondestructive material from the sample (laser ablation). Second was measuring deformations around the hole by high-resolution digital holographic interferometry. They carried out experiments measuring in-plane stresses at the curved strip geometry specimen with the hole drilling and milling techniques. Thus, these technique along with laser proved that, without contacting the workpiece, it was possible to analyse the residual stress.

Yoshida et al. (2016) used acoustic-elasticity and optical interferometry for the analysis of residual stresses. They used Electronic Speckle Interferometry (ESPI) analysis to understand the dynamic behavior of the specimen under the influence of residual stresses. By combining both acoustic elasticity and optical interferometry, they measured residual stresses quantitatively. Using this nondestructive method two-dimensional data was collected.

Wyant (2002) compared the three conventional white light interferometry with laser light interferometry. The scattering plate of the interferometer was appeared in the zero-order fringe, and it was automatically obtained by Wyant. The author also obtained good interferograms using a white light interferometer.

Schelkens et al. (2019) investigated the conversion of two-dimensional jpeg by using the JPEG Pleno method. This method allows to produce three-dimensional images from the data obtained from two-dimensional images. Jamal, Ahmadi, Khanzadeh, and Malekzadeh (2019) developed a Convolutional Neural Network (CNN) for avoiding moire artefacts data extracted from experimental work. This CNN method allowed to maintain the signal accuracy and image resolution. CNN used in the post-processing stage via the image learning process. Jacquot (2008) reviewed the background and development of the speckle interferometry. He found that in-line and out-of-plane measurements and their producers were discussed clearly.

Pedrini, Pfister, and Tiziani (1993) studied double-pulse ESPI by microsecond interval. They studied the fringes and explored the results of the fringes, which was very helpful for vibrational analysis. Advantages of this method like, it reduce the time delay for film processing and reconstruction of the hologram. This can helped to increasing the resolution by modifying the speckle size. Kemper and Von Bally (2008) studied the digital holography technique for live cell application, for which they created multifocal imaging using the numerical focus method. By using these simultaneous and marker-free approaches with nondestructive test, Bally studied morphology analysis.

Stetson and Brohinsky (1985) studied the electro-optic process and reuse of the data recorded by the TV camera. They estimated the computational time for reusing the data by using five-phase steps to form a hologram. Saleem, Wildman, Huntley, and Whitworth (2003) discussed the strain distribution in biscuits by using speckle interferometry. Apart from the normal industrial application, this study focused on changing the moisture content in the biscuits. The researchers measured the holographic expansion coefficient, and stored their data for simulating the biscuit behavior.

Takayama (1983) discussed the application of the holographic technique to analyse shock waves. This analysis covered the application in shock wave propagation in aerofoil. They conducted their complete analysis under the dynamic gas area. Zhang (1998) studied the in-plane displacement field by using the basic concepts of Leendert and ESPI method. Zhang The placed a charge-coupled device (CCD) camera as an integral part of the instrument to provide space for light interfacing. Zhang combined this study with the blind hole drilling method to investigate the concentration of residual stress.

Xu, Peng, Miao, and Asundi (2001) studied the in-situ measurement of the microstructure using a microscopic holography system. The system was developed mainly to measure the deformation of the cantilever in the Micro Electro Mechanical System (MEMS).

The literature review above evidences that stress measurement techniques have been slowly developed. Also, the relationship between their developments in technical aspects needs a clear view.

MAIN FOCUS OF THE CHAPTER

The main theme of this chapter has consolidated the works in stress measurement techniques. The following sections will give the outline these techniques. Also it provide details with regard to their recent advancement. This will give researchers to get an idea of how to determine stress in a component by selecting the proper method. For example, considering composite structure automation on stress measurement is difficult, thus it is better to work without the aid of automation. This chapter is going through the limitations of the process with respect to similar conditions. This will provide the reader with additional information to select or maintain the workpiece condition to get the best results.

Types of Optical Methods for Measuring The Stress

1. Moire interferometry.
2. DIC.
3. ESPI.
4. Holographic interferometry.
5. Other interferometry approaches.

Major Parts

Light is the heart of the optical microscope system. Fluorescent light source, laser light source, and white light source are some types of light. Partially reflected and fully reflected mirrors are also used in the OM. For ESPI, the sensor is the most important part and secondly the CCD camera was the most important one.

MOIRE INTERFEROMETRY

Moire interferometry was developed in 1980. It allows to measure stresses in structural applications, cyclic loading equipment, welded joints, and vibrational parts by identifying their surface displacement. In addition, it allows to measure polycrystalline materials, fracture mechanism for materials, and components produced by layer, such as additive manufacturing. This method has the advantage of measuring in-plane stress, too. Also, it can be coupled with the hole drilling process to identify strain in a given component. In the hole drilling method, the sequence of holes to be drilled must be identified first. Later the fringes are aimed for study. This chapter will investigate the fringes which are obtained from this drilled area. This technique can also be combined with a strain gauge to identify the magnitude of stress. This process gives noble sensitivity for determining displacements. Figure 1 shows the setup of moire interferometry. A cross-line grating is applied to the area to be studied. However, the condition or restriction of this OM is the object must be flat. For this case, a flat specimen is chosen with gratings. In this instrument, laser light is used as a source. Later this Laser Beam (LB) is divided as two portions such LB1 and LB2 as shown in figure 1. The diffracted lights appear in the z-axis as a reflected one. When forces are applied, the displacement of the object, as well as the gratings, are disturbed. Thus, the supplied light is diffracted to the camera for recording or identification. This will give the fringe as a pattern from axis x. LB3 and LB4 give the displacement in the vertical axis (i.e., y axis). When LB1 and

Figure 1. Schematic representation of the setup of moire interferometer

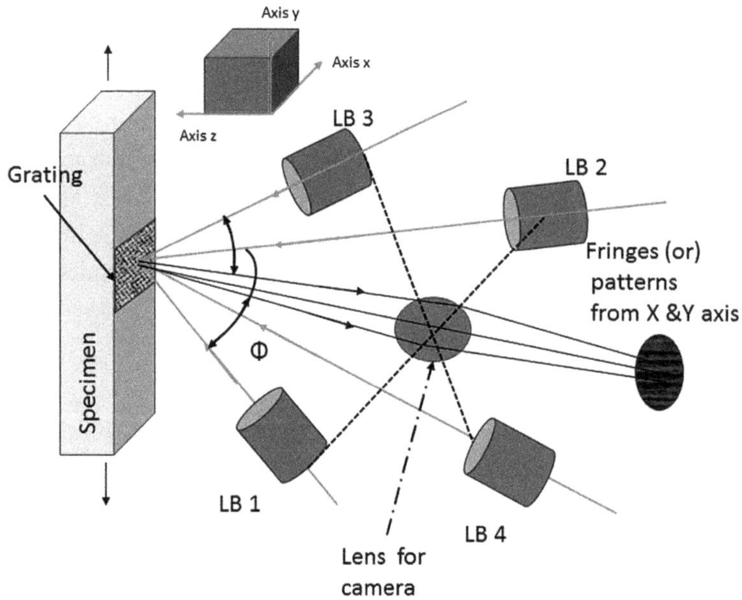

LB2 are jammed, the patterns are recorded for the y axis. Similarly, LB3 and LB4 are jammed after that x-axis patterns will be recorded (refer figure 1). In moire interferometer, by adjusting the instrument, it is possible to produce equally spaced fringes. By changing the angle, it is conceivable to produce the same fringes at no-load condition. The same will be produced under tensile or compression load condition. These fringes appear parallel to the reference gratings.

Implementation of Moire Interferometry

This section starts with an example assume that Moire interferometry is going to measure the deformation of an object under mechanical and thermal loading condition. This combined loading discussion is very important, because experimental arrangements are quite different for different materials. It can vary with size, shape, and loading condition (thermal or structural). Thus, depending on the cases, moire interferometry can be mounted with separate experimental setup or built as an integral part. However, generally, the specimen is mounted with the fixture that is related with the interferometer. In a chamber, the specimens are placed to avoid contamination from environmental hazards. For thermal loading, the specimen must be placed inside the oven, so that the hazards will not affect the specimen. For mechanical loading, the specimen should not be affected by the vibration caused by the moving parts in it. Once the specimen is free from the hazards and other external factors, moire interferometry can perform well without the errors. Now it can enchant the possibility of real-time measurement readings. The following example introduces a condition: If a thin chip made with different materials is going to operate with a range of temperature changes (e.g., the room temperature is25°C) to 100°C, different materials can experience different strain (thermal expansion) with respect to temperature changes. These changes can be effectively measured by a moire interferometer. In case of a constant time interval for measuring the changes in infringes, the consolidated fringes will give an idea of the thermal stress and strain introduced in a given component.

Another example is a bi-thermal loading condition on a specimen. This is just the reverse order of the previous condition. In the previous condition, the temperature is raised by the room condition, so the readings and analysis are carried out while the temperature rises. For this second case (i.e., a bi-thermal loading condition on a specimen), the assumption is the specimen is heated to a certain level (elevated temperature) and, while it cools down to room temperature, the interferometer is allowed to read the changes (i.e., strains). This measurement is very useful in the composite manufacturing process. The fabrication of composites will produce heat due to the exothermal reaction caused by the resin and hardener mixture. The temperature will go to the peak and slowly cool to room temperature. The changes are monitored continuously by the interferometer and recorded. The rise in temperature during curingmay produce dimensional changes. As a result, the final product may have a strong deviation. For real-time measurement, it is advisable to identify if the mould has any dimensional changes on that operating temperature condition. If so, it is necessary to observe the changes which occurred with the mould and the composite specimen. This calculation will give accurate perdition in dimensional changes. This interferometer can also be engaged with materials behavior under cryogenic conditions, and effect of heat treatment in materials.

Hole-Drilling Method

As mentioned before, the hole drilling method can combine with the moire interferometry to measure strain accurately (Furgiuele, Pagnotta, & Poggialini (1991), Nicoletto et al. (1988) and Ribeiro, Monteiro,, Lopes, & Vaz (2011)) In addition, it is possible to use computational techniques to measure strain correctly by the data obtained from the experiments (Wu, Lu, & Han (1998)). In this process, a hole up to a certain depth will be drilled, and then the patterns of more fringes appear (Figure 2). In the hole drilling method, radial displacement can be written as follows:

$$R_u\,(r.\,\Phi) = M\,(\sigma_x + \sigma_y) + N\,(\sigma_x - \sigma_y)\,[\cos 2\Phi + 2\,\tau_{xy}\,\sin 2\Phi] \tag{1}$$

where M and N are constants and can be obtained from the finite element method analysis. This equation contains the parameters E, γ, and h/d ratio (i.e., hole depth to diameter ratio) and finally radial position to the whole radius (R_{xy}/H_r). The displacement in X_u and Y_u are obtained from X and Y directions, and can relate to the order of fringes as X_N and Y_N.

$X_u = (1/2\,F_g)X_N$ and $Y_u = (1/2\,F_g)Y_N$, whereas F_g is the frequency of the gratings. The radial displacement can be expressed by using R_x and R_y, as follows:

$$R_u\,(r,\,\Phi) = X_u\,(X_i,\,Y_i)\cos\Phi_i + Y_u\,(X_i,\,Y_i)\sin\Phi_i\,(X_i^2 + Y_i^2 = r^2) \text{ and } \Phi_i = \text{tan-1}\,(Y_i/X_i) \tag{2}$$

where Φ_i represents angle relations between the fringes which are likely be 0°, 45°, and 90°. The matrix below allows to correlate residual stress:

$$[X_n\,(X_i,Y_i)\;Y_n\,(X_i,Y_i)]\begin{bmatrix} cos\Phi \\ \sin\Phi \end{bmatrix} = 2\,F_g\,[\text{P} + \text{Q}\cos 2\Phi_i\;\;\text{P-Q}\cos 2\Phi_i\;\;2\text{B}\sin 2\Phi_i]\begin{bmatrix} \sigma x \\ \sigma y \\ \sigma z \end{bmatrix} \tag{3}$$

Figure 2. More fringe obtained in the x-axis direction

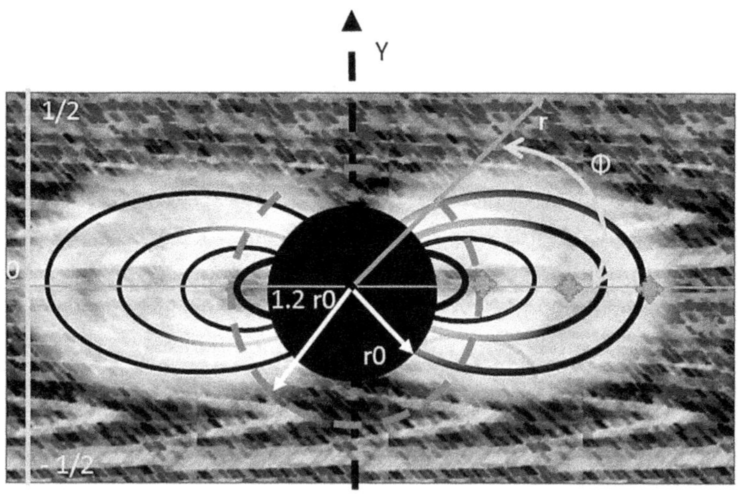

Fringe order X_n and Y_n can be identified for an angle of Φ_i and radial location for Figure 2. The fringes are started from the centre of the hole and increasein progression of, for example, 0.5, 1.0, and 1.5. This approach can be used for composites (including laminated) also to identify their stress distribution.

Microscopic Moire Interferometry

Microscopic moire interferometry, too, allows to measure surface strains from small sectioning. A grating is placed on the surface of the sample with photolithography. This grating generally has a pitch of 5 microns for large strain measurements. It can be much smaller, for some special cases, to measure elastic strain. As the author discussed before, in this apparatus also two LBs are used to illuminate the surface of the specimen. Two diffracted beams from the grating are collected and interfered. This action produces a fringe pattern. By using a grid, two perpendicular gratings collect the fringe pattern. These three components allow to explore and measure the surface strain. This method is used to measure the strain produced in the MEMS equipment (e.g., cantilever beams and grips). These devices are in the microlevel, so a micro moire interferometer is required. Bimetallic strips in the sensors can also be measured by using micro moire interferometry.

Curved Surface: A Special Case

Normally, moire interferometry is used for a flat surface, as the author discussed before. However, in some important cases, this interferometer can be also used for a curved surface. Importantly, this technique has some limitations while measuring the curved sections. A curved surface includes a cylindrical profile or other irregular geometry that can be measured by this method. Thus, initially, it is necessary to break the part into small segments. Then, the measurement can be taken for each of those segments.

DIGITAL IMAGE CORRELATION

DIC produces a 3D image of the surface, which needs to be studied (Sutton, (2008)). For three-dimensional image studies, a white light source and two high-quality cameras are needed. The concept of these image techniques is to compare the initial condition and deformed condition after a certain time or interval. At first, the image is captured, digitalized, and stored in a particular location. Later, the second image is recorded and digitalized with refer to that location. Both images are be compared and correlated, and the changes are noted. The x, y, and z axis are taken as a coordinating system, at the beginning. Subsequently, the 3D coordinates are continuously tracked by the images. Compared to the previous technique, this DIC holds the advantage of being free from some restrictions, such assurface conditions (i.e., no need of maintaining flat surface). Software packages such as ISI-SYS, La Vision, and DANTES are availed on the market to do this work better.

Digital Image Correlation with Hole Drilling

DIC can also be adopted with the hole drilling method by capturing the initial image of hole size and taking this image as a reference for future use (Nelson, Makino, & Schmidt, 2006). This method can be implemented with a few restrictions, such as little surface preparation (e.g., surface coatings). For a number of given anglesfor the radius, r/r0 can generate the displacement coordinates,

$$R_{ui} = (P+Q\cos 2\Phi_i)\,\sigma_x + (P-Q\cos 2\Phi_i)\,\sigma_y + (2P\sin 2\Phi_i)\,\tau_{xy} \tag{4}$$

The second method of this hole drilling is by image correlations. After surface preparation, a hole is drilled on the object to release residual stress. Due to the absence of stress, the new coordinates will be different. The difference between the initial and final coordinates is taken for analysis. This method can use an optimized process with analytical knowledge, too.

The process of DIC with hole drilling has some advantages, such as normal white light can be used for the stress measurement instrument, and it can be used for contour and irregular profiles, too. DIC can give displacement data in addition to the above mentioned results. One drawback is the calibration of the instrument before the experiment is needed, because it uses x, y, and z coordinate values as a reference.

The above-mentioned hole drilling method is also suitable for macrolevel measurement. Even smaller geometry (e.g., shapes in micro and nano-level) can be measured by using this method. The concept is the same, but the object is small, so the images must be captured by a Scanning Electron Microscope (SEM). However, a focused ion beam is used for making holes at micro or nano level. While removing the materials for making a hole or slot the material, burr may be deposited in the nearby surface. For this method, it is good to take the microstructural changes at that portion. Later computational approaches are also included in this analysis. This DIC method allows to measure small deflection, as well, by reversing the idea that "certain deflection is caused by how much stress arises" in the object. This deflection may raise because of the surface modification process.

Implementation of Digital Image Correlation

DIC techniques can have a good interface with modern technology. The results from the DIC can readily be reported to finite element methods. Later, using these details, it is possible to build and predict the model behavior under a desired condition. This model may be two-dimensional or three-dimensional. This can be applied to direct monitoring of components in the industry (e.g., expansion rods, sensors, and thermostats). The extended application of this DIC measures the changes in dry textiles under different loading conditions. Another application is the tensile behavior of various components under tensile and bending test. In polymer composites, if the matrix is formed by general-purpose resin after the cure, it becomes more brittle. During the tensile test, attention is needed to monitor the initiation of cracks. DIC is be the best choice to monitor this and extract the data for future finite element analysis.

ELECTRONIC SPECKLE INTERFEROMETRY

ESPI is used to measure in-plane stress, displacement, and surface vibration. The extend application may be used to measure localized necking in metal works, too. This method uses the diffracted light beam from the workpiece and later analyzes the reflected light beams. As other methods, recording devices are used to store the data. The minimum sparkle size is determined by the camera resolution. The fringe pattern before and after deformation is be compared, and results may be evaluated immediately. Instead of a holographic plate, a CCD camera is placed in ESPI. This method can measure the vibrational behavior of objects, as well. The practical application of this ESPI is material testing and quality control process in the industry, because ESPI can be used for in-plane and out-of-plane measurement. For analyzing the full components, preprocessing and postprocessing have to be carried out. In the postprocessing stage, the collected two-dimensional data are merged and developed as three-dimensional elements. The curved surface can be measured by extracting the data from various views. With the help of today's technology, it is possible to measure this complicated structure within a few seconds. The results can be stored in a computer for future use.

Figure 3. Electronic speckle pattern interferometry with the hole drilling method

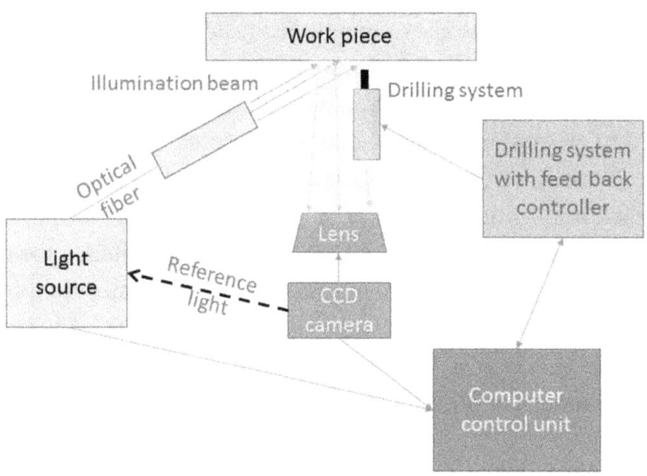

Electronic Speckle Pattern Interferometry with the Hole Drilling Method

In this process, a blind hole is drilled, and a small amount of material is removed from the stressed portion. The depth of the drill and other parameters associated with this process can determined by the customer. Figure 3 sows this experimental setup with the hole drilling attachment. As a result of this action, a new area will be sharing the stress the objects raise. This change in stress may be smaller, this measurement is called as ESPI with hole drilling method. The camera is used to measure and identify the changes continuously for verification.

Practical Application of Electronic Speckle Pattern Interferometry

A few practical applications with the usage of ESPI are listed below. In space vehicles, components such as a central cylinder, deck plates, solar panel substrates, and antenna reflectors are made of aluminium/ composite with honeycomb sandwich construction. The propellant tanks of a spacecraft are made of titanium alloy. It is very significant to quantify the whole field displacement/slope of the components of the above space vehicle under static and dynamic environments, because the production of those components is specially manufactured. Thus, conducting destructive tests are not recommended; the best choice is nondestructive testing. A more suitable process is ESPI only,due to the internal pressure, the minute deformation in the structure of the propeller tank can easily identify by ESPI. This process can also extend to identify the fatigue life of steel structure and welded components.

HOLOGRAPHIC INTERFEROMETRY

Holographic interferometry provides quantitative information for a slight change in small surface displacement (nano or micro) (Kreis (2006); Sharpe & william (2008); Vandenrijt & Georges, (2010)). In holographic interferometry, a low power laser is used as a source that is reflected by the object, and is passed through the lens. The laser is used because it has the property of being highly coherent. Figure 4 illustrates the principle of this technique. The reference light is needed to produce the hologram location.

Figure 4. Schematic diagram of electronic speckle pattern interferometry holographic technique

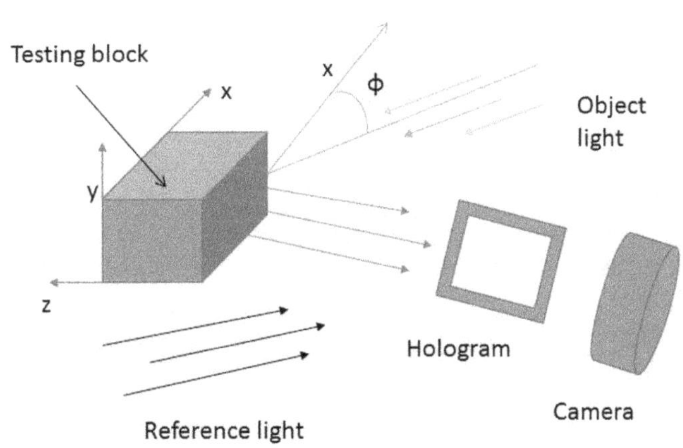

Both digital and analogue holographic methods are used. In the analogue method, after the processing, the storage device is erased (Toal (2011)). Once the information are collected, the holograms are re-produced at any time by using the light source. This hologram will give the replica of the workpiece. If changes are taken in the workpiece, the path of the hologram will deviate in digital holographic interferometry, instead of the camera, the hologram is stored electronically (i.e., CCD and image sensors), and images before and after the surface displacement are compared and analyzed.

The Hole Drilling Method with Electronic Speckle Pattern Interferometry

This hole drilling method is an extended application of the ESPI system (Antonov, (1983); Bass, Schmitt, & Ahrens (1986); Khanna, He & Agrawal (2001); McDonach, McKelvie, MacKenzie, & Walker (1983); Nelson & McCrickerd (1986)). This is a common technique and it is good option, in comparison with all the other techniques which are reviewed in this chapter. Initially, the hole is formed (see the hole location in Figure 5). The coordinates have to be determined in terms of radius and angle. The displacement to the blind hole formed can get from the following matrix (Makino, Nelson, Fuchs, & Williams, 1996):

$$\begin{Bmatrix} ru \\ r\Phi \\ zu \end{Bmatrix} = \begin{bmatrix} P+QCos2\Phi & P-QCos2\Phi & 2QSin2\Phi \\ RSin2\Phi & -RSin2\Phi & -2RCos2\Phi \\ U+VCos2\Phi & U-VCos2\Phi & 2VSin2\Phi \end{bmatrix} \begin{Bmatrix} \sigma x \\ \sigma y \\ \tau xy \end{Bmatrix} \quad (5)$$

The radial and tangential directions of the in-plane displacement were taken as r_{un} and Φ_u (Figure 6). The displacement in the Z direction is taken as Z_u. As in all other techniques, if the displacement happens, the fringes change from their usual pattern.

Figure 5. Representation of fringes produced by the hole drilling method

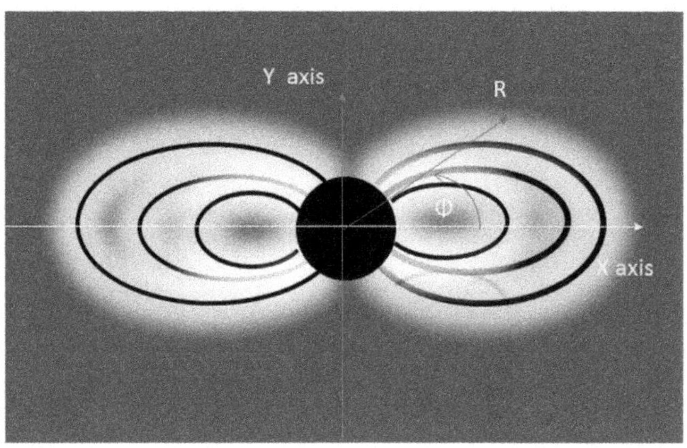

Figure 6. Stress distribution in the hole drilling method

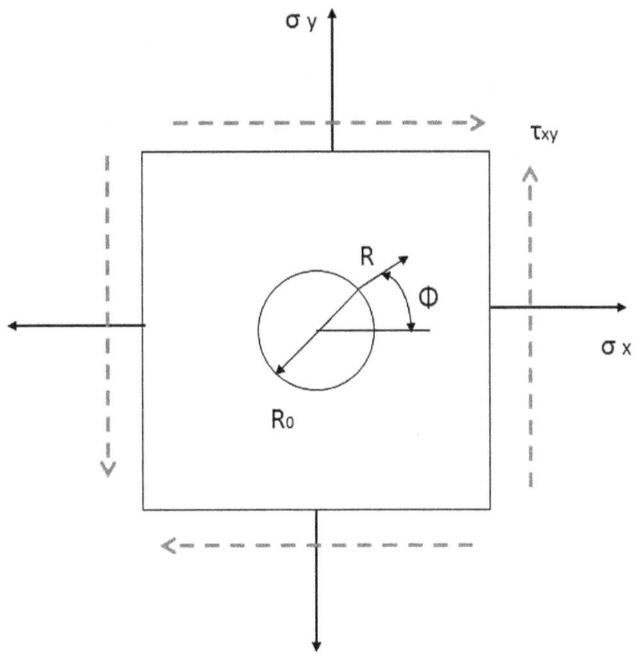

OTHER INTERFEROMETRY APPROACHES

Apart from the above measurement techniques, a little more discussion is needed to cover all the methods. Stereography and interferometry strain rosette are other interferometry methods. A short discussion of these methods is given below.

Stereography

Stereography is a nondestructive testing method to find the defects caused by deformation. As interferometers, where laser light is used for processing, the reflected light is passed through the birefringent element, lens, and polariser. Then, the light reaches the camera for further processing (e.g., a recording). Here, the initial condition and final condition are identified. Finally, their difference is also calculated. A few steps (e.g., reporting, sizing, and localising) are followed for processing. Some attractive advantages of stereography are it isnonconduct and quite sensitive, allows real-time inspection, and it is a nondestructive method, which encourages its adoption in more fields.

Interferometry Strain Rosette

Interferometry strain rosette is based on the microindentation depressed on the specimen surface. This depression allows to measure the three in-plane strains. This process is superior to other methods (e.g., resistance strain gauge rosette), because it is noncontact, this method allows to handle complex geometry with steep gradients. Further, this method is suitable to measure the objects at a higher temperature. Also in the worst environment condition, this method is the best choice.

OTHER METHODS FOR MEASURING RESIDUAL STRESSES

Other methods for measuring residual stresses are (deep hole method, the contour method, diffraction techniques by using X-ray and neutron, sound of acoustics and magnetic methods (i.e., the magnetic Barhausen noise method the ultrasonic method)) briefly illustrated below.

The deep hole method is classified as a semidestructive method (DeWald & Hill, (2003); Leggatt, Smith, Smith, & Faure, (1996)). In this method, a hole is drilled, then a trepanning operation is carried out around it. The diameter of the hole after the trepanning must be measured. The changes on the hole dimensions give an idea of the residual stress present. This deep hole method is the best solution for isotropic materials, carbon steels, aluminium, stainless-steel materials, and workpieces having thick sections. As the name implies, it is a sectioning method, so a part of "thin" material is removed from the specimen. The deformation of the portion is measured for analyses (Shadley, Rybicki, E. F., & Shealy, (1987); Tebedge, Alpsten, & Tall et al. (1973)). By using this deformation, the residual stress formed in the above-mentioned materials is measured. The contour method was introduced in 2000 for measuring residual stress present in two-dimensional (2D) specimens (Prime & Gonzales, 2000). This method can cover T-joint weld configurations in welding, quenched thick plates in heat treatment areas,and forgings. This deep hole method is quite easy and covers most of the manufacturing sectors. The initial step is preparing the sample from the specimen. The next step is measuring the contour of the sample. The final step is extracting the data and find the amount of stress by their analysis.

The next major division for measuring residual stresses is diffraction techniques by using X-ray and neutron. These methods have a similar concept, but the penetration power of neutron diffraction is much better than the penetration power of the X-ray. For example, stainless steel of 25 mm can be examined by a neutron, not by the X-ray, because penetrating such a high thickness is very difficult by X-ray (Kim, Kim, & Lee, (2009)).

The other part of the nondestructive method may be covered by the sound of acoustics and magnetic methods. The magnetic Barhausen noise method is widely used in stress measurement at surface level. This method covers ferromagnetic materials (Altpeter, Dobmann, Kröning, Rabung, & Szielasko, 2009). In the ultrasonic method, sound waves are passed from the transducer to another end or by using the same transducer by itself. By using recorded sound, residual stresses in the components are compared and finally calculated.

FUTURE RESEARCH DIRECTIONS

Over time, stress measurement has been improved for various components with different conditions. Slowly, the noncontact type of stress measurement devices has been developed and has reached its saturation level. Currently, the finite element result analysis is becoming popular. Moreover, the analysis of image correlating techniques is also being developed. Nevertheless, the accuracy of stress measurement instruments must be improved, in the future. In particular, the pixel for image correlation techniques must be improved for getting good results. Future research should be focused on interferometry and other measurement techniques.

CONCLUSION

The above-discussed methods are followed by many industries for providing the solution to scientific problems. Various methods are available to measure residual stresses, but nondestructive methods have good advantages over other methods. The author discussed the usage of interferometry and the final conclusion are listed as follows:

- The combination of measurement technique with the hole drilling method and how both techniques are combined and interfaced.
- Moire interferometry, which can give quantitative results about the object. For tiny zones, also high spatial resolution can be produced.
- The latest techniques (e.g., ESPI), with their detailed working process.
- The use and advantages of other techniques (e.g., stereography) in comparison with other processes.
- The holographic method for measuring residual stress for objects; using reference light to produce a hologram, as the latest technique for stress measurement; finally, residual stress measurement for a block.
- Other nondestructive methods.

ACKNOWLEDGMENT

This research received no specific grant from any funding agency in public, commercial or not-for-profit sectors.

REFERENCES

Altpeter, I., Dobmann, G., Kröning, M., Rabung, M., & Szielasko, S. (2009). Micro-magnetic evaluation of micro residual stresses of the IInd and IIIrd order. *NDT & E International*, *42*(4), 283–290. doi:10.1016/j.ndteint.2008.11.007

Antonov, A. A. (1983). Development of the method and equipment for holographic inspection of residual-stresses in welded structures. *Welding Production*, *30*(12), 41–43.

Baldi, A. (2014). Residual stress measurement using hole drilling and integrated digital image correlation techniques. *Experimental Mechanics*, *54*(3), 379–391. doi:10.100711340-013-9814-6

Bass, J. D., Schmitt, D., & Ahrens, T. J. (1986). Holographic in situ stress measurements. *Geophysical Journal International*, *85*(1), 13–41. doi:10.1111/j.1365-246X.1986.tb05170.x

Chen, Y. H., Chen, X., Xu, N., & Yang, L. (2014). *The digital image correlation technique applied to hole drilling residual stress measurement* (No. 2014-01-0825). SAE Technical Paper. doi:10.4271/2014-01-0825

DeWald, A. T., & Hill, M. R. (2003). Improved data reduction for the deep-hole method of residual stress measurement. *Journal of Strain Analysis for Engineering Design*, *38*(1), 65–77. doi:10.1243/030932403762671908

Furgiuele, F. M., Pagnotta, L., & Poggialini, A. (1991). Measuring residual stresses by hole-drilling and coherent optics techniques: A numerical calibration. *Journal of Engineering Materials and Technology*, *113*(1), 41–50. doi:10.1115/1.2903381

Jacquot, P. (2008). Speckle interferometry: A review of the principal methods in use for experimental mechanics applications. *Strain*, *44*(1), 57–69. doi:10.1111/j.1475-1305.2008.00372.x

Jamal, F., Ahmadi, F., Khanzadeh, M., & Malekzadeh, S. (2019). Application of image processing in optical method, moire deflectometry for investigating the optical properties of zinc oxide nanoparticle. *Optic and Laser Technology*. doi:10.13140/RG.2.2.23059.32804

Kemper, B., & Von Bally, G. (2008). Digital holographic microscopy for live cell applications and technical inspection. *Applied Optics*, *47*(4), A52–A61. doi:10.1364/AO.47.000A52 PMID:18239699

Khanna, S. K., He, C., & Agrawal, H. N. (2001). Residual stress measurement in spot welds and the effect of fatigue loading on redistribution of stresses using high sensitivity moiré interferometry. *Journal of Engineering Materials and Technology*, *123*(1), 132–138. doi:10.1115/1.1286218

Kim, S. H., Kim, J. B., & Lee, W. J. (2009). Numerical prediction and neutron diffraction measurement of the residual stresses for a modified 9Cr–1Mo steel weld. *Journal of Materials Processing Technology*, *209*(8), 3905–3913. doi:10.1016/j.jmatprotec.2008.09.012

Kreis, T. (2006). *Handbook of holographic interferometry: Optical and digital methods*. John Wiley & Sons.

Leggatt, R. H., Smith, D. J., Smith, S. D., & Faure, F. (1996). Development and experimental validation of the deep hole method for residual stress measurement. *Journal of Strain Analysis for Engineering Design*, *31*(3), 177–186. doi:10.1243/03093247V313177

Lord, J. D., Penn, D., & Whitehead, P. (2008). The application of digital image correlation for measuring residual stress by incremental hole drilling. In Applied Mechanics and Materials (Vol. 13, pp. 65-73). Trans Tech Publications. doi:10.4028/www.scientific.net/AMM.13-14.65

Makino, A., Nelson, D. V., Fuchs, E. A., & Williams, D. R. (1996). Determination of biaxial residual stresses by a holographic-hole drilling technique. *Journal of Engineering Materials and Technology*, *118*(4), 583–588. doi:10.1115/1.2805960

Martínez-García, V., Wenzelburger, M., Killinger, A., Pedrini, G., Gadow, R., & Osten, W. (2014). Residual Stress Measurement with Laser-Optical and Mechanical Methods. *Advanced Materials Research*, 996.

McDonach, A., McKelvie, J., MacKenzie, P., & Walker, C. A. (1983). Improved moiré interferometry and applications in fracture mechanics, residual stress. and damaged composites. *Experimental Techniques*, *7*(6), 20–24. doi:10.1111/j.1747-1567.1983.tb01766.x

Nelson, D. V., Makino, A., & Schmidt, T. (2006). Residual stress determination using hole drilling and 3D image correlation. *Experimental Mechanics*, *46*(1), 31–38. doi:10.100711340-006-5859-0

Nelson, D. V., & McCrickerd, J. T. (1986). Residual-stress determination through combined use of holographic interferometry and blind-hole drilling. *Experimental Mechanics*, *26*(4), 371–378. doi:10.1007/BF02320153

Nicoletto, G. (1988). Theoretical fringe analysis for a coherent optics method of residual stress measurement. *Journal of Strain Analysis for Engineering Design*, *23*(4), 169–178. doi:10.1243/03093247V234169

Pedrini, G., Pfister, B., & Tiziani, H. (1993). Double pulse-electronic speckle interferometry. *Journal of Modern Optics*, *40*(1), 89–96. doi:10.1080/09500349314550111

Prime, M., & Gonzales, A. (2000). The contour method: Simple 2-D mapping of residual stresses. *6th International Conference on Residual Stress Proceedings*, *1*, 617-624.

Ribeiro, J., Monteiro, J., Lopes, H., & Vaz, M. (2011). Moire interferometry assessement of residual stress variation in depth on a shot peened surface. *Strain*, *47*, e542–e550. doi:10.1111/j.1475-1305.2009.00653.x

Rossini, N. S., Dassisti, M., Benyounis, K. Y., & Olabi, A. G. (2012). Methods of measuring residual stresses in components. *Materials & Design*, *35*, 572–588. doi:10.1016/j.matdes.2011.08.022

Saleem, Q., Wildman, R. D., Huntley, J. M., & Whitworth, M. B. (2003). A novel application of speckle interferometry for the measurement of strain distributions in semi-sweet biscuits. *Measurement Science & Technology*, *14*(12), 2027–2033. doi:10.1088/0957-0233/14/12/001

Schelkens, P., Ebrahimi, T., Gilles, A., Gioia, P., Oh, K. J., Pereira, F., & Pinheiro, A. M. (2019). JPEG Pleno: Providing representation interoperability for holographic applications and devices. *ETRI Journal*, *41*(1), 93–108. doi:10.4218/etrij.2018-0509

Shadley, J. R., Rybicki, E. F., & Shealy, W. S. (1987). Application guidelines for the parting out step in a through thickness residual stress measurement procedure. *Strain*, *23*(4), 157–166. doi:10.1111/j.1475-1305.1987.tb00640.x

Sharpe, J., & William, N. (Eds.). (2008). *Springer handbook of experimental solid mechanics.* Berlin: Springer. doi:10.1007/978-0-387-30877-7

Stetson, K. A., & Brohinsky, W. R. (1985). Electrooptic holography and its application to hologram interferometry. *Applied Optics*, *24*(21), 3631–3637. doi:10.1364/AO.24.003631 PMID:18224099

Sutton, M. A. (2008). Digital image correlation for shape and deformation measurements. In W. N. Sharpe (Ed.), *Springer handbook of experimental solid mechanics* (pp. 565–600). Berlin: Springer. doi:10.1007/978-0-387-30877-7_20

Takayama, K. (1983, October). Application of holographic interferometry to shock wave research. In *Industrial Applications of Laser Technology* (Vol. 398, pp. 174–181). International Society for Optics and Photonics. doi:10.1117/12.935372

Tebedge, N., Alpsten, G., & Tall, L. (1973). Residual-stress measurement by the sectioning method. *Experimental Mechanics*, *13*(2), 88–96. doi:10.1007/BF02322389

Toal, V. (2011). *Introduction to holography.* CRC Press. doi:10.1201/b11706

Vandenrijt, J. F., & Georges, M. P. (2010). Electronic speckle pattern interferometry and digital holographic interferometry with micro-bolometer arrays at 10.6 µm. *Applied Optics*, *49*(27), 5067–5075. doi:10.1364/AO.49.005067 PMID:20856279

Walker, D. (2001). Residual stress measurement techniques. *Advanced Materials & Processes*, *159*(8), 30–33.

Wu, Z., Lu, J., & Han, B. (1998). Study of residual stress distribution by a combined method of Moire interferometry and incremental hole drilling, Part I: Theory. *Journal of Applied Mechanics*, *65*(4), 837–843. doi:10.1115/1.2791919

Wyant, J. C. (2002, July). White light interferometry (Vol. 4737). SPIE.

Xu, L., Peng, X., Miao, J., & Asundi, A. K. (2001). Studies of digital microscopic holography with applications to microstructure testing. *Applied Optics*, *40*(28), 5046–5051. doi:10.1364/AO.40.005046 PMID:18364784

Yoshida, S., Sasaki, T., Usui, M., Sakamoto, S., Gurney, D., & Park, I. K. (2016). Residual stress analysis based on acoustic and optical methods. *Materials (Basel)*, *9*(2), 112. doi:10.3390/ma9020112 PMID:28787912

Zhang, J. (1998). Two-dimensional in-plane electronic speckle pattern interferometer and its application to residual stress determination. *Optical Engineering (Redondo Beach, Calif.)*, *37*(8), 2402–2410. doi:10.1117/1.602007

ADDITIONAL READING

Berkovic, G., & Shafir, E. (2012). Optical methods for distance and displacement measurements. *Advances in Optics and Photonics*, *4*(4), 441–471. doi:10.1364/AOP.4.000441

Huang, X., Liu, Z., & Xie, H. (2013). Recent progress in residual stress measurement techniques. *Guti Lixue Xuebao*, *26*(6), 570–583.

Kishimoto, S. (2012). Electron moiré method. *Theoretical and Applied Mechanics Letters*, *2*(1), 011001. doi:10.1063/2.1201101

Li, K. (1997). Application of interferometric strain rosette to residual stress measurements. *Optics and Lasers in Engineering*, *27*(1), 125–136. doi:10.1016/S0143-8166(95)00014-3

Ljunggren, C., Chang, Y., Janson, T., & Christiansson, R. (2003). An overview of rock stress measurement methods. *International Journal of Rock Mechanics and Mining Sciences*, *40*(7-8), 975–989. doi:10.1016/j.ijrmms.2003.07.003

Ma, Q., & Clarke, D. R. (1993). Stress measurement in single-crystal and polycrystalline ceramics using their optical fluorescence. *Journal of the American Ceramic Society*, *76*(6), 1433–1440. doi:10.1111/j.1151-2916.1993.tb03922.x

Pedrini, G., Martínez-García, V., Weidmann, P., Singh, A., & Osten, W. (2016). Optical methods for the analysis of residual stresses and measurement of displacements in the nanometric range. In *Proceedings of the 2016 IEEE 14th International Conference on Industrial Informatics (INDIN)* (pp. 570-575). IEEE. 10.1109/INDIN.2016.7819227

Ribeiro, J., Monteiro, J., Vaz, M., Lopes, H., & Piloto, P. (2009). Measurement of residual stresses with optical techniques. *Strain*, *45*(2), 123–130. doi:10.1111/j.1475-1305.2008.00421.x

Rossini, N. S., Dassisti, M., Benyounis, K. Y., & Olabi, A. G. (2012). Methods of measuring residual stresses in components. *Materials & Design*, *35*, 572–588. doi:10.1016/j.matdes.2011.08.022

Servin, M., Quiroga, J. A., & Padilla, M. (Eds.). (2014). *Fringe pattern analysis for optical metrology: theory, algorithms, and applications*. Germany: John Wiley & Sons. doi:10.1002/9783527681075

Suterio, R., Albertazzi, A., & Cavaco, M. A. M. (2003, June). Preliminary evaluation: The indentation method combined with a radial interferometer for residual stress measurement. In *SEM Ann. Conf. Exposition Exp. Appl. Mech Bethel: Society for ExperimentalMechanics; 2003* (pp. 115-121).

Withers, P. J., & Bhadeshia, H. K. D. H. (2001). Residual stress. Part 1–measurement techniques. *Materials Science and Technology*, *17*(4), 355–365. doi:10.1179/026708301101509980

KEY TERMS AND DEFINITIONS

Charge-Coupled Device: A charge-coupled device (CCD) is a device for the movement of electrical charge, usually from CCDs containing grids of pixels are used in digital cameras, optical scanners, and video cameras as light-sensing devices.

Deformation: Changes in an object's shape or form due to the application of a force or forces.

Electronic Speckle Pattern Interferometry: Electronic speckle pattern interferometry, also known as TV Holography, is a technique which uses laser light, together with video detection, recording and processing to visualize static and dynamic displacements of components with optically rough surfaces.

Interferometry: It is a family of techniques in which waves, usually electromagnetic waves, are superimposed causing the phenomenon of interference in order to extract information.

Laser: A device that stimulates atoms or molecules to emit light at particular wavelengths and amplifies that light, typically producing a very narrow beam of radiation. The emission generally covers an extremely limited range of visible, infrared, or ultraviolet wavelengths.

Michelson Interferometer: The Michelson interferometer produces interference fringes by splitting a beam of monochromatic light so that one beam strikes a fixed mirror and the other a movable mirror. When the reflected beams are brought back together, an interference pattern results.

Microelectromechanical Systems: It is a technology that in its most general form can be defined as miniaturized mechanical and electromechanical elements (i.e., devices and structures) that are made using the techniques of microfabrication.

Nondestructive Method: It is a wide group of analysis techniques used in science and technology industry to evaluate the properties of a material, component, or system without causing damage.

Optical Microscope: The optical microscope often referred to as the light microscope, is a type of microscope that commonly uses visible light and a system of lenses to magnify images of small objects. Optical microscopes are the oldest design of the microscope and were possibly invented in their present compound form in the 17th century.

Stress: Stress is defined as the resistance force acting per unit cross-section area of the body. It is also defined as the ratio of applied load to the cross-section area of the body.

Chapter 8
Optical X-Ray Diffraction Data Analysis Using the Williamson-Hall Plot Method in Estimation of Lattice Strain-Stress

Manikandan Padinjare Kunnath
Federation University, Australia

Malaidurai Maduraipandian
https://orcid.org/0000-0002-1882-154X
Indian Institute of Technology, Danbad, India

ABSTRACT

Lattice stress and strain was analysed with estimated crystalline size of the synthesised ZnFe2O4 nanoparticles from x-ray diffraction data using Williamson-Hall (W-H) method. This very peculiar method was used to analyse the other physical parameters such as strain, stress, and energy density. Values calculated from the W-H method include uniform deformation model, uniform deformation stress model, and uniform deformation energy density model. These are very useful methods to label each data point on the Williamson-Hall plot according to the index of its reflection. Particularly, the root mean square value of strain was calculated from the interplanar distance using these three models. The three models have given different strain values by reason of the anisotropic nature of the nanopartcles. The average grain size of ZnFe2O4 nanoparticles estimated from FESEM image, Scherrer's formula, and W-H analysis is relatively correlated.

DOI: 10.4018/978-1-7998-1690-4.ch008

INTRODUCTION

Spinel oxide type AB_2O_4, where A and B represent two different ionic comparable cations, are a class of chemically and thermally stable materials that are suitable for a wide-ranging application such as the catalyst and magnetic materials. In the spinel structure, oxygen ions form a cubic closed structure, and cations A and B occupy two different crystallographic sites: tetrahedral and octahedral sites (Ashtaputre et al., 2005; Gupta, 1990; Harbour & Hair, 1979; Mitra, Chatterjee, & Maiti, 1998). In the distribution of cations, A and B, these two sites are influenced by the combination and nature of the two cations and are strongly dependent on the preparation and processing conditions. Spin cation distribution has been given a lot of attention because it allows an understanding of the correlations between the structure and properties such as color, diffusivity, magnetic behavior, and optical properties, which are heavily based on better occupation with these two metal sites (Hotchandani & Kamat, 1992).In the class of nanomaterials, zinc ferrite nanocrystalline spinel ($ZnFe_2O_4$) is known as Zinc iron brown, commonly used as a catalyst, color filter for automotive lamps, and a pigment layer on luminescent materials due to their optical properties, thermal, chemical, peculiar stability, and photochemistry (Cullity & Stock, 2001; Ramakanth, 2007; Suryanarayana, 2004; Ungár, 2007). In recent years, much work has been done on the preparation of nanoscale $ZnFe_2O_4$ for optical properties . A variety of methods, such as combustion (Warren & Aver bach, 1950), Pechini method (Suryanarayana & Norton, 1998), sol-gel (Wasa, Kitabatake, & Adachi, 2004), and micro-emulsion (Zhang, Zhang, Xu, & Ji, 2006) have been successfully performed for the preparation of $ZnFe_2O_4$ nanoparticles.

XRD diffraction peak broadening of the nanoparticle is an important parameter to estimate the three main patterns, such as crystalline size, stress, and strain (Qin & Szpunar, 2005). Crystallization of the perfect powder crystals should extend in all directions with infinity lattice points. But according to the crystallography defined that no crystals can be perfect because of their finite size. This deviation of the flawless nanocrystals can be detected from the line broadening of the individual x-ray diffraction peaks. The two main properties extracted from peak width analysis are (a) crystallite size and (b) lattice strain. Crystallite size is a measure of the size of a coherently diffracting domain. The crystallite size of the particles is not generally the same as the particle size due to the presence of polycrystalline aggregates (Ramakanth, 2007). The lattice strain is the measurement of the dispersion of the lattice constants arising from the deformation of the crystal, that is, the displacement of the lattice, and this is the main source of the other strain, such as the three junctions of the grain boundaries, assembling liabilities, coherency stresses. . X-ray line broadening is used to examine the distribution, lattice displacements, reducing crystal sizes, and mixed composites, in addition to mechanical milling powders, induces greater strain (Suryanarayana & Norton, 1998). As already stated, X-ray diffraction grading analysis is a simple tool for estimating the size of the crystal and mainly the lattice strain, as well as the characteristics of the crystal. The values of the crystallite size and lattice strain obtained by using a pseudo-Voigt function, Rietveld purification, and Warren-Averbach analysis from the X-ray diffraction data(Warren & Averbach, 1952) . Scherrer first introduced the idea that a crystallite size also examined that a single crystal comprised several grains or particles. Grain had a crystal within themselves, and the size of the crystalline particles differed from that of the grain. Lattice strain can be affected by the displacement of lattice location and point defects due to crystal defect (Dorf, 2003). Two types of lattice strains in the crystals (i.e., uniform and nonuniform strains) that cause the amplitude of the peak to be found in the same crystal lattice (Gul, Maqsood, Naeem, & Ashiq, 2010). Some methods to find deformation of the sized materials have represented by Scherrer (Salavati-Niasari, Davar, & Emadi, 2010), Williamson-Hall (W-H) (Bersuker,

1997), strain–size plot (SSP), and Warren-Averbach methods (Maurizio et al., 2010). The Scherer and W-H methods set the Full-width Half Maximum (FWHM) values and integrated integral line broadening peaks, these are indicating from the Fourier coefficient profile in the Warren-Averbach method (Mimani, 2001). But in the Warren-Averbach method is mathematically complex (Irfan, Racik K, & Anand, 2018). Among these, the Williamson-Hall (W-H) analysis is a simplified, integrated Full-width method where the peak width of both the size-induced and the strain-induced broadening is decomposed into a function of 2θ .

In this present work, a comparative estimation of the average particle size of $ZnFe_2O_4$ nanoparticles obtained from direct examined FESEM image measurements, and the peak width broadening of powder X-ray diffraction (XRD) is reported.

The strain accompanying the $ZnFe_2O_4$ samples prepared by co-precipitation at 500°C due to lattice deformation was calculated by the modified form of W-H, i.e., the uniform deformation model (UDM). Other unmodified models, such as the uniform deformation stress model (UTSM) and the uniform deformation energy density model (UTETM), have provided an idea of the stress-strain relationship and strain as a function of energy density U. In UDM, this model assumes that the isotropic nature of the crystal is considered, while the USTM and UDEM crystals also assumed that the crystals have an anisotropic nature. The strain accompanying with the anisotropic nature of the simple cubic crystal is compared and conspired with the strain follow-on from the d spacing. We have reported $ZnFe_2O_4$ nanoparticles have been synthesized by the co-precipitation method in this paper. X-ray peak profile analysis was performed by origin software to determine the size of the crystals and the lattice strain of the $ZnFe_2O_4$ nanoparticles based on the UTM, USTM, and UDETM models. This work has been addressing the importance of the W-H models and the SSP method in the tuning of the crystal size and strain parameters of the $ZnFe_2O_4$ nanoparticles.

Sample Preparation

Zinc ferrite powder was prepared by two steps. Firstly, the aqueous solutions of zinc nitrate $Zn(NO_3)_2\bullet 6H_2O$ and ferric nitrate $Fe(NO_3)_3\bullet 2H_2O$ were added to the aqueous solution of sodium hydroxide NaOH with vigorous stirring. After they reacted for 2 h, the obtained precipitates were then washed with deionized water and dried at 100°C. Secondly, the precursor powder was sintered to form $ZnFe_2O_4$ crystallization under air for 2 hours at 500°C temperature.

Characterization

XRD and FESEM (Figure 1) were used to obtain the textural parameters like size, shape, and crystal structure in order to understand the enhanced properties of as-prepared and annealed $ZnFe_2O_4$ nanoparticles. The precipitation precursor was investigated by energy dispersive spectroscopy (EDS). X-ray diffraction (XRD) with Cu-Kα radiation was used for phase analysis. The surface morphology and size of the grains were studied by scanning electron microscope (SEM). It was used to examine the morphology of the annealed $ZnFe_2O_4$ nanoparticles. Crystallite size and lattice strain were determined using Scherrer's formula and W-H analysis.

Figure 1. FESEM image of the ZnFe2O4

XRD Analysis

The XRD pattern of the $ZnFe_2O_4$ powder, as shown in Figure 2 is having a, or other $ZnFe_2O_4$ phases are detected, indicating that the pure $ZnFe_2O_4$ nanoparticles are crystalline in nature. The peaks intensity is sharp and narrow, confirming that the sample is of high quality with good crystallinity and fine grain size. Using XRD data, lattice parameters were calculated $a = c = 5.2063Å$).

CRYSTALLINE SIZE AND STRAIN

W–H Method

Uniform Deformation Model (UDM)

In addition to getting influenced by the size of crystallites, the X-ray diffraction patterns are also influenced by the lattice strain and lattice defects in many cases. W-H analysis is a simple integral breath method. It clearly differentiates the armature size and strain-induced deformation peak considering the broadening of the peak width as a function of 2θ. The broadening of peaks is evidence of the occurrence of grain refinement and large strain associated with the powder. The instrumental broadening (β_{hkl}) was corrected, corresponding to each diffraction peak of $ZnFe_2O_4$ material using the relation:

$$\beta_{hkl} = \left[(\beta_{hkl})^2_{Measured} - (\beta_{hkl})^2_{Instrumental} \right]^{1/2} \quad (1)$$

The average nanocrystalline size was calculated using Debye-Scherrer's formula:

Figure 2. The XRD pattern of the $ZnFe_2O_4$ powder

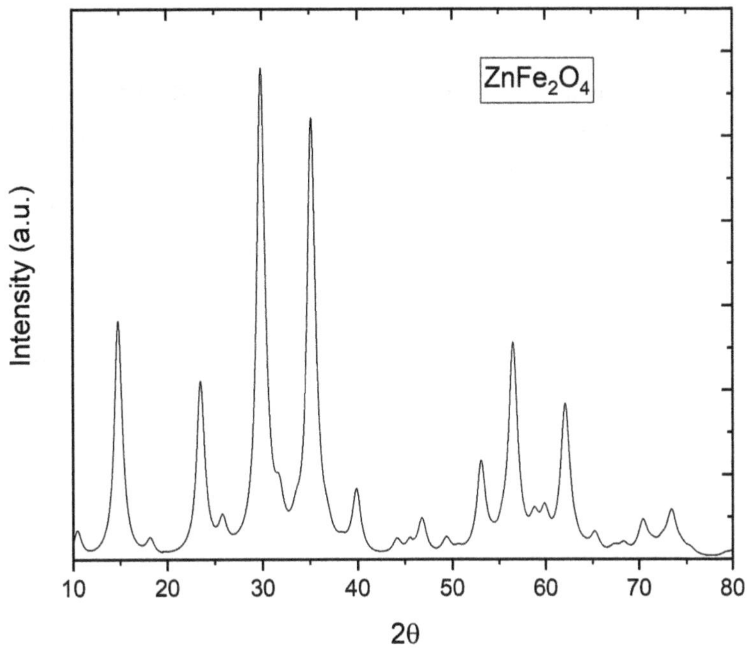

$$D = \frac{K\lambda}{\beta_{hkl}\cos\theta} \tag{2}$$

where D = crystalline size, K = shape factor (0.9), and λ= wavelength of CuK_α radiation. From the calculations, the average crystalline size of the $ZnFe_2O_4$ nanoparticles is 27nm. The strain-induced in powders due to crystal imperfection and distortion was calculated using the formula:

$$\varepsilon = \frac{\beta_{hkl}}{4\tan\theta} \tag{3}$$

From Equations 2 and 3, it was confirmed that the peak width from crystallite size varies as $1/\cos\theta$ strain varies as $\tan\theta$. If the particle size and strain contributions to line broadening are independent of each other and both have a Cauchy-like profile, the observed line breadth is simply the sum of Equations2 and 3.

$$\beta_{hkl} = \frac{K\lambda}{D\cos\theta} + 4\varepsilon\tan\theta \tag{4}$$

By rearranging the above equation, we get

$$\beta_{hkl}\cos\theta = \frac{K\lambda}{D} + 4\varepsilon\sin\theta \tag{5}$$

The above equations are known as W-H equations. The graph is drawn with 4sinθ along the *x*-axis and β_{hkl} cosθ along the *y*-axis for as-prepared $ZnFe_2O_4$ nanoparticles, as shown in Figure 2.

From the linear fit to the data, the crystalline size was estimated from the *y*-intercept, and the strain, from the slope of the fit. Equation 5 represents the UDM. Here, the strain was assumed to be uniform in all crystallographic directions, consequently considering the isotropic nature of the crystal. Also, the material properties are independent of the direction along which they are measured. The uniform deformation model for $ZnFe_2O_4$ nanoparticles is shown in Figure 3. In Equation5, the h homogeneous isotropic nature of the crystal is considered. However, in many cases, the assumption of homogeneity and isotropy is not satisfied. Moreover, all the constants of proportionality related to the stress-strain relation are no longer independent when the strain energy density *u* is taken into account. According to Hooke's law, the energy density *u* (energy per unit volume) as a function of strain is $u=\epsilon^2E_{hkl}/2$. Therefore, Equation (6) can be modified to the form, where *u* is the energy density (energy per unit volume):

$$\beta_{hkl}\cos\theta = \frac{K\lambda}{D} + 4\sin\theta\,(2u/E_{hkl})^{1/2} \tag{6}$$

The uniform deformation energy density (UDEDM) can be obtained from the slope of the line plotted between β_{hkl} cosθ and 4sinθ $(2u/E_{hkl})^{1/2}$. The lattice strain can be obtained by knowing the E_{hkl} values of the sample. W-H equations are modified assuming UDEDM and the corresponding plot are shown in Figure 4. From Equations6 and 8, the energy density and the stress can be associated with UDSM and UDEDM, but approaches are different, based on the assumption of uniform deformation stress, according to Equation6. The assumption of uniform deformation energy is as per Equation8, even though both models consider the anisotropic nature of the crystallites. From Equations6 and 8, the deformation stress and deformation energy density are related $u=\epsilon^2E_{hkl}/2$.

It may be noted that though both Equations 6 and 8 are considered in the anisotropic nature of the elastic constant, they are essentially different. This is because, in Equation 3, it is assumed that the deformation stress has the same value in all crystallographic directions allowing *u* to be anisotropic, while Equation 8 is developed assuming the deformation energy to be uniform in all crystallographic directions treating the deformation stress to be anisotropic. Thus, from Williamson-Hall plots using Equations6 and 8, a given sample may result in different values for lattice strain and crystallite size. For a given sample, Williamson-Hall plots may be plotted using Equations1 to 8, and the most suitable model may be chosen as the one, which results in the best fit of the experimental data. A comparison of the three evaluation procedures for the nanocrystalline $ZnFe_2O_4$ sample is possible from the analysis of Figure 4. The scattering of the points away from the linear expression is lesser for Figures 3and 4. Further, the average crystallite sizes are estimated from the *y*-intercept of the graphs shown in Figures 3, and 4, i.e., 36, and 36nm, respectively

Uniform Stress Deformation Model (USDM)

Uniform stress deformation and uniform deformation energy density were considered the anisotropic nature of Young's modulus of the crystal is more realistic. According to Hooke's law, within the elastic limit, there exists a linear proportionality relation between the stress (σ) and strain (ε) i.e. $\sigma = E\,\varepsilon$

Where E is the elasticity modulus or Young's modulus. This equation is an approximation that is valid for the significantly small strain. Hence, it is assumed that the lattice deformation stress is uniform in the second term of equation signifying USDM and is replaced by $\varepsilon = (\sigma/E)$. Here, the stress is proportional to the strain, with the constant of proportionality being the modulus of elasticity or Young's modulus, denoted by E_{hlk}. In this approach, the Williamson-Hall equation is modified by substituting the value of in Equation5; we get

$$\beta_{hkl}\cos\theta = \frac{K\lambda}{D} + 4\sin\theta\sigma / E_{hkl} \tag{7}$$

E_{hkl} is Young's modulus in the direction perpendicular to the set of the crystal lattice plane (hkl). The uniform stress can be calculated from the slope line plotted between $4\sin\theta/E_{hkl}$ and $\beta_{hkl}\cos\theta$ and the crystallite size D, from the intercept, as shown in Figure 3. The slope of the straight line between $4\sin\theta/E_{hkl}$ and $\beta_{hkl}\cos\theta$ gives the uniform stress, and the crystallite size D easily determined from the intercept (Figure 3). Youngs modulus E_{hkl} is related to their elastic compliances S_{ij} as (Balzar & Ledbetter, 1993)

$$E_{hkl} = \frac{\left[h^2 + (h+2k)^2/3 + (al/c)^2\right]^2}{S_{11}\left(h^2 + (h+2k)^2/3 + S_{33}(al/c)^4 + (2S_{13} + S_{44})(h^2 + ((h+2k)^2/3)(al/c)^2)\right)} \tag{8}$$

Young's modulus E_{hkl} for samples with a cubic crystal phase is related to their elastic compliances S_{ij} (for $ZnFe_2O_4$, $S_{11} = 6.07$, $S_{12} = -3.83$ and $S_{44} = 7.22\ TPa^{-1}$) as:

$$\frac{1}{E_{hkl}} = S_{11} - 2(S_{11} - S_{12} - 0.5S_{44})(m_1^2 m_2^2 + m_2^2 m_3^2 + m_3^2 m_1^2) \tag{10}$$

where, $m_1 = h\,(h^2+k^2+l^2)^{-0.5}$, $m_2 = k\,(h^2+k^2+l^2)^{-0.5}$, and $m_3 = l\,(h^2+k^2+l^2)^{-0.5}$

Equation7 represents USDM. Plotting the values of $\beta_{hkl}\cos\theta$ as a function of $4\sin\theta/E_{hkl}$, the uniform deformation stress can be calculated from the slope of the line and lattice strain. USDM for annealed $ZnFe_2O_4$ nanoparticles is shown in Figure 3. The Young's modulus E_{hkl} value for cubic $ZnFe_2O_4$ NPs was calculated to be 2.44 TPa. Figure 3 shows the plot of $\beta_{hkl}\cos\theta$ values as a function of $4\sin(\theta)/E_{hkl}$, and the uniform deformation stress can be calibrated from the slope.

Uniform Deformation Energy Density Model (UDEDM)

The following model that can be used to find the energy density of a crystal is called the Uniform Deformation Energy Density Model (UDEDM). Previously, it was assumed that crystals are homogeneous

Figure 3. USDM plot for $ZnFe_2O_4$-NPs calcined at 900°C

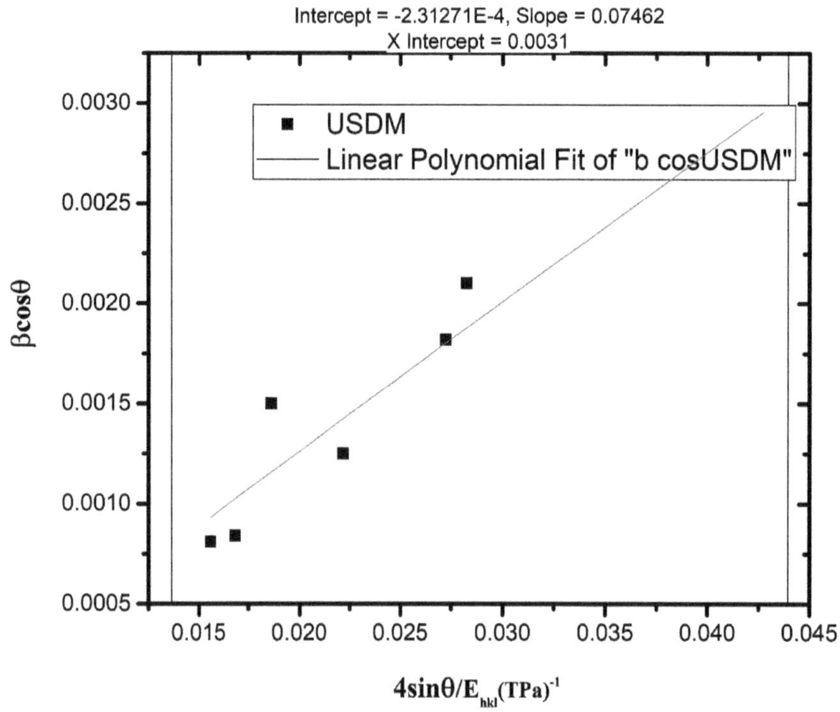

and isotropic. Although, in many cases, the assumption of homogeneity and isotropy is not justified. In addition, the proportional constants for the strain–stress relationship are not widely independent when studying the deformation energy density (u). For an elastic system that follows Hooke's law, the energy density (unit energy) can be calculated from the relation $u = (\varepsilon^2 E_{hkl})/2$. Thus, Equation (9) can be rewritten according to the energy and strain relationship, that is,

$$\beta_{hkl}\cos\theta = \frac{K\lambda}{D} + 4\sin\theta\,(2u/E_{hkl})^{1/2} \tag{11}$$

The plot of $\beta_{hkl}\cos\theta$ versus $4\sin\theta(2u/E_{hkl})^{1/2}$ is shown in Figure 4. The anisotropic energy density (u) is estimated from the slope and the crystallite size (D) of the Y-intercept.

SSP Method

The W–H plot showed that the line broadening was isotropic. This indicates that the diffraction domains were isotropic, and there was also a microstrain involvement. However, in the case of isotropic line broadening, it is possible to acquire a better calculation of the size–strain considering that an average ''size–strain plot" (SSP), which has relatively less weight gain, is given to high-angle reflections, where accuracy is generally lower. In this approach, the "crystalline dimension" profile is assumed to be described by a Lorentz function and the "strain profile" of a Gaussian function. As a result, we have:

Figure 4. UDEDM plot for $ZnFe_2O_4$-NPs calcined at 900°C

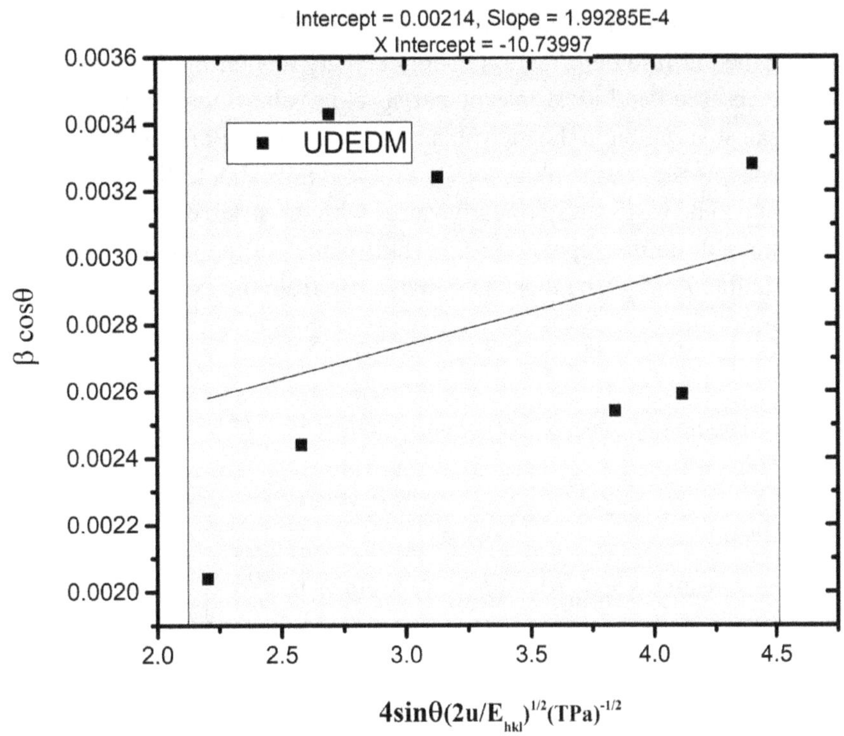

$$(d_{hkl}\beta_{hkl}\cos\theta)^2 = \left(\frac{K}{Dv}\right)(d_{hkl}^2\beta_{hkl}\cos\theta) + \left(\frac{\varepsilon}{2}\right)^2 \quad (12)$$

where K is a constant dependent on the shape of the particles, for spherical particles, it is given as 03/04. In Figure 5, like W–H methods, the term $(d_{hkl}\beta_{hkl}\cos\theta)^2$ is plotted with respect to $d^2_{hkl}\beta_{hkl}\cos\theta$ for all $ZnFe_2O_4$-NPs orientation peaks. In this case, the particle size is determined by the slope of the linearly fitted data, and the intercepted root yields strain. The size of crystallites (D) varies with the result of the calcination temperature obtained by the Scherrer formula, W–H (UDM, USDM, and UD-EDM) models and SSP methods are shown in Figure 5. It can be observed that the values of the average crystallite size obtained from the UDEDM, UDM, and UDSM are in accordance with the results of the TEM analysis. Thus, it may be concluded that these models are more pragmatic in the present case. The values of crystallites obtained from the three models are in good agreement with the values obtained from Scherrer's formula and TEM. The results we obtained from our experiments are more accurate than the reported literature. As far as authors are concerned, a detailed study using these models on the $ZnFe_2O_4$-synthesized sample annealed at 500°C is not reported yet. This study throws some more light and reveals the importance of models in the determination of particle size of $ZnFe_2O_4$ nanomaterials. We suggest that these three models are the best models for the evaluation of the crystallite size of $ZnFe_2O_4$ nanoparticles. This agrees with the results of Rosenberg et al. that for metallic samples with cubic structures, the uniform deformation energy model is suitable. In our case, all three models are found worthy of the determination of crystallite size.

The root means square (RMS) lattice strain, $\langle \varepsilon_{rms} \rangle = (2/\pi)^{1/2} (\Delta d / d_0)$ estimated (Mote, Purushotham, & Dole, 2012) from the observed variation in the interplanar spacing values are plotted against those estimated using the uniform deformation energy density model, E_{hkl} as shown in Figure 5. Here, d and $d_0$represent the observed and ideal interplanar spacing values, respectively. Figure 5 shows the plot of RMS strain against variation in the interplanar spacing for the uniform deformation energy density model. Theoretically, if the strain values agree, all the points should lie on a straight line with an angle of 45° to the x-axis. The RMS strain linearly varies with the strain calculated from the interplanar spacing, which attributed to no discrepancy on the (hkl) planes in the nanocrystalline nature. Lattice strain in the nanocrystalline $ZnFe_2O_4$ nanoparticles may arise from the excess volume of grain boundaries due to dislocations.

Table 1. Geometric parameters of the $ZnFe_2O_4$ nanoparticles

Scherrer's method	**W-H method**								
	UDM		**USDM**			**UDEDM**			
***D* (nm)**	***D* (nm)**	**$\varepsilon \times 10^{-4}$**	***D* (nm)**	**Σ (MPa)**	**$\varepsilon \times 10^{-4}$**	***D* (nm)**	***U* (kJm^3)**	**Σ (MPa)**	**$\varepsilon \times 10^{-4}$**
27	35	0.00131	36	166	0.00130	36	67	100.5	0.00113

Figure 5. Size–strain plot of $ZnFe_2O_4$-NPs calcined at 900°C

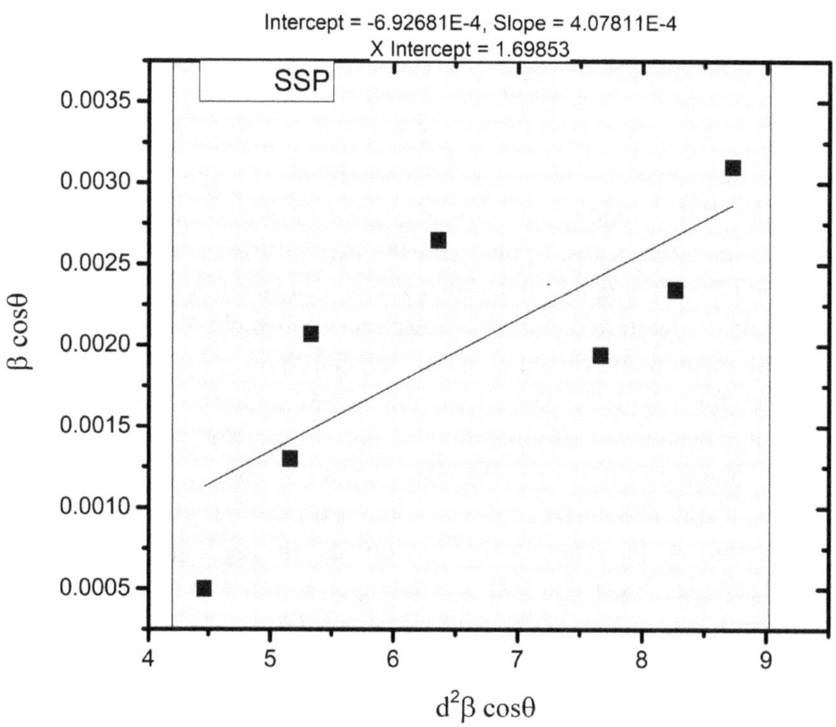

Young's modulus (*E*) is calculated using Equation7 for $ZnFe_2O_4$-NPs and is approximately 127GPa, which agrees with the bulk $ZnFe_2O_4$. Summarization of the geometric parameters of $ZnFe_2O_4$-NPs obtained from Scherrer's formula is given in the table 1, various modified forms of W-H analysis, and TEM results. By comparing the values of average crystallite size obtained from UDM, UDSM, and UDEDM, it was found that the values are almost alike, suggesting that the incorporation of strain in various forms has a very small consequence on the average crystallite size of $ZnFe_2O_4$nanoparticles.

However, the average crystallite size obtained from Scherrer's formula and W-H analysis shows a small disparity; this is because of the difference in averaging the particle size distribution. The values of strain from each model are calculated by considering Young's modulus E_{hkl} to be 127GPa. The average crystallite size and the strain values obtained from the graphs plotted for various forms of W-H analysis, i.e., UDM, UDSM, and UDEDM, were found to be precise, comparable, and fair, as their entire preferred high-intensity points lay close to the linear fit.

Micrographs measured by the FESEM instrument are given in Figure 5. It can be seen that the crystalline grains are distributed between several hundred nanometers and several micrometers, and the grains grow surprisingly well with increasing temperature, indicating that over-warming is beneficial for Zinc ferrite formation and crystallization.

CONCLUSION

$ZnFe_2O_4$ nanoparticles were synthesized and collected by the coprecipitation process and characterized by particle XRD and Feschem analysis. The size enhancement of ZnFe2O4 nanoparticles due to the crystal size and strain defined from the powder XRT analysis was analyzed by the Scherrer formula. The size and strain contributions to the line expansion were analyzed by the Williamson and Hall method with uniform dispersion pressure, uniform dispersion pressure, and uniform dispersion energy density models.

The uniform dispersion energy and density model makes the strain model very accurate. The modified W – H plot was developed and adopted to determine the crystal size and the width induced by the strain of the lattice decay. W-H analysis based on the UTM, UTSM, and UTETM models was very helpful in quantifying the crystal size and strain. With the assumption of a cubic anisotropic crystalline nature, the RMS lattice strain differs from the strain calculated from UTSM and UTETM. The FESEM image of the annual $ZnFe_2O_4$ nanoparticles reveals the nature of the nanocrystal, and their particle size is found to be 50nm. W-H has been developed and established to determine the size of the crystals and the longitudinal induced decay due to the deformation of the network. UDM-based U-DMS, UDSM, and UDEDM analyzes are useful for dimensional estimation and calculation of crystals. The size and strain of crystals estimated by XRT powder measurements are in good agreement with the FESEM results. The elastic properties of the Young Sij (Ehkl) module was estimated by the values of the lattice (h, k, l) plane. The three modified forms of the W-H analysis were used to determine the value of strain, stress, and energy density with a specific approximation, and therefore, it is desirable to define the crystal completeness of these models. The value of crystal size calculated from the W-H analysis is in agreement with the average crystal size measured from the FESEM.

REFERENCES

Ashtaputre, S. S., Deshpande, A., Marathe, S., Wankhede, M., Chimanpure, J., Pasricha, R., & Kulkarni, S. (2005). Synthesis and analysis of ZnO and CdSe nanoparticles. *Pramana*, *65*(4), 615–620. doi:10.1007/BF03010449

Balzar, D., & Ledbetter, H. (1993). Voigt-function modeling in Fourier analysis of size-and strain-broadened X-ray diffraction peaks. *Journal of Applied Crystallography*, *26*(1), 97–103. doi:10.1107/S0021889892008987

Bersuker, I. B. (1997). Limitations of density functional theory in application to degenerate states. *Journal of Computational Chemistry*, *18*(2), 260–267. doi:10.1002/(SICI)1096-987X(19970130)18:2<260::AID-JCC10>3.0.CO;2-M

Cullity, B. D., & Stock, S. R. (2001). *Elements of X-ray Diffraction* (Vol. 3). Prentice Hall.

Dorf, R. C. (2003). *CRC Handbook of Engineering Tables*. CRC Press. doi:10.1201/9780203009222

Gul, I., Maqsood, A., Naeem, M., & Ashiq, M. N. (2010). Optical, magnetic and electrical investigation of cobalt ferrite nanoparticles synthesized by co-precipitation route. *Journal of Alloys and Compounds*, *507*(1), 201–206. doi:10.1016/j.jallcom.2010.07.155

Gupta, T. K. (1990). Application of zinc oxide varistors. *Journal of the American Ceramic Society*, *73*(7), 1817–1840. doi:10.1111/j.1151-2916.1990.tb05232.x

Harbour, J. R., & Hair, M. L. (1979). Radical intermediates in the photosynthetic generation of hydrogen peroxide with aqueous zinc oxide dispersions. *Journal of Physical Chemistry*, *83*(6), 652–656. doi:10.1021/j100469a003

Hotchandani, S., & Kamat, P. V. (1992). Photoelectrochemistry of semiconductor ZnO particulate films. *Journal of the Electrochemical Society*, *139*(6), 1630–1634. doi:10.1149/1.2069468

Irfan, H., Racik, K. M., & Anand, S. (2018). Microstructural evaluation of CoAl2O4 nanoparticles by Williamson–Hall and size–strain plot methods. *Journal of Asian Ceramic Societies*, *6*(1), 54–62. doi:10.1080/21870764.2018.1439606

Maurizio, C., El Habra, N., Rossetto, G., Merlini, M., Cattaruzza, E., Pandolfo, L., & Casarin, M. (2010). XAS and GIXRD study of Co sites in CoAl2O4 layers grown by MOCVD. *Chemistry of Materials*, *22*(5), 1933–1942. doi:10.1021/cm9018106

Mimani, T. (2001). Instant synthesis of nanoscale spinel aluminates. *Journal of Alloys and Compounds*, *315*(1-2), 123–128. doi:10.1016/S0925-8388(00)01262-7

Mitra, P., Chatterjee, A. P., & Maiti, H. S. (1998). ZnO thin film sensor. *Materials Letters*, *35*(1-2), 33–38. doi:10.1016/S0167-577X(97)00215-2

Mote, V., Purushotham, Y., & Dole, B. (2012). Williamson-Hall analysis in estimation of lattice strain in nanometer-sized ZnO particles. *Journal of Theoretical and Applied Physics*, *6*(1), 6. doi:10.1186/2251-7235-6-6

Qin, W., & Szpunar, J. (2005). Origin of lattice strain in nanocrystalline materials. *Philosophical Magazine Letters*, *85*(12), 649–656. doi:10.1080/09500830500474339

Ramakanth, K. (2007). *Basics of X-ray Diffraction and its Application.* IK.

Salavati-Niasari, M., Davar, F., & Emadi, H. (2010). Hierarchical nanostructured nickel sulfide architectures through simple hydrothermal method in the presence of thioglycolic acid. *Chalcogenide Letters*, *7*(12), 647–655.

Suryanarayana, C. (2004). *Mechanical Alloying and Milling*. Marcel Dekker.

Suryanarayana, C., & Norton, M. G. (1998). *Practical aspects of X-ray diffraction X-Ray Diffraction.* Springer.

Ungár, T. (2007). Characterization of nanocrystalline materials by X-ray line profile analysis. *Journal of Materials Science*, *42*(5), 1584–1593. doi:10.100710853-006-0696-1

Warren, B., & Averbach, B. (1950). The effect of cold-work distortion on X-ray patterns. *Journal of Applied Physics*, *21*(6), 595–599. doi:10.1063/1.1699713

Warren, B., & Averbach, B. (1952). The separation of cold-work distortion and particle size broadening in X-ray patterns. *Journal of Applied Physics*, *23*(4), 497–497. doi:10.1063/1.1702234

Wasa, K., Kitabatake, M., & Adachi, H. (2004). *Thin film materials technology: sputtering of control compound materials*. Springer Science & Business Media.

Zhang, J. M., Zhang, Y., Xu, K.-W., & Ji, V. (2006). General compliance transformation relation and applications for anisotropic hexagonal metals. *Solid State Communications*, *139*(3), 87–91. doi:10.1016/j.ssc.2006.05.026

Chapter 9
Deformation Assessment of Stainless Steel Sheet Using a Shock Tube

Saibal Kanchan Barik
Indian Institute of Technology, Guwahati, India

Niranjan Sahoo
Indian Institute of Technology, Guwahati, India

Nikki Rajaura
Indian Institute of Technology, Guwahati, India

ABSTRACT

In the present study, a high-velocity sheet metal forming experiment has been performed using a hemispherical end nylon striker inside the shock tube. The striker moves at a high velocity and impacts the sheet mounted at the end of the shock tube. Three different velocity conditions are attained during the experiment, and it helps to investigate the forming behavior of the material at different ranges of velocity conditions. Various forming parameters such as dome height, effective strain distribution, limiting strain, hardness, and grain structure distribution are analysed. The dome height of the material increases monotonically with the high velocity. The effective-strain also follows the similar variation and a biaxial stretching phenomenon is observed. The comparative analysis with the quasi-static punch stretching process illustrates that the strain limit is increased by 40%-50% after the high-velocity forming. It is because of the inertial effect generated on the material during the high-velocity experiment, which stretches the sheet further without strain localization.

DOI: 10.4018/978-1-7998-1690-4.ch009

INTRODUCTION

Austenitic stainless steel has been more desirable for different structural applications, particularly in the field of automotive industries due to its relatively high strength, high formability and increased resistance to corrosion (Campos, Butuc, Gracio, Rocha, & Duarte, 2006), (Chen, Gau, & Lee, 2009). The application of the SS 304L grade steel sheet is vital in many industries. However, its plastic deformation behavior is affected mainly by its various material properties such as strain hardening properties, strain rate sensitivity, and anisotropic ratio (Makkouk et al. 2008), (Talyan, Wagoner, & Lee, 1998). The existing forming properties of the material can be enhanced either by conducting the forming process at higher-temperature or at a higher-strain rate (Stachowicz, Trzepieciński, & Pieja, 2010), (Lichtenfeld, Van Tyne, & Mataya, 2006). To avoid the martensitic transformation and omit the annealing process, the worm forming is useful but, it adds the cost of the forming process and makes the environment unpleasant for the people work on it (Kim, Huh, Bok, & Moon, 2011). Thus, the plastic deformation at a high strain rate becomes more popular for the last two decades (Bronkhorst et al. 2006), (Kim et al. 2011).

In order to characterize the material properties, a quasi-static tensile test and split-Hopkinson pressure bar test are used for so long (Lee, & Lin, 2001), (Gilat, Kuokkala, Seidt, & Smith, 2017). Though the engineering stress and strain data are quite accurate to predict the material properties at the different regimes of strain rate, the uni-axially obtained results restrict their usage during real-time application. It may not be wise to interpolate the uni-axial results for the calculation of the bi-axial properties of the material. Therefore, several types of research have been employed in the field of quasi-static as well as dynamic loading conditions to characterize the bi-axial material properties (Acharya et al. 2019).

Many researchers have used quasi-static bulge testing technology since 1940, and it has become an established experimental technique for the determination of forming limit diagrams of the sheet material (Koç, Billur, & Cora, 2011), (Kaya, 2016). The forming limit diagram (FLD) proposed by Keeler (Keeler, 1989) is one of the major characterization techniques for the prediction of failure of the sheet material and it has been used widely by the sheet metal forming industries to evaluate the formability of the material by minimizing the shop floor trials. There are several standard experimental techniques are also available for the material forming behavior analysis such as Erichsen cup test, limiting dome height (LDH) test and Fukui conical cup test, etc (Logesh, Raja, & Velu, 2015), (Meuleman, Siles, & Zoldak, 1985), (Gerdeen, & Daudi, 1983). These all hydraulic operated experimental facilities enable us to perform the test only in the lower magnitude of strain rate. However, various modification on the simple bulge test has been performed to study the material properties at a higher range of strain rate. Broomhead and Grieve (Broomhead, & Grieve, 1982) used a drop hammer rig to apply pressure loading which helped to determine the FLD of low carbon steel with a strain rate up to 70 s^{-1}. In order to obtain the multi-axial material properties under higher strain rate conditions, Grolleau et al. (Grolleau, Gary, & Mohr, 2008) modified the conventional Split Hopkinson Pressure Bar (SHPB) apparatus as a dynamic bulging device to perform biaxial tests on the material at higher strain rate but, in a limitation, several complexities have been observed in the experimental set up during the experiment (Ramezani, & Ripin, 2010).

The high energy rate forming (HERF) processes like electro-hydraulic forming (EHF), electromagnetic forming (EMF) and explosive forming (EF) have been developed to generate dynamic loading on the material (Rohatgi et al. 2012) (Oliveira, Worswick, Finn, & Newman, 2005), (Mynors, & Zhang, 2002). All the HERF processes generate the impulsive wave, which hits and transmits through the material. It results in the generation of inertial forces, which reduces the load in the necking area and widens the forming region, and it turns into an increase in the total elongation. Generally three influencing factors should be accounted for this kind of hyper-plasticity effect such as, inertia stabilization (Balanethiram, Hu, Altynova, & Daehn, 1994), tool/ sheet impact (Oliveira et al. 2005) and the changes in the constitutive behaviour (Verleysen, Peirs, Van Slycken, Faes, & Duchene, 2011). Along with this, the additional benefit of HERF experiments can be the reduced spring back due to the impulsive nature of the pressure wave which deforms the sheet. Besides several advantages of these advanced manufacturing processes, the major limitations of these processes are higher capital cost, the requirement of skilled personnel, complexity in the instrumentation and difficulties in handling.

A shock tube has been introduced in many experimental studies to investigate the material properties under dynamic loading conditions. The shock wave generated inside the shock tube is generally a planer shock wave and it can be easily controlled and measured. This facilitates its application in the variety of studies to understand the dynamic properties of both the sheet and the composite materials (Wang, & Shukla, 2010), (Tekalur, Bogdanovich, & Shukla, 2009). The experimental studies by Stoffel et al. (Stoffel, Schmidt, & Weichert, 2001) and Kumar et al. (Kumar, LeBlanc, & Shukla, 2011) have focused on the dynamic response of the metallic plates when subjected to different levels of blast and shock loading. In another study, Stoffel (Stoffel, 2007) compared the mechanical response of the dynamically deformed plates with the quasi-statically deformed plates by simulation using the visco-plastic model. Justusson et al. (Justusson, Pankow, Heinrich, Rudolph, & Waas, 2013) used a shock tube to extract the bi-axial rate-dependent mechanical properties of the thin homogeneous material and demonstrated it as a dynamic bulge testing device to obtain the material properties in an intermediate to high range of strain rate. The microstructural and textural response of the thin Al sheets was studied by Ray et al. (Ray, Jagadeesh, & Suwas, 2015) and the change in grain size was not that significant. However, the intra-granular misorientation was evident. A similar type of study has been demonstrated by Bisht et al. (Bisht, Kumar, Subburaj, Jagadeesh, & Suwas, 2019) where the variation in crystallographic orientation has been studied on a pure copper sheet. The path of the texture evolution showed deformation bands and deformation twins which are strain-rate dependent.

The dynamic properties of the material can be studied at a higher range of strain rate by applying the impulsive forces by a rigid body. Rusinek et al. (Rusinek, Rodríguez-Martínez, Arias, Klepaczko, & López-Puente, 2008) used the Hopkinson tube to perform the direct impact of a hemispherical projectile on the mild steel at different velocities. Different failure modes were observed and the perforation process was simulated by the application of 3D analysis using ABAQUS/Explicit FE code (Arias, Rodríguez-Martínez, & Rusinek, 2008). A gas gun has been used in the various study to drive a projectile to impact on a metallic plate and the impact situation was simulated to analyze the perforation behavior of the different shape of the projectiles and the factors that influence the deformation behavior of the target plate (Gupta, Iqbal, & Sekhon, 2006).

From the authors' knowledge, it has not been observed that the shock tube can be used to drive a projectile at a certain velocity in a controlled manner. Thus, in the present work, a hemispherical projectile made of nylon is used as a punch that travels down the tube at a high velocity with the help of the pressure-induced due to the shock wave and hits the target material mounted at the end of the shock tube. Along with this, fewer research works have focused on the variation of microstructural configuration after the projectile deforms the material. Therefore, the current study focuses on the variation of the microstructure and the hardness of the material after deformation has been observed.

Stainless steel of 304L grade material is used in the automotive and industrial sectors very commonly. Thus, an SS 304L sheet of 1 mm thickness has been used during the current study to analyze the variation in the forming parameters when it is deformed under high-velocity forming using the shock tube. The hemispherical projectile moves at a high-velocity after receiving energy from the shock wave induced mass motion and deforms the sheet kept at the end. It depicts the multi-axial stretching of the material at a high-strain rate, which gives a clear indication of the rise in the forming parameters after the comparison with the quasi-static punch stretching process. The forming parameters such as dome height, effective strain distribution, strain limit analysis, microstructural configuration and hardness variation after deformation are analyzed after deformation.

EXPERIMENTAL METHODOLOGY

Material Specification

During the present experimental investigation, the forming behavior of SS 304 L sheet of 1 mm thickness has been studied. All the specimens are cut of dimension 150 × 150 mm^2 and have a free area to deform of diameter 110 mm. In order to obtain the mechanical properties of the sheet material, tensile samples are cut along 0^0, 45^0 and 90^0 to the rolling direction as per ASTM E8 and tested at a cross-head speed of 1 mm/min using a universal testing machine (UTM) of 200 kN capacity. The anisotropic coefficients of the material along all the three directions are also obtained as per ASTM E517. All the mechanical properties are demonstrated in table 1. Along with this, the chemical composition of the material is obtained by EDX analysis, as shown in table 2. Circular grids of dimension 1.924 ± 0.002 mm are printed on the opposite surface to the loading direction on the sheet as shown in Figure 1 to measure the effective strain generated on the material after deformation.

Figure 1. Circular grids printed on the surface of the specimen for the strain measurement

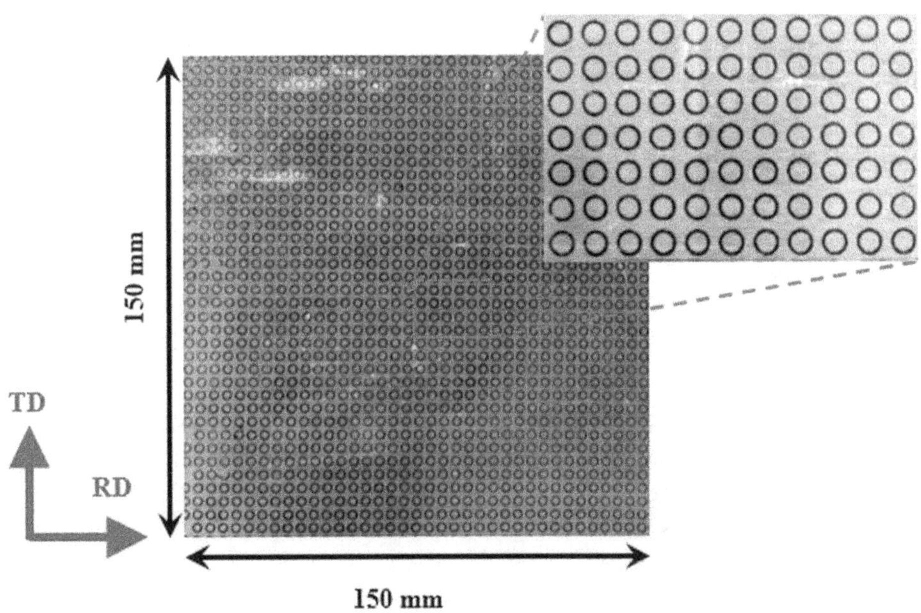

Table 1. Mechanical properties of SS 304L sheet

Material SS 304L (1mm)	Mg	Cu	Si	Fe	Mn	Cr	Al
	2.68	0.10	0.92	0.31	0.13	0.32	Balance

Table 2. Chemical composition of SS 304L sheet (Weight percentage)

Material	RD	σ_{ys} (MPa)	*UTS* (MPa)	*n*	*K* (MPa)	e_u* (%)	e_t* (%)	*r*
SS 304L (1 mm)	0°	351 ± 3	807 ± 3	0.35	1533 ± 5	55.5 ± 1.4	58.5 ± 1.2	1.22
	45°	357 ± 3	778 ± 3	0.34	1522 ± 4	56.6 ± 1.8	60.5 ± 1.8	1.15
	90°	353 ± 4	778 ± 2	0.34	1451 ± 4	55.7 ± 1.2	59.4 ± 1.1	1.18

RD: Rolling direction; σ_{ys}: yield strength; *UTS*: ultimate tensile strength; *n*: strain hardening coefficient; *K*: strength coefficient; e_u: uniform elongation; e_t: total elongation; *r*: plastic strain ratio. * At 25 mm gauge length.

Experimental Facility

The shock tube is a pneumatically controlled device which is divided into two sections, such as a high-pressure driver section and a driven section. Both the sections are divided by a metallic or plastic diaphragm. When the pressure difference reaches the critical limit, the diaphragm ruptures and the sudden release of the gas generates a shock wave. The impulsive nature of the shock wave can be used to study the dynamic behavior of the material at different loading conditions. The evolved shock wave has sufficient strength to drive a rigid body at a high-velocity inside the tube. This facilitates the use of a hemispherical striker as a punch inside the shock tube, which moves at a high velocity and strikes the sheet material kept at the end of the shock tube. This experimental condition develops a dynamic loading environment that deforms the material at a high-velocity.

The shock tube used in the present work is 4 m long consisting of 2 m driver section and 2 m driven section as shown in Figure 2. All the tubes have the same inner diameter of 55 mm and having a thickness of 11 mm. Different layers of Mylar sheet of thickness 0.1 mm have been used as a diaphragm to obtain the different magnitude of shock wave during the experiment. The shock waves of different magnitude drive the hemispherical striker at different velocities. The hemispherical striker used during the experiment is a Nylon rod of overall length 95 mm and having diameter 54.7 mm as shown in Figure 3. The sheet material is kept in between two flanges as shown in Figure 2 and a free area to deform of diameter 110 mm is kept open for free deformation. Both the flanges are clamped properly to avoid slipping out of the testing sample during the experiment. Two pressure transducers (PCB Piezotronic; USA; Model 113B22) of sensitivity 14.62 mv/MPa have been mounted on the shock tube to measure the incident, reflected and the magnitude of shock Mach number (M_s) during the experiments. The pressure transducers are connected to a signal conditioner before it is connected to a Tektronix oscilloscope. The oscilloscope used during the experiment has model number Tektronix MDO 3024 which is having a maximum sampling rate of 2.5 GS/s.

Figure 2. Shock tube experimental set-up

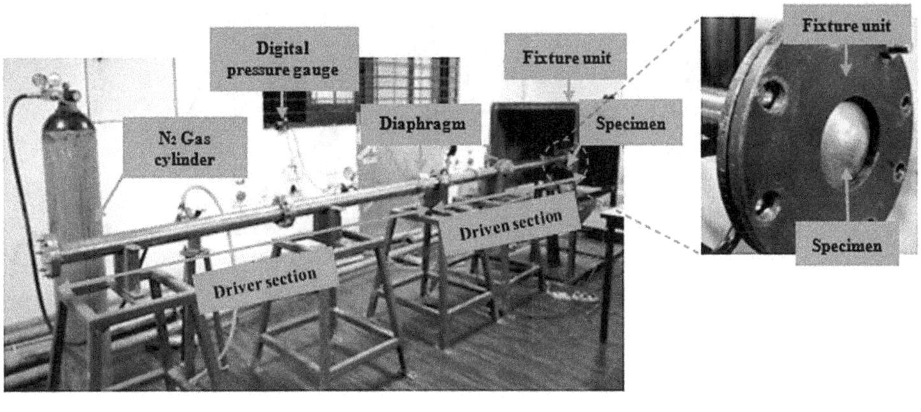

Figure 3. The dimension of the striker used during the experiment

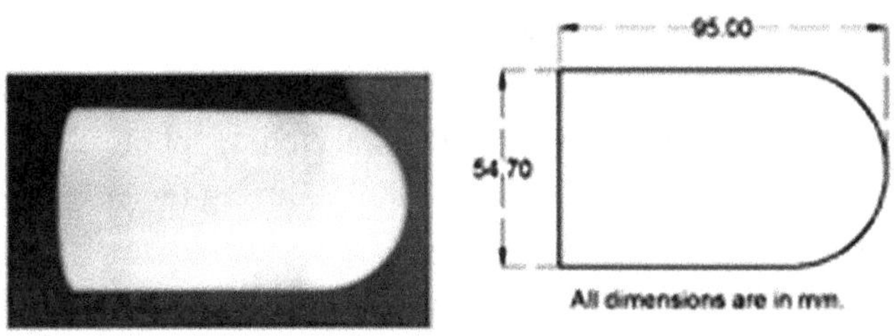

In order to measure the velocity of the striker, infrared sensors are connected on the driven section of the tube as depicted in Figure 4. The infrared sensor is an electronic instrument that is used to sense the opaque bodies. It consists of an IR emitter and an IR receiver. The infrared emitter is generally a light-emitting diode (LED) that emits infrared radiation, whereas the infrared receiver detects the radiation from an IR transmitter. When the infrared signal bounces back from the surface of an object, the signal is received by the infrared receiver. Both the sensors are connected to the LM 358M integrated circuit. The circuit is connected to a 5V DC power supply. The voltage output from the circuit is connected to the oscilloscope to acquire the results. During the current experiment, both the IR emitter and the receiver are mounted on the driven section of the shock tube at a distance of 130 mm from the end of the shock tube and both are facing towards each other. The emitted infrared radiation falls directly on the receiver and when the striker moves at a high velocity and crosses between both the sensors, there is an interruption in the signal and it results in a sudden voltage drop at the output. When the striker passes out from the sensors, again the receiving diode receives the infrared radiation and results in a further rise in output voltage. This time interval of voltage fluctuation becomes very crucial during the calculation of the striker velocity by dividing the time of voltage drop to its self-length. All the experiments are conducted thrice to ensure repeatability during the experiment.

Figure 4. IR sensors mounted on the shock tube to measure the velocity of the striker

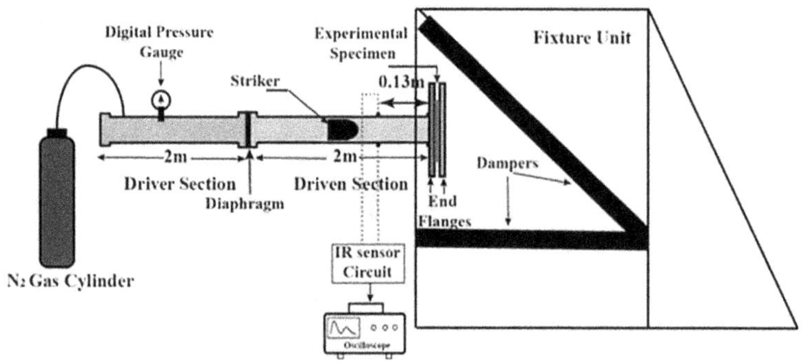

Strain Distribution Analysis

The variation of strain after deformation is a major parameter to measure the formability of the material. Higher uniformity of the strain distribution results in higher limit strains leading to better formability. Several techniques have been developed to measure the effective stress and strain generated on the material during deformation such as screen printing, electrochemical etching, photo-chemical etching, serigraphy and laser marking (Lee, & Hsu, 1994), (Ozturk et al. 2009). During the current experiment, the screen-printing method was preferred because of the accuracy of the grid geometry with a lower capital cost. The diameter of the grids is measured using a Nikon AZ100 Macroscope having a precision of 0.001 mm. The average grid diameter printed on the surface of the samples is having a diameter of 1.924 ± 0.005 mm. After deformation, the circular grids are changed to elliptical grids and the major and minor diameter of the grids located on the deformed region are measured. Hill's 1948 yield criterion has been used during the calculation of the effective strain (Hosford, & Caddell, 2011).

$$\bar{\varepsilon}^{2} = \frac{(G+H)}{(FG+FH+GH)^{2}}\left[F^{2}(G+H)\varepsilon_1^2 + G^{2}(F+H)\varepsilon_2^2 + H^{2}(F+G)\varepsilon_3^2\right] \quad (1)$$

where, $\varepsilon_1 = \ln {}^{d_1}\!/_{d_0}$ = Major strain, $\varepsilon_2 = \ln {}^{d_2}\!/_{d_0}$ = Minor strain, $\varepsilon_3 = \ln {}^{t}\!/_{t_0}$ = thickness strain, $\bar{\varepsilon}$ = effective strain, $F=r_{RD}$ = Anisotropic coefficient across rolling direction, $G=r_{TD}$ = Anisotropic coefficient across transverse direction, $H=r_{RD}\, r_{TD}$.

The limit of deformation has been set by the velocity of the striker because it is not possible to stop the test exactly at the necking in such a high-velocity experiment. During this bi-axial high-velocity forming process, the generated limiting strain is compared with the conventional punch stretching process in which the samples are stretched up to the initiation of necking. The circular grids located near the necking location are considered for the analysis during the calculation of failure strain in both the forming processes and compared both the results. The height of deformation is measured using a coordinate measuring machine having a precision of 0.001 mm.

RESULTS AND DISCUSSION

Calibration of Pressure Generated in the Shock Tube

During the experiment, a hemispherical striker has been used as a punch and it is kept inside the driven section of the shock tube at a distance of 800 mm from the end of the shock tube as shown in Figure 5. The shock wave is generated in the shock tube by bursting different layers of Mylar diaphragm. Two pressure transducers are mounted at a distance of 385 mm and 885 mm from the end of the shock tube. Three different pressure bursting conditions are generated during the experiment and the pressure signals obtained from the experiment are plotted in the Figure. 6. The magnitude of incident shock pressure reflected shock pressure and the shock Mach number (M_s) generated during the experiments are depicted in Table 3. Different pressure magnitude provides different loading conditions and it helps to accelerate the rigid striker at a different velocity, which allows studying the rate-dependent material forming parameters of SS 304L steel sheets.

Figure 6 shows the pressure variation with and without striker at different bursting pressures, which illustrates the dissipation of pressure energy during the experiment to drive the striker kept inside the shock tube. The pressure signals obtained from the pressure transducers without striker shows the actual pressure variation inside the shock tube. The first step of pressure variation is because of the incident shock wave. When the incident shock wave hits the end flange of the shock tube and returns back, the pressure wave turns in to reflected shock wave which induces higher pressure and temperature than the incident one. When the striker is kept inside the shock tube, the shock wave induced mass motion hits the striker and reflects back as depicted from the pressure variation data shown in Figure 6. The higher

Figure 5. Schematic diagram of the shock tube with pressure transducers

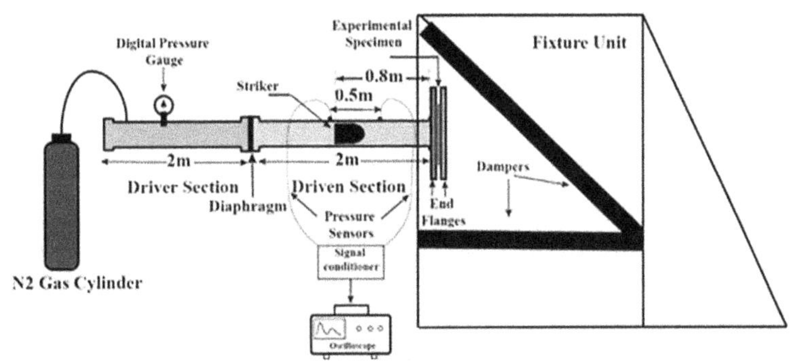

Table. 3. Average pressure generated during the experiments

Average burst pressure (bar)	Average incident pressure (bar)	Average reflected pressure (bar)	Average Shock Mach number (M_s)
11.5 ± 1.1	3.35 ± 0.5	11.2 ± 0.4	1.61 ± 0.02
17.8 ± 1.1	3.61 ± 0.6	15.25 ± 0.5	1.71 ± 0.02
24.2 ± 1.1	4.29 ± 0.4	19.95 ± 0.5	1.92 ± 0.02

Figure 6. Pressure-time history for the bursting pressure (a) 11.5 ± 1.1 bar (b) 17.8 ± 1.1 bar (c) 24.2 ± 1.1 bar

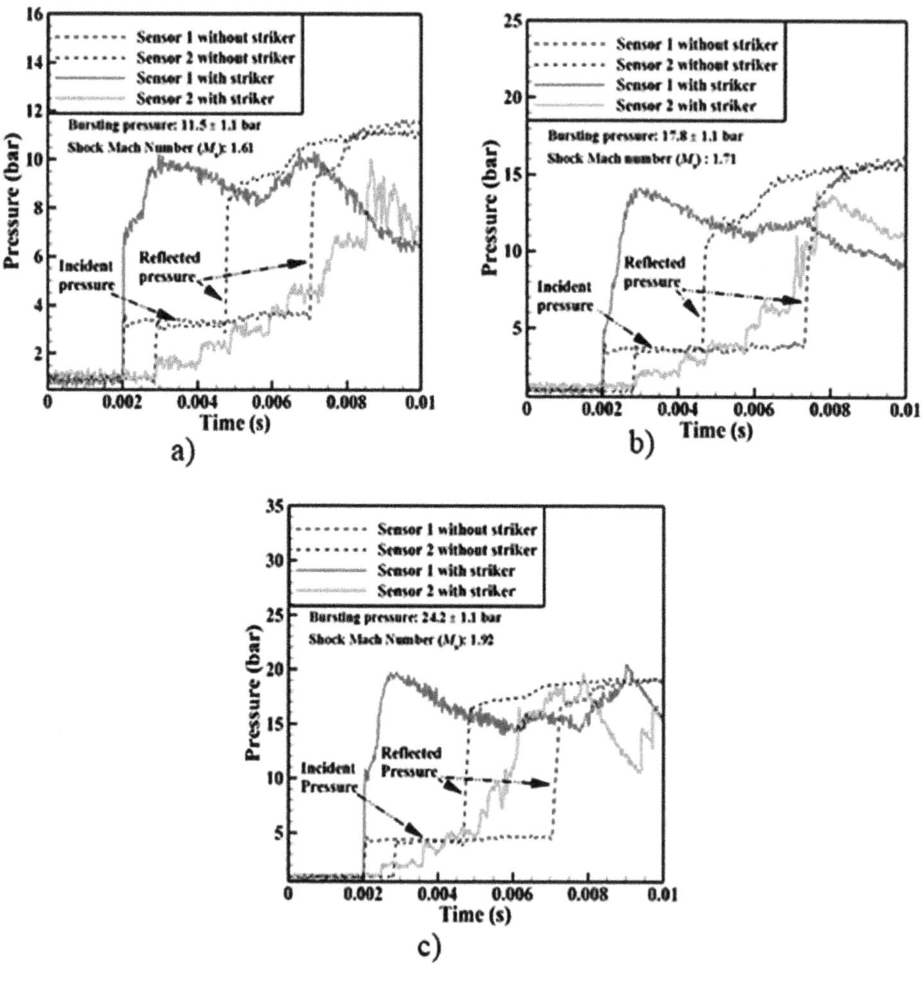

pressure condition of the reflected shock wave helps to keep the momentum of the striker. The sensor 1 shows the actual magnitude of pressure hitting the striker whereas sensor 2 shows the dissipation rate of the pressure energy.

Calibration of the Velocity of the Striker

The shock wave-induced mass motion accelerates the motion of the striker and keeps the momentum of the strike before it hits the specimen kept at the end of the shock tube. The IR sensors mounted on the shock tube sense the striker when it passes between them and give a voltage output, as illustrated in Figure 7. The time interval of voltage drop can be divided with the self-length of the striker and the final velocity of the striker hitting the sheet can be found out. Table 4. Specifies the velocity of the striker at different experimental conditions taken for the analysis.

Figure 7. Signal obtained from the IR sensor for the bursting pressure 11.5 ± 1.1 bar

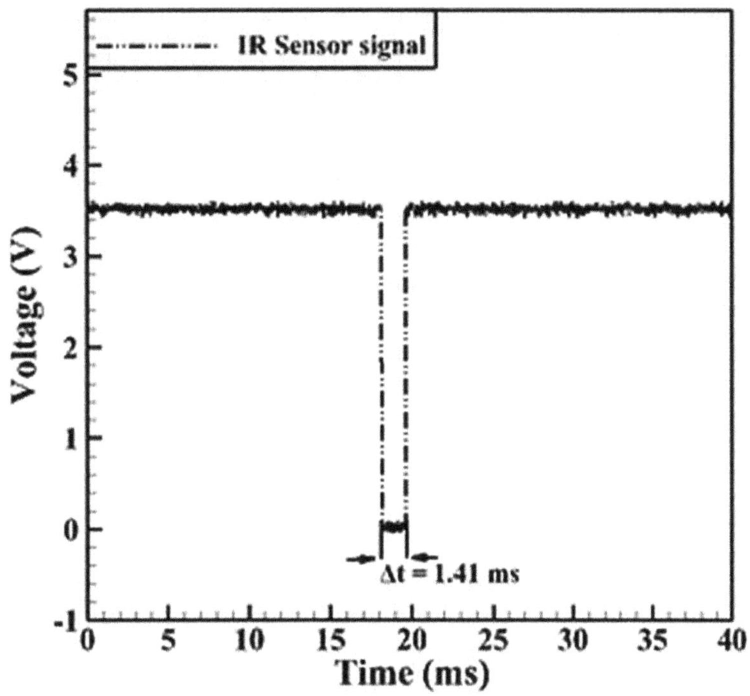

Table 4. Velocity of the striker obtained from the IR sensor

Avg. Bursting pressure (bar)	Avg. Shock Mach number (M_s)	Time *(ms)*	Velocity of the striker *(m/s)*
11.5 ± 1.1	1.61 ± 0.02	1.41 ± 0.06	67.37 ± 1.48
17.8 ± 1.1	1.71 ± 0.02	1.20 ± 0.08	78.62 ± 1.85
24.2 ± 1.1	1.92 ± 0.02	1.05 ± 0.07	90.47 ± 1.67

Free Forming Deformation Analysis

The sheet material is mounted rigidly between the two flanges during the experiment and the sheet is allowed for free deformation. In order to analyze the forming behavior, the sheet material is deformed at a different velocity of the striker conditions and the height of the deformation is measured and plotted in Figure 8.

The results illustrate that the sheet is deformed uniformly in the exposed area. The height of deformation increases monotonically at higher velocity. The high-velocity forming phenomenon stretches the sheet uniformly because of the inertial effect. The inertial forces are stabilized by generating additional tensile stresses on the material, which helps to stretch the material even more. The deformation profile in both X-axis and Y-axis signifies the sheets are deformed bi-axially. The deformed sheets at three different velocity conditions are illustrated in Figure 9. The deformation profile of the sheets signify the uniform deformation of the sheet, but the slope of deformation near the striker hitting zone is slightly sharper in

Figure 8. Deformation along (a) X-axis and (b) Y-axis at three different velocity Conditions

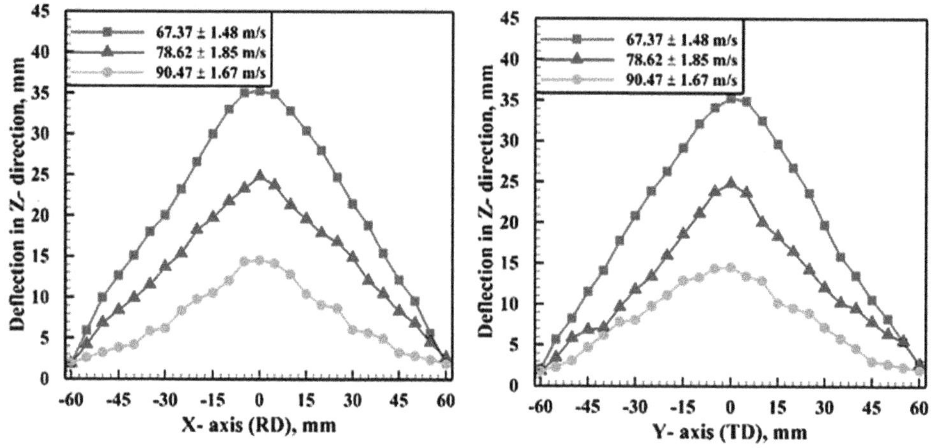

Figure 9. Deformed sheets at three different velocity conditions

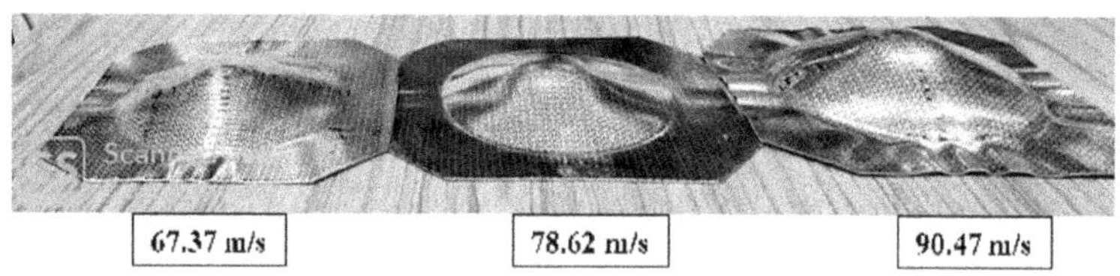

all the three cases. It accentuates the stretching of the material is more around the striker hitting zone, but because of the inertial stabilization, the sheet has stretched further without strain localization. Due to the lack of the gripping force, wrinkles appear on the gripping zone during deformation, but it does not affect the magnitude of deformation.

Effective Strain Distribution Analysis

The diameter of the circular grids printed on the sheets is measured along both the X-axis and Y-axis and the effective strain at each grid point is calculated by using Hill's 1948 yield criterion. The results plotted in Figure 10 illustrates the strain variation in uniform without strain localization. Higher uniformity in the strain distribution accentuates higher limiting strains which lead to better formability. The limit of deformation is decided by the velocity of the striker hitting the specimen. The tests are performed until the safe strains are obtained because it is not possible to stop the test exactly at the necking at such high-velocity experiments. The results depict three different velocity variations of the striker, which results in characterizing the material forming process within the safe limit. The frictional constraint between the sheet and the striker can be ignored as the effective strain variation in both the X-axis and Y-axis doesn't show any evidence of strain localization even if the striker-hitting areas have a larger magnitude of strain.

The percentage of rising in strain generated on the material increases monotonically with the rise in velocity without any tensile failure. However, in the general limiting dome height test, the evolved limiting strain for SS 304L sheet of 1 mm thickness is remaining in the range of 0.3 ± 0.02 whereas, the present high-velocity forming experiment induces safe strain in the range of 0.4 ± 0.02. Several reasons

Figure 10. Effective strain distribution along (a) X-axis and (b) Y-axis for three velocity conditions

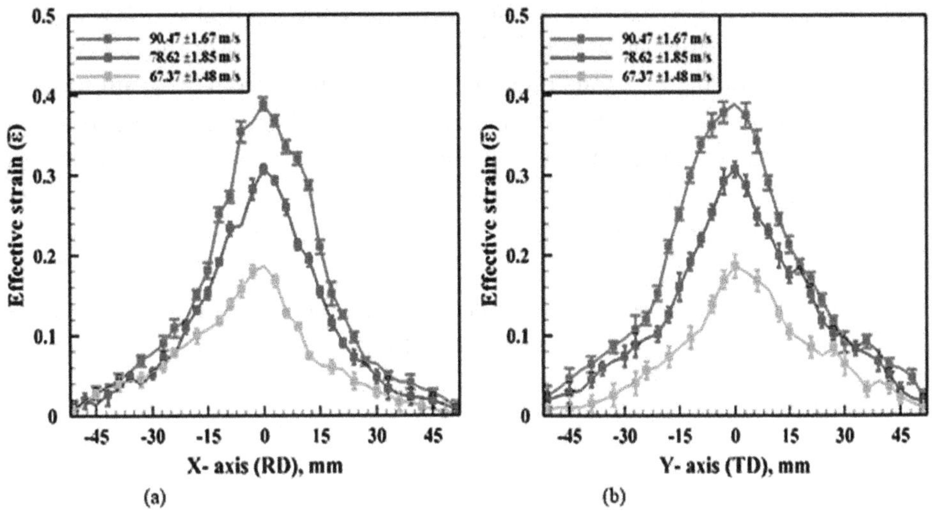

or agreements have been postulated for the rise in the forming limit of the material during the high-velocity experiments. The major important factor is the inertial stabilization which prolonged the forming limits of the material during the deformation. Almost in all quasi-static hydraulic forming experiments, the local velocity gradient in the local sites varies simultaneously, which results in the formation of onset necking. On the other hand, during the high-velocity forming process, the local velocity gradient is minimized by generating additional tensile stresses outside the necking region. This helps to deform the material further before it fails due to necking and results to generate higher strain as compared to the conventional forming processes.

Hardness and Grain Size Measurement

The variation of hardness after deformation is measured along the thickness direction for all the three samples, as depicted in Figure 11, and it is observed that the sheets are strain hardened after deformation. The hardening behavior of the material increases significantly after plastic deformation. The rate of strain hardening rises with respect to the rise in the velocity of the forming. This substantial increase in the hardness is attributed due to the change in the constitutive behavior of the material. A similar observation has been illustrated by Huang et al. (Huang, Matlock, & Krauss, 1989). During material deformation analysis during the electromagnetic forming process and quoted that the dislocation wall strengthening is the dominant mechanism which increases the strain hardening behavior of the material. During the current forming process, the hardness variation of the sheet after deformation varies uniformly like the effective strain, which is also become good evidence of no further strain localization.

To further investigate the plastic deformation mechanism at different loading conditions, the microstructure of the deformed sheets at the central location has been evaluated where the deformation is severe. Figure 12 shows the grain boundary configuration of the deformed sheets. The results are compared with the grain structure of the material before deformation which is an evenly distributed equiaxed grain pattern having the average grain diameter of 11 μm. After the deformation, the grains are stretched apparently along the rolling direction and small grains are generated as depicted in Figure 10.

Similar observation has been observed during the high strain rate forming of SS 304 sheet by Chen et al. (Chen, Ruan, Wang, Chan & Wang, 2011) and reported that the different strain rate of deformation produces different volume of dislocation which glides to martensite transformation and then leads to deformation twinning with increasing strain rate. In the present study, from the grain structure analysis, it is observed that the high-velocity of deformation nucleates small grains, and the heavy distortion in the grain structure may lead to martensite transformation. The mechanical properties are the embodiment of the microstructure. The increase in the hardness along the thickness direction is attributed due to the misorientation of the grain structure which strengthens the material after deformation (Hedayati, Najafizadeh, Kermanpur, & Forouzan, 2010).

The grain structure of the material after deformation under the quasi-static punch stretching test is compared with the parent material, and it is illustrated in Figure 13. From the comparison of the results with the high-velocity experiments (as shown in Figure 12), it is observed that the grains of the material is stretched after deformation, but the percentage of nucleation of grains is lesser than the high-velocity deformation. This may be attributed due to the localized deformation near the punch due to the friction during the quasi-static stretching process. This elucidates smaller and localized misorientation of the grain structure.

Strain Limit Analysis

In order to analyze the strain limit of the SS 304L sheet material under bi-axial loading at both quasi-static and high-velocity experiments, the sheets are stretched up to failure and the limiting major strain and minor strain magnitude are calculated and compared in Figure 14. From the results, it can be concluded that all the points during high-velocity experiments lie in the 39.7% - 68.8% (major strain) and 38.5% -

Figure 11. Variation of hardness on the deformed location at different loading conditions

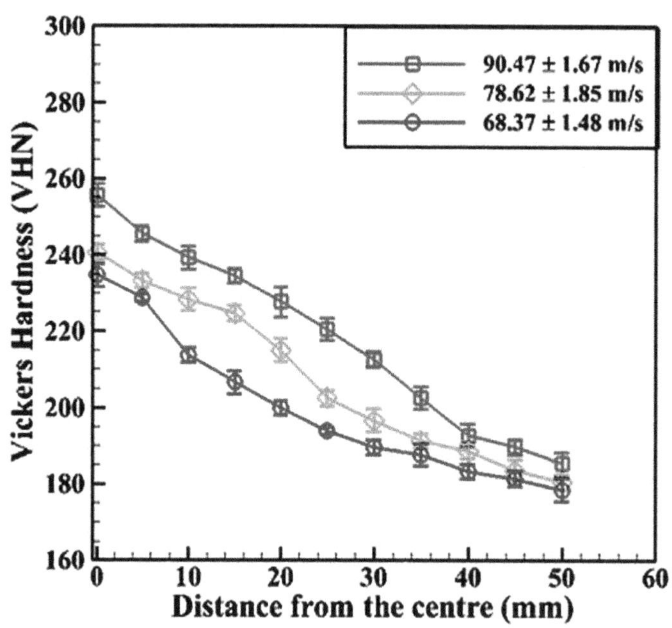

Figure 12. Grain structure configuration of (a) the parent sheet and the deformed sheets at (b) V = 68.37 m/s, (c) V = 78.62 m/s and (d) V = 90.47 m/s

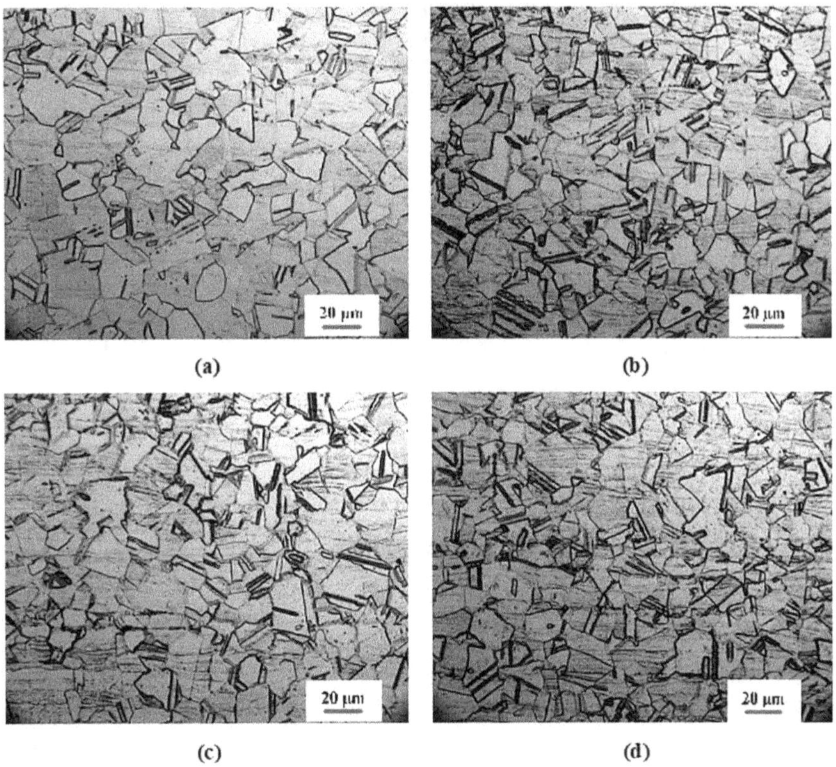

Figure 13. The grain structure of (a) the parent material and (b) quasi-statically deformed sheet at 5 mm/min

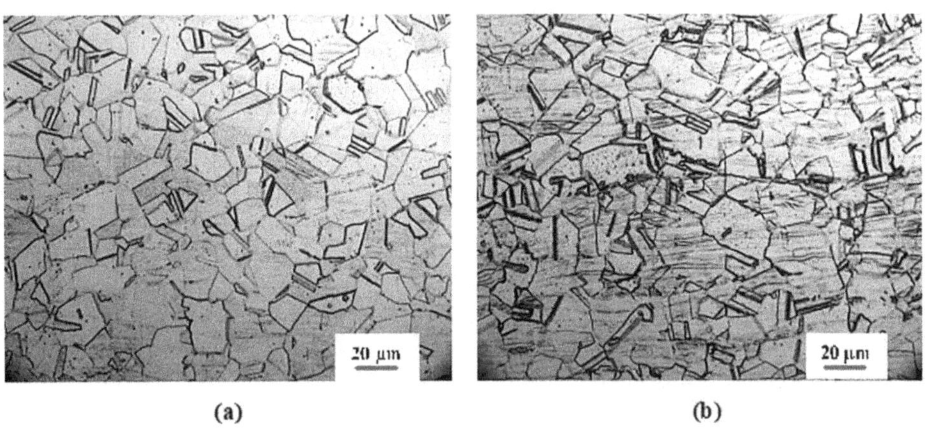

Figure 14. Limiting strain comparison for quasi-static and high-velocity experiments

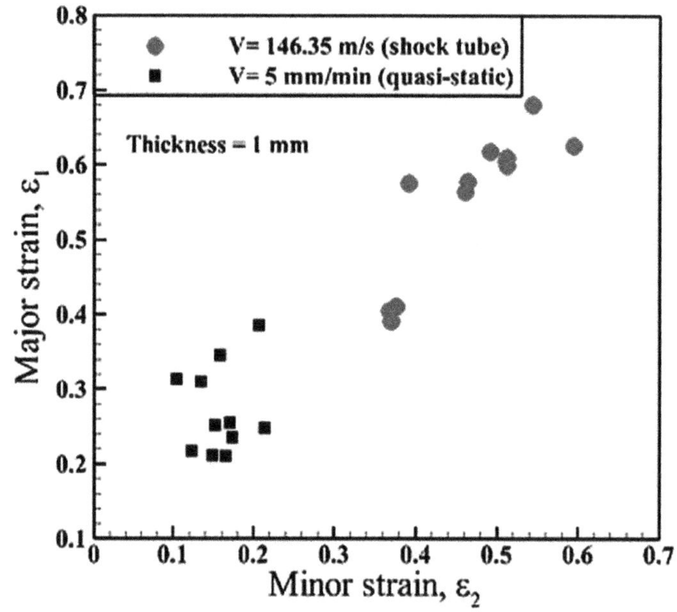

62.4% (minor strain) strain range. However, in the case of quasi-static experiments, the limiting strains are lying in the range of 21.6% - 41.5% of major strain and 12.8% - 22.8% of minor strain respectively. A significant improvement in formability is observed during the comparative analysis.

The most dominant phenomenon during the high-velocity forming experiments is the inertial stabilization (Balanethiram et al. 1994). The inertial forces developed during these processes help to stretch the material without a sharp velocity gradient in between the two adjacent points on the material surface, which further encourages the material to stretch without any strain localization. Along with this, the higher dislocation density developed on the material after deformation results into a higher tendency of multi-slip motion of dislocation and helped to deform the material even more which is absent in quasi-static stretching processes (Balanethiram et al. 1994; Liu, Yu, & Li, 2012).

CONCLUSION

A high-velocity forming experiment has been performed using a shock tube, where the pressure generated due to the shock wave drives a hemispherical end projectile at a high velocity. The projectile acts like a punch and deforms the SS 304L sheet of 1 mm thickness at a high-velocity. From the experimental analysis, the following conclusions are derived:

1. The deformation profile after deformation signifies the uniform bi-axial stretching of the material. With the rise in the velocity, the material deforms monotonically without strain localization. The deformation along both the X-axis and Y-axis signifies the biaxial deformation of the material.

2. The effective strain distribution analysis demonstrates the uniform stretching of the material. The material stretching is more on the striker hitting zone, but this does not result in strain localization at any point. The uniform distribution of strain confirms that the friction is less between the striker and the sheet, and the deformation is larger near the pole region.
3. The hardness distribution of the deformed sheets confirms that the sheets are hardened due to the strain hardening effect. The grain structure configuration of the deformed sheets near to the pole region accentuates the nucleation of the new grains and misorientation of the existing equiaxed grains of the base material after the high-velocity deformation which is the main cause of the rise in the hardness of the material after deformation.
4. The limiting strains of the material after the deformation are compared with the quasi-static experimental results and concluded that the limit strains of the material are increased by 40% -50% during the bi-axial high-velocity experiments. The significant improvement in the strain limit is because of the inertial stabilization, which deforms the sheet further without strain localization.

REFERENCES

Acharya, S., Moitra, A., Bysakh, S., Nanibabu, M., Krishanan, S. A., Mukhopadhyay, C. K., & Ghosh, K. S. (2019). Effect of high strain rate deformation on the properties of SS304L and SS316LN alloys. *Mechanics of Materials*, *136*, 103073. doi:10.1016/j.mechmat.2019.103073

Arias, A., Rodríguez-Martínez, J. A., & Rusinek, A. (2008). Numerical simulations of impact behavior of thin steel plates subjected to cylindrical, conical and hemispherical non-deformable projectiles. *Engineering Fracture Mechanics*, *75*(6), 1635–1656. doi:10.1016/j.engfracmech.2007.06.005

Balanethiram, V. S., Hu, X., Altynova, M., & Daehn, G. S. (1994). Hyperplasticity: Enhanced formability at high rates. *Journal of Materials Processing Technology*, *45*(1-4), 595–600. doi:10.1016/0924-0136(94)90404-9

Bisht, A., Kumar, L., Subburaj, J., Jagadeesh, G., & Suwas, S. (2019). Effect of stacking fault energy on the evolution of microstructure and texture during blast assisted deformation of FCC materials. *Journal of Materials Processing Technology*, *271*, 568–583. doi:10.1016/j.jmatprotec.2019.04.029

Bronkhorst, C. A., Cerreta, E. K., Xue, Q., Maudlin, P. J., Mason, T. A., & Gray Iii, G. T. (2006). An experimental and numerical study of the localization behavior of tantalum and stainless steel. *International Journal of Plasticity*, *22*(7), 1304–1335. doi:10.1016/j.ijplas.2005.10.002

Broomhead, P., & Grieve, R. J. (1982). The effect of strain rate on the strain to fracture of a sheet steel under biaxial tensile stress conditions. *Journal of Engineering Materials and Technology*, *104*(2), 102–106. doi:10.1115/1.3225043

Campos, H. B., Butuc, M. C., Gracio, J. J., Rocha, J. E., & Duarte, J. M. F. (2006). Theoretical and experimental determination of the forming limit diagram for the AISI 304 stainless steel. *Journal of Materials Processing Technology*, *179*(1-3), 56–60. doi:10.1016/j.jmatprotec.2006.03.065

Chen, A. Y., Ruan, H. H., Wang, J., Chan, H. L., Wang, Q., Li, Q., & Lu, J. (2011). The influence of strain rate on the microstructure transition of 304 stainless steel. *Acta Materialia*, *59*(9), 3697–3709. doi:10.1016/j.actamat.2011.03.005

Chen, C. H., Gau, J. T., & Lee, R. S. (2009). An experimental and analytical study on the limit drawing ratio of stainless steel 304 foils for microsheet forming. *Materials and Manufacturing Processes*, *24*(12), 1256–1265. doi:10.1080/10426910903129786

Gerdeen, J. C., & Daudi, A. (1983). Formability Limits of Relatively Thick Sheet Metal as Determined by the Fukui Conical Cup Test. *Journal of Engineering for Industry*, *105*(4), 276–281. doi:10.1115/1.3185899

Gilat, A., Kuokkala, V. T., Seidt, J. D., & Smith, J. L. (2017). Full-field measurement of strain and temperature in quasi-static and dynamic tensile tests on stainless steel 316L. *Procedia Engineering*, *207*, 1994–1999. doi:10.1016/j.proeng.2017.10.1057

Grolleau, V., Gary, G., & Mohr, D. (2008). Biaxial testing of sheet materials at high strain rates using viscoelastic bars. *Experimental Mechanics*, *48*(3), 293–303. doi:10.100711340-007-9073-5

Gupta, N. K., Iqbal, M. A., & Sekhon, G. S. (2006). Experimental and numerical studies on the behavior of thin aluminum plates subjected to impact by blunt-and hemispherical-nosed projectiles. *International Journal of Impact Engineering*, *32*(12), 1921–1944. doi:10.1016/j.ijimpeng.2005.06.007

Hedayati, A., Najafizadeh, A., Kermanpur, A., & Forouzan, F. (2010). The effect of cold rolling regime on microstructure and mechanical properties of AISI 304L stainless steel. *Journal of Materials Processing Technology*, *210*(8), 1017–1022. doi:10.1016/j.jmatprotec.2010.02.010

Hosford, W. F., & Caddell, R. M. (2011). *Metal forming: mechanics and metallurgy*. Cambridge University Press. doi:10.1017/CBO9780511976940

Huang, G. L., Matlock, D. K., & Krauss, G. (1989). Martensite formation, strain rate sensitivity, and deformation behavior of type 304 stainless steel sheet. *Metallurgical Transactions. A, Physical Metallurgy and Materials Science*, *20*(7), 1239–1246. doi:10.1007/BF02647406

Justusson, B., Pankow, M., Heinrich, C., Rudolph, M., & Waas, A. M. (2013). Use of a shock tube to determine the bi-axial yield of an aluminum alloy under high rates. *International Journal of Impact Engineering*, *58*, 55–65. doi:10.1016/j.ijimpeng.2013.01.012

Kaya, S. (2016). Nonisothermal warm deep drawing of SS304: FE modeling and experiments using servo press. *International Journal of Advanced Manufacturing Technology*, *83*(5-8), 1047–1056. doi:10.100700170-015-7620-2

Keeler, S. P. (1989). On the Origins of the FLD. *The Minerals, Metals & Materials Society, Forming Limit Diagrams: Concepts, Methods, and Applications*, 3-8.

Kim, S. B., Huh, H., Bok, H. H., & Moon, M. B. (2011). Forming limit diagram of auto-body steel sheets for high-speed sheet metal forming. *Journal of Materials Processing Technology*, *211*(5), 851–862. doi:10.1016/j.jmatprotec.2010.01.006

Koç, M., Billur, E., & Cora, Ö. N. (2011). An experimental study on the comparative assessment of hydraulic bulge test analysis methods. *Materials & Design*, *32*(1), 272–281. doi:10.1016/j.matdes.2010.05.057

Kumar, P., LeBlanc, J., & Shukla, A. (2011). Effect of curvature on shock loading response of aluminum panels. In Dynamic Behavior of Materials (vol. 1, pp. 369-374). doi:10.1007/978-1-4614-0216-9_51

Lee, R. S., & Hsu, Q. C. (1994). Image-processing system for circular-grid analysis in sheet-metal forming. *Experimental Mechanics*, *34*(2), 108–115. doi:10.1007/BF02325707

Lee, W. S., & Lin, C. F. (2001). Effects of prestrainning on the impact response and twinning structure of 304L stainless steel. *Materials Transactions*, *42*(10), 2080–2086. doi:10.2320/matertrans.42.2080

Lichtenfeld, J. A., Van Tyne, C. J., & Mataya, M. C. (2006). Effect of strain rate on stress-strain behavior of alloy 309 and 304L austenitic stainless steel. *Metallurgical and Materials Transactions. A, Physical Metallurgy and Materials Science*, *37*(1), 147–161. doi:10.100711661-006-0160-5

Liu, D. H., Yu, H. P., & Li, C. F. (2012). Comparative study of the microstructure of 5052 aluminum alloy sheets under quasi-static and high-velocity tension. *Materials Science and Engineering A*, *551*, 280–287. doi:10.1016/j.msea.2012.05.018

Logesh, K., Raja, V. B., & Velu, R. (2015). Experimental investigation for characterization of formability of epoxy based fiber metal laminates using Erichsen cupping test method. *Indian Journal of Science and Technology*, *8*(33), 1–7. doi:10.17485/ijst/2015/v8i33/72244

Makkouk, R., Bourgeois, N., Serri, J., Bolle, B., Martiny, M., Teaca, M., & Ferron, G. (2008). Experimental and theoretical analysis of the limits to ductility of type 304 stainless steel sheet. *European Journal of Mechanics. A, Solids*, *27*(2), 181–194. doi:10.1016/j.euromechsol.2007.04.004

Meuleman, D. J., Siles, J. L., & Zoldak, J. J. (1985). *The limiting dome height test for assessing the formability of sheet steel* (No. 850005). SAE Technical Paper.

Mynors, D. J., & Zhang, B. (2002). Applications and capabilities of explosive forming. *Journal of Materials Processing Technology*, *125*, 1–25. doi:10.1016/S0924-0136(02)00413-2

Oliveira, D. A., Worswick, M. J., Finn, M., & Newman, D. (2005). Electromagnetic forming of aluminum alloy sheet: Free-form and cavity fill experiments and model. *Journal of Materials Processing Technology*, *170*(1-2), 350–362. doi:10.1016/j.jmatprotec.2005.04.118

Ozturk, F., Dilmec, M., Turkoz, M., Ece, R. E., & Halkaci, H. S. (2009, June). Grid marking and measurement methods for sheet metal formability. *5th International Conference and Exhibition on Design and Production of MACHINES and DIES/MOLDS*, 41-49.

Ramezani, M., & Ripin, Z. M. (2010). Combined experimental and numerical analysis of bulge test at high strain rates using split Hopkinson pressure bar apparatus. *Journal of Materials Processing Technology*, *210*(8), 1061–1069. doi:10.1016/j.jmatprotec.2010.02.016

Ray, N., Jagadeesh, G., & Suwas, S. (2015). Response of shock wave deformation in AA5086 aluminum alloy. *Materials Science and Engineering A*, *622*, 219–227. doi:10.1016/j.msea.2014.10.010

Rohatgi, A., Stephens, E. V., Davies, R. W., Smith, M. T., Soulami, A., & Ahzi, S. (2012). Electrohydraulic forming of sheet metals: Free-forming vs. conical-die forming. *Journal of Materials Processing Technology*, *212*(5), 1070–1079. doi:10.1016/j.jmatprotec.2011.12.014

Rusinek, A., Rodríguez-Martínez, J. A., Arias, A., Klepaczko, J. R., & López-Puente, J. (2008). Influence of conical projectile diameter on perpendicular impact of thin steel plate. *Engineering Fracture Mechanics*, *75*(10), 2946–2967. doi:10.1016/j.engfracmech.2008.01.011

Stachowicz, F., Trzepieciński, T., & Pieja, T. (2010). Warm forming of stainless steel sheet. *Archives of Civil and Mechanical Engineering*, *10*(4), 85–94. doi:10.1016/S1644-9665(12)60034-X

Stoffel, M. (2007). Experimental validation of anisotropic ductile damage and failure of shock wave-loaded plates. *European Journal of Mechanics. A, Solids*, *26*(4), 592–610. doi:10.1016/j.euromechsol.2006.12.002

Stoffel, M., Schmidt, R., & Weichert, D. (2001). Shock wave-loaded plates. *International Journal of Solids and Structures*, *38*(42-43), 7659–7680. doi:10.1016/S0020-7683(01)00038-5

Talyan, V., Wagoner, R. H., & Lee, J. K. (1998). Formability of stainless steel. *Metallurgical and Materials Transactions. A, Physical Metallurgy and Materials Science*, *29*(8), 2161–2172. doi:10.100711661-998-0041-1

Tekalur, S. A., Bogdanovich, A. E., & Shukla, A. (2009). Shock loading response of sandwich panels with 3-D woven E-glass composite skins and stitched foam core. *Composites Science and Technology*, *69*(6), 736–753. doi:10.1016/j.compscitech.2008.03.017

Verleysen, P., Peirs, J., Van Slycken, J., Faes, K., & Duchene, L. (2011). Effect of strain rate on the forming behaviour of sheet metals. *Journal of Materials Processing Technology*, *211*(8), 1457–1464. doi:10.1016/j.jmatprotec.2011.03.018

Wang, E., & Shukla, A. (2010). Analytical and experimental evaluation of energies during shock wave loading. *International Journal of Impact Engineering*, *37*(12), 1188–1196. doi:10.1016/j.ijimpeng.2010.07.003

Chapter 10
Micromotion Analysis of a Dental Implant System

R. Manimaran
SRM Institute of Science and Technology, India

Vamsi Krishna Dommeti
https://orcid.org/0000-0003-0633-9825
SRM Institute of Science and Technology, India

Emil Nutu
University Politehnica of Bucharest, Romania

Sandipan Roy
SRM Institute of Science and Technology, India

ABSTRACT

The objective of project is to reduce the micromotion of novel implant under the static loads using function of uniform design for FE analysis. Integrating the features of regular implant, a new implant model has been done. Micromotion of the novel implant was obtained using static structural FE analysis. Compared to the existing International team for implantology implants, the micromotion of the novel implant model was considerably decreased by static structural analysis. Six control factors were taken for achieving minimizes the micromotion of novel dental implant system. In the present work, uniform design technique was used to create a set of finite element analysis simulation: according to the uniform design method, all FE analysis simulation; compared to the original model, the micromotion is 0.01944mm and micromotion of improved design version is 0.01244mm. The improvement rate for the micromotion is 35.02%.

DOI: 10.4018/978-1-7998-1690-4.ch010

INTRODUCTION

Nowadays, the implant system can easily and immediately replace missing teeth or damaged teeth. Dental implants act like teeth or become permanent parts of the human mandible. Immediate loading of the dental implant system has newly gained fame due to several factors including pain, healing time, aesthetic appearance, and psychological benefits to the patient. The biological process between implants and bone is called osseointegration. The primary stability is the main role of the successive surgical process. Achieving primary stability is an important factor for successful osseointegration, such as bone quality, design of the implant, and surgical methods.

Researchers consider primary stability a requirement of successful osseointegration(Branemark, 1977). Bone quality, implant shape, and surgical technique influence the primary stability of dental implants. Implant design is the relevant parameter for gaining primary stability (Chong, Khocht, Suzuki, & Gaughan, 2009). The taper implant model provides better primary stability, and it has shown better stability through the healing process(O'Sullivan, Sennerby, & Meredith, 2004). X. Li and Dong (2017) investigated the influence of various neck designs on implant stress distribution. In this study, the researcher investigated the adult mandible with various implant neck structures and different loading conditions. Researchers indicated that various neck morphologies with v-shaped micro threads of implant influence the stress circulation around the implant-bone surfaces. Moreover, researchers observed the overall stress at the cortical bone area and around the implant neck. Researchers examined the FE analysis and investigated the load distribution of various thread profiles and material properties of the dental implant around the mandibular bone. Several researchers investigated Zirconium dental implants and observed in their study that they have high mean stress at their implant bodies and low mean stress around their bones compared to TI implants,(Shafi, Kadir, Sulaiman, Kasim, & Kassim, 2013).

Additionally, researchers investigated the influence of different thread profiles with different osseointegration conditions on stress distribution in a peri-implant bone. A study showed that high von misses stress was distributed in the mesiodistal direction and maximum stresses was distributed at the cortical bone area. In the majority of models, the degree of osseointegration increases when von misses stresses gradually increases in the supporting structure (Mosavar, Ziaei, & Kadkhodaei, 2015). Moreover, researchers carried out FE analysis to identify the length of the posterior mandible with low bone quality and the optimum ranges of the implant diameter. A diameter of four mm and a length of 12mm is the best grouping for implant in mandibles with low bone quality (T. Li et al., 2011). Using the finite element analysis method to assess the maximum von misses' stress, researchers studied jawbones with various thread widths and heights that had immediately loaded implants. Type 2 bones, thread height of more than 0.44mm and width of 0.19 to 0.23 mm is better for biomechanical properties for immediately loaded implants(Ao et al., 2010). Researchers performed both experimental investigations and finite element studies on stress distribution around dental bone-implant interface, and they carried out stress evaluation on the implants fitted with a bone model that had three thread profiles. The researchers performed photoelastic stress analysis, which has been validated by FEA results. Researchers observed the high-stress value on the reverse buttress thread implant profile in the apical region. They observed the smallest values on the v thread implant. Therefore, the v-thread implant profile is best compared with the other two because it has the highest shear stress values (Dhatrak, Shirsat, Sumanth, & Deshmukh, 2018).

Furthermore, researchers can expect a high success rate for fully and partially edentulous patients in mandible bones, and type 1, 2, and 3 bones give the best strength. Type IV bone has a thin cortical and low cancellous density. Failure of the implant can be decreased using a pre-surgical determination method of type IV bone. Research has revealed that external hexogen implants lower biomechanical behavior compared to Morse taper implant systems, which lack primary stability and show increased levels of micromotion. Further, the highest micro motion leads to a failed osteointegration process between the implant and bone (Macedo et al., 2017).The authors present the optimization method for single piece zirconium ceramic type of dental implant by finite element simulation under dynamics load, and they carried out experimental confirmation using the fatigue test.

The uniform design method determines micromotion of the dental implant, and the objective function of the statistical micro-motion results show that an optimized implant is 27.58μm less than 72.62μm for original statistical micro motion and that the safety factor also increased. Therefore, the new procedure gives a dental implant with an optimized shape. Additionally, researchers investigate fatigue behaviors and dynamic behavior of dental implants with various loading conditions compared with other works. The maximum stress value is under the limits of yield strength of prosthetic and abutment screws(Kayabaşı, Yüzbasıoğlu, & Erzincanlı, 2006). Researchers have investigated stress distribution and micro gap formation and at the connections of various types of the abutments and implants. Non-cylindrical abutments give a steady locking mechanism that decreased micromotion, and octagonal and hexagonal abutments deliver linked patterns of stress distribution and micromotion. Conical shape abutment delivered the highest amount of micromotion, whereas the tri-lobe connection gives the lowest amount of micromotion because of a polygonal shape. FGBM implants develop the mechanical behaviors of dental implants system concerning deformation and stress as well asFGBM parameter, which plays vital role in creating a better biomechanical function (Kayabaşı et al., 2006). Additionally, researchers used new experimental system to directly measure the implant displacement as the importance of occlusal loading. Considerable differences in micromotion resulted from inserting placing implants in bone with different densities (Karl, Graef, & Winter, 2015). Researchers carried out various contact connections between bone and implant simulation using three different types of implicit biomechanical models, bone deformation, and stress at the implant-bone interface. Next, researchers calculated implant deformation under an axial load 200N. Moreover, researchers conducted a micro CT-based 3D finite element computational study to asses strain magnitude around the bone structure under axially loaded, and they also investigated micromotion between implant and bone interface(Limbert et al., 2010).

Researchers investigated the micromotion analysis of bone density, implant configuration, and cortical bone thickness in immediately loaded mandible full-arch dental implant system restorations. Eliminating cantilever load during the healing time should successfully decrease the risk of extreme micromotion in patients with minimum-density cancellous bone and thin cortical bone (Sugiura, Yamamoto, Horita, Murakami, & Kirita, 2018).Additionally, researchers introduced a uniform design concept application for experimental simulation methods to reduce the micromotion of an international team for implantology dental implant systems under the dynamic loads .Also, researchers modeled new implant by mixing the characteristics of the traditional international team for implantology and nanotite implants, and they analyzed experimental simulation for the primordial model and the micromotion of the fresh edition implant was 45.11um. Later, applying the uniform design method, the micromotion decreased to 31.37um. The micro-motion development percentage was 30.5%. This statistic shows that uniform design method was a great use to decreases micromotion for fresh dental implants under dynamic loads (Cheng, Lin, Jiang, & Lee, 2015).

Furthermore, researchers have studied the dynamic FE analysis of the Zimmer implant with three different types of the dental implant system under the dynamic chewing loads. From the study by Semados, it is clear that the implant system is the minimum micromotion. The micromotion of the original Semados implant system with the Zimmer implant model was 33.39um. After researches applied the uniform design concept, the micromotion decreased to 22.22um. Moreover, the development rate of micromotion was 33.45%. This statistic shows that the uniform design method was a greatly helpful tool to decrease the micromotion for the Semados dental implant system with the Zimmer implant model(Cheng, Lin, & Jiang, 2016). In the implant-bone interface, reliability and the stability of the implant influence. Additionally, several authors have studied porous dental implants with computational techniques by varying the percentage reduction of the volume to achieve the optimum porosity for implants (Roy, Dey, Khutia, Chowdhury, & Datta, 2018; Roy, Panda, Khutia, & Chowdhury, 2014).

In this study, the authors applied a uniform design method to minimizing the micromotion of the novel dental implant system under static load using FE analysis. Furthermore, the authors compared the micro-motions for novel dental implants and studied them by static finite element analysis techniques using a uniform design concept and a new dental implant system that has been suggested.

METHODOLOGY

Combining the aspect of universal implants, the authors modeled a fresh implant. The authors modeled anew3D dental implant system using solid work software, as shown in Figure 1. New implant dimensional characteristics are: (A) diameter of an implant at top; (B) diameter of an implant at bottom; (C) depth of thread; (D) pitch thread; and(E) as given in Table 1. Figure 1 shows the important dimensions' characteristics of integrated dental implants system and elements of dental implant systems.

The authors took implant material considered as Ti and bone properties from (Cheng et al., 2015). A martial property of the Finite element model is given in Table 2.

To achieve primary stability for immediate loading conditions, the authors completed modeling through nonlinear frictional contact elements, which allowed minor displacement between bone-implant interfaces. The authors kept the frictional coefficient at 0.3 and it was completely fixed in all directions on the distal and mesial surface for the boundary condition of 3D, models as shown in Figure 2. Additionally, the authors considered the 120N load(Kasani et al., 2019) on the implant axial direction, as shown in Figure3. After applying all boundary conditions, the authors carried out all material properties and loading conditions simulation. After this step, the authors completed simulation with the help of FEA software, ANSYS. The extracted results of FE analysis and they listed in Table 3.

Table 1. Geometrical properties of the dental implant

Diameter/ Length of Implant at Top A (mm)	Implant Diameter of Body D (mm)	Depth of Thread C (mm)	Diameter of Implant at Bottom B (mm)	Pitch of Thread E (mm)
4/11	3	0.25	2	0.5

Table 2. Geometrical properties of the dental implant

Material	Density(mg *mm^3)	Young's Modulus E (MPa)	Poisson Ratio
Implant	4.5	110000	0.35
Cortical bone	2.4	13000	0.3
Cancellous bone	1.1	345	0.3

Figure 1. (a) Dental implant (b) Assembly model

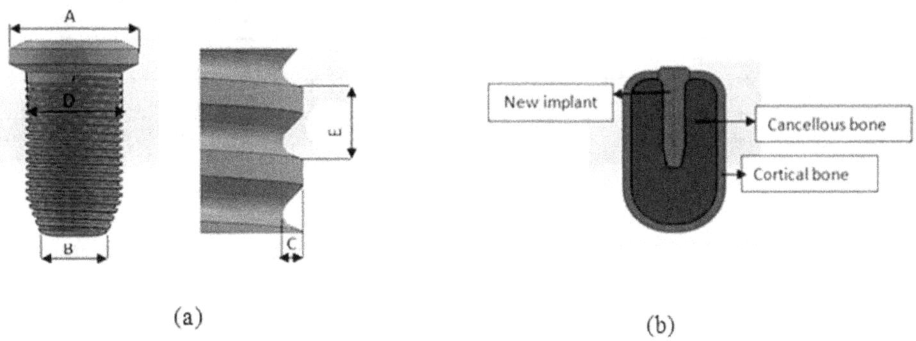

Figure 2. Nomenclature with a dental implant system

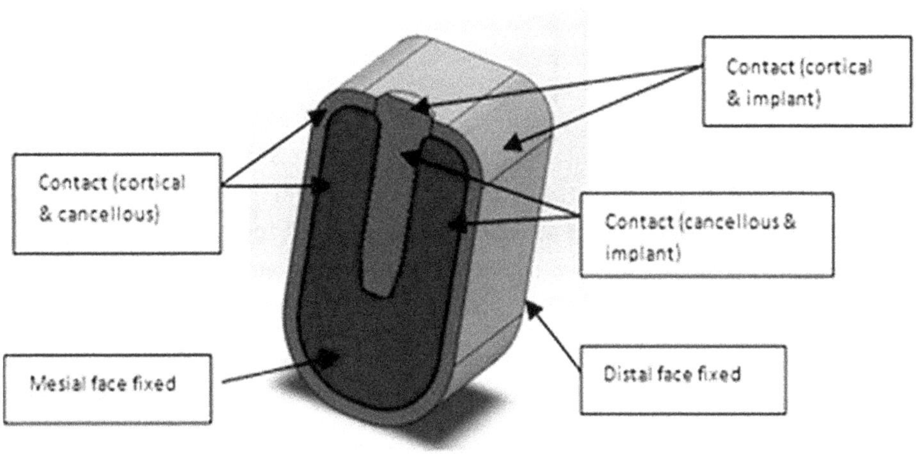

Figure 3. Load applied

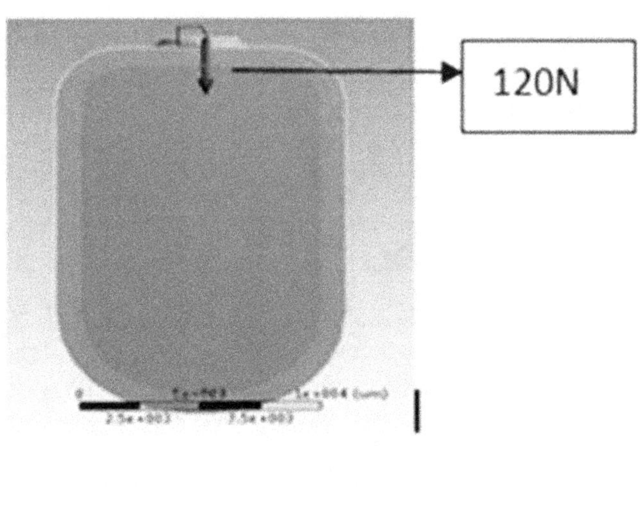

Table 3. Deformation of different elements for a new implant system

Load	120N Static Load		
Mesh element size(mm)	1	0.95	0.90
Element	37696	40156	44000
Total Deformation(mm)	0.019147	0.019714	0.01995

From the analysis results, it is evident that the micromotion of the integrated implant and bone system is 0.019147mm and 0.017356mm, respectively. FE analysis of the integrated implant system is shown in Figure 4, and FE analysis of the bone system is shown in Figure5. After that, the authors completed the convergence study for the element with mesh sizes 0.95 and 0.9, respectively. Table 3 gives the element size. Therefore, the authors selected the integrated version of the dental implant system for the enhanced design of the dimensions.

RESULTS

In this study incorporated a dental implant system that has three sets of dimension control factors: the pitch of thread, depth of thread, and a diameter of the body. The three material control factors are a density of cancellous bone, density of cortical bone, and Young's modulus of cancellous bone. The control factor for the design range is shown in Table 4. Since all control factors and design gaps also constant space, the design points are infinite in the continuous design space, and the valuation of all

Figure 4. FE analysis of implant

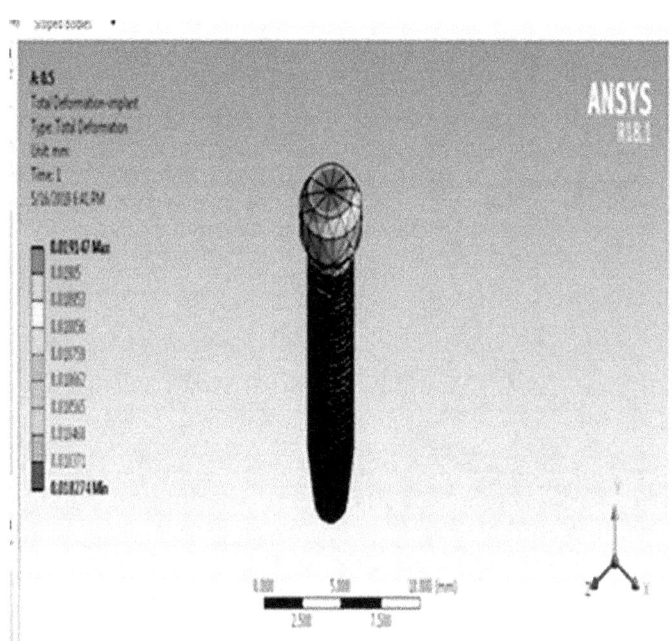

Figure 5. FE analysis of bone

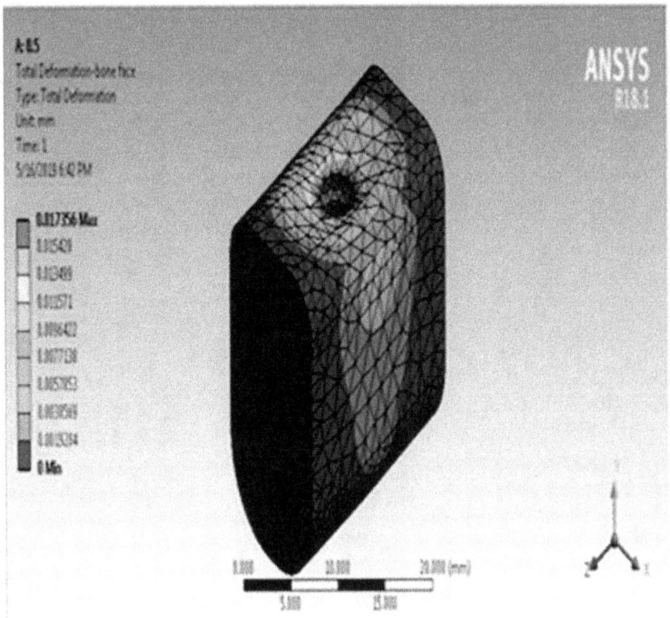

design points is not possible. Therefore, the authors applied uniform design experiments in this study, and theyproposed X.Uniform design is given in Table 5. In this study, the authors divided each control factorinto 16 levels to the limitations of computational resources.

The authors utilized the uniform design Table U (16) to form 16 experiments, as shown in Table 6. For each new dental implant system, the authors used solid works software to construct the 3D model, and they used ANSYS workbench 18.1 to calculate the reaction of the new dental implant model under static loads. The minimum micro motion happens at the tenth analysis, so that the tenth experiment is found as the better edition of the design. The improved edition of the implant system causes the micromotion of 0.01244mm, which means the primary stability of the implant has improved by varying the depth of thread, diameter, and thread pitch of implant in Figure 6.Moreover, Figure 8 shows that the micromotion of the novel implant is less than an integrated implant. Various micro-motion results are given in Table 7. Additionally, Figure 7 shows FE analysis results of the bone system for an improved version.

Table 4. Design range for the control factor

Control Factor	Lower Bond	Basic	Upper Bond
Thread pitch(mm)	0.3	0.5	0.7
Diameter(mm)	2.5	3	3.7
Thread depth(mm)	0.2	0.25	0.45
Density-cancellous(mg*mm^3)	970	1100	1210
Density-cortical(mg*mm^3)	2.12	2.4	2.36
Young's modulus cancellous (mPa)	200	345	650

Figure 6. Improved version FE analysis of implant

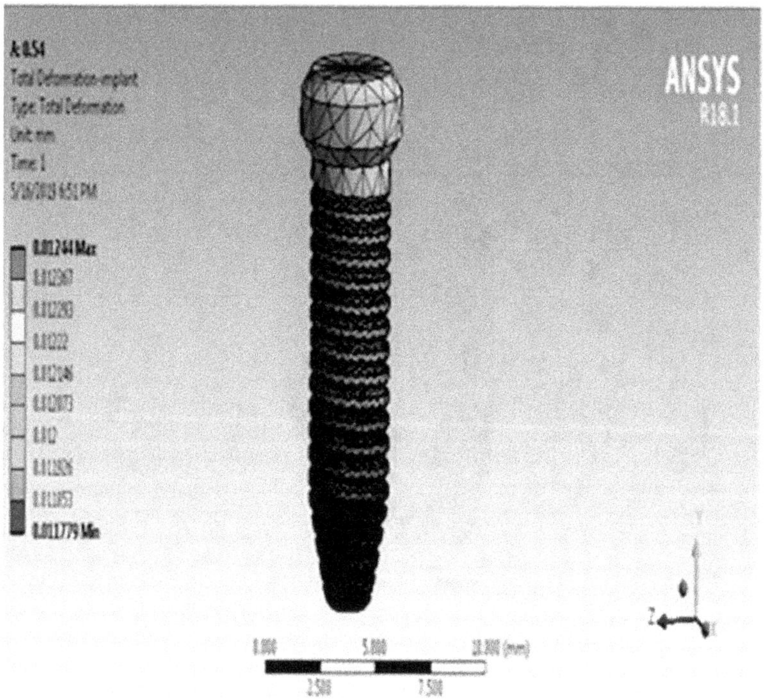

Figure 7. Improved version FE analysis bone

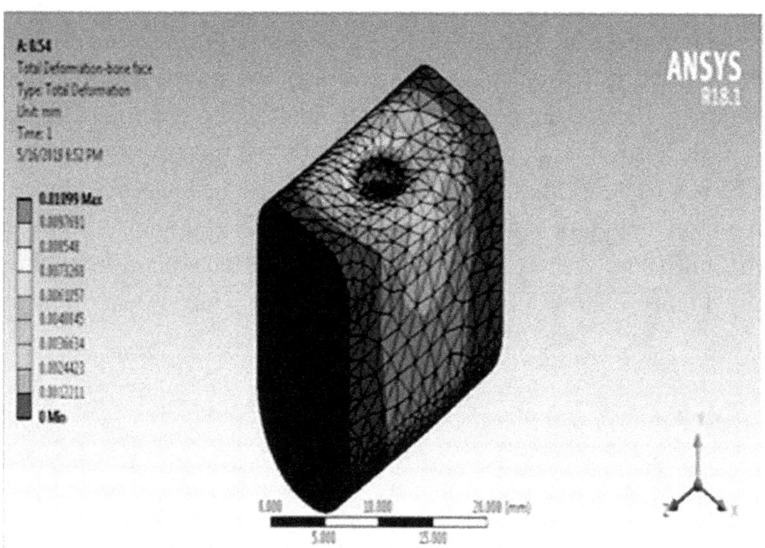

Table 5. Uniform Table U16 (16^12)

Uniform Table U16(16^12)												
No.	1	2	3	4	5	6	7	8	9	10	11	12
1	1	2	4	5	6	8	9	10	13	14	15	16
2	2	4	8	10	12	16	1	3	9	11	13	15
3	3	6	12	15	1	7	10	13	5	8	11	14
4	4	8	16	3	7	15	2	6	1	5	9	13
5	5	10	3	8	13	6	11	16	14	2	7	12
6	6	12	7	13	2	14	3	9	10	16	5	11
7	7	14	11	1	8	5	12	2	6	13	3	10
8	8	16	15	6	14	13	4	12	2	10	1	9
9	9	1	2	11	3	4	13	5	15	7	16	8
10	10	3	6	16	9	12	5	15	11	4	14	7
11	11	5	10	4	15	3	14	8	7	1	12	6
12	12	7	14	9	4	11	6	1	3	15	10	5
13	13	9	1	14	10	2	15	11	16	12	8	4
14	14	11	5	2	16	10	7	4	12	9	6	3
15	15	13	9	7	5	1	16	14	8	6	4	2
16	16	15	13	12	11	9	8	7	4	3	2	1

Table 6. Results from FEA

Exp No	Thread Pitch (mm)	Density-Cancellous Bone (mg*mm^3)	Thread Depth (mm)	D(mm)	Density-Cortical Bone (mg*mm^3)	Young's Modulus Cancellous Bone (mpa) (Mpa)	Micromotion(um)
1	0.3	0.1018	0.29	3.22	2.328	620	68.62
2	0.33	0.1082	0.39	2.66	2.28	560	37.48
3	0.35	0.1146	0.2	3.46	2.232	500	67.982
4	0.38	0.121	0.3	2.9	2.184	440	20.31
5	0.41	0.1002	0.4	3.7	2.136	380	15.003
6	0.43	0.1066	0.22	3.14	2.36	320	30.84
7	0.46	0.113	0.32	2.58	2.312	260	24.85
8	0.49	0.1194	0.42	3.38	2.264	200	23.99
9	0.51	0.986	0.24	2.82	2.216	650	15.07
10	0.54	0.105	0.34	3.62	2.168	590	12.44
11	0.57	0.1114	0.44	3.06	2.12	530	15.031
12	0.59	0.1878	0.25	2.5	2.344	470	19.99
13	0.62	0.9709	0.35	3.3	2.296	410	17.673
14	0.65	0.1034	0.45	2.74	2.248	350	20.381
15	0.67	0.1098	0.27	3.54	2.2	290	14.066
16	0.7	0.1162	0.37	2.98	2.152	230	26.305

Table 7. Micro-motion results

Type of Model	Total Deformation of Implant System (mm)	Deformation of Bone Landmark (mm)	Implant Micro Motion (mm)
Integrated Version	0.019147	0.017356	0.001791
Novel Version	0.01244	0.01099	0.00145

Figure 8. Implant micro-motion of novel Vs integrated version

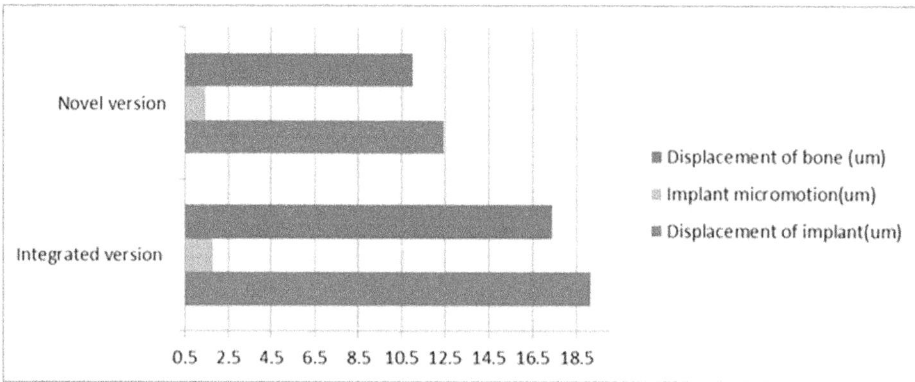

CONCLUSION

Within the limitations of the study, the authors modeled the dental implant system by incorporating the features of the traditional International Team for Implantology implants and micromotion of the novel dental implant system. The integrated implant micromotion decreased from 0.00179mm to 0.00145mm for the novel design. The micromotion of the integrated dental implant system also decreased from 0.019147mm to 0.01244mm for the primordial design, as shown in Table 6, which presents the results of experiment number 10. In the present study, the development percentage is evaluated as 35.02% which provides better stability with lesser micromotion of the implant system. From the outcome values uniform design tool is the strength to decrease the micromotion of dental implant structure.

REFERENCES

Ao, J., Li, T., Liu, Y., Ding, Y., Wu, G., Hu, K., & Kong, L. (2010). Optimal design of thread height and width on an immediately loaded cylinder implant: A finite element analysis. *Computers in Biology and Medicine*, *40*(8), 681–686. doi:10.1016/j.compbiomed.2009.10.007 PMID:20599193

Branemark, P.-I. (1977). Osseointegrated implants in the treatment of the edentulous jaw. Experience from a 10-year period. *Scandinavian Journal of Plastic and Reconstructive Surgery. Supplementum*, 16. PMID:356184

Cheng, Y.-C., Lin, D.-H., & Jiang, C.-P. (2016). Micromotion Improvement and Applications for Abutment-Implant System by Uniform Design and Kriging Interpolation. *International Journal of Bioscience, Biochemistry, Bioinformatics*, *6*(2), 41–49. doi:10.17706/ijbbb.2016.6.2.41-49

Cheng, Y.-C., Lin, D.-H., Jiang, C.-P., & Lee, S.-Y. (2015). Design improvement and dynamic finite element analysis of novel ITI dental implant under dynamic chewing loads. *Bio-Medical Materials and Engineering*, *26*(s1), S555–S561. doi:10.3233/BME-151346 PMID:26406049

Chong, L., Khocht, A., Suzuki, J. B., & Gaughan, J. (2009). Effect of implant design on initial stability of tapered implants. *The Journal of Oral Implantology*, *35*(3), 130–135. doi:10.1563/1548-1336-35.3.130 PMID:19579524

Dhatrak, P., Shirsat, U., Sumanth, S., & Deshmukh, V. (2018). Finite element analysis and experimental investigations on stress distribution of dental implants around implant-bone interface. *Materials Today: Proceedings*, *5*(2), 5641–5648.

Karl, M., Graef, F., & Winter, W. (2015). *Determination of Micromotion at the Implant Bone Interface–An: Vitro*. Academic Press.

Kasani, R., Attili, B. K. R. S., Dommeti, V. K., Merdji, A., Biswas, J. K., & Roy, S. (2019). Stress distribution of overdenture using odd number implants–A Finite Element Study. *Journal of the Mechanical Behavior of Biomedical Materials*, *98*, 369–382. doi:10.1016/j.jmbbm.2019.06.030 PMID:31326699

Kayabaşı, O., Yüzbasıoğlu, E., & Erzincanlı, F. (2006). Static, dynamic and fatigue behaviors of dental implant using finite element method. *Advances in Engineering Software*, *37*(10), 649–658. doi:10.1016/j.advengsoft.2006.02.004

Li, T., Hu, K., Cheng, L., Ding, Y., Ding, Y., Shao, J., & Kong, L. (2011). Optimum selection of the dental implant diameter and length in the posterior mandible with poor bone quality–A 3D finite element analysis. *Applied Mathematical Modelling*, *35*(1), 446–456. doi:10.1016/j.apm.2010.07.008

Li, X., & Dong, F. (2017). Three-dimensional finite element stress analysis of uneven-threaded ti dental implant. *International Journal of Clinical and Experimental Medicine*, *10*(1), 307–315.

Limbert, G., van Lierde, C., Muraru, O. L., Walboomers, X. F., Frank, M., Hansson, S., ... Jaecques, S. (2010). Trabecular bone strains around a dental implant and associated micromotions—A micro-CT-based three-dimensional finite element study. *Journal of Biomechanics*, *43*(7), 1251–1261. doi:10.1016/j.jbiomech.2010.01.003 PMID:20170921

Macedo, J., Pereira, J., Faria, J., Pereira, C., Alves, J., Henriques, B., ... López-López, J. (2017). Finite element analysis of stress extent at peri-implant bone surrounding external hexagon or Morse taper implants. *Journal of the Mechanical Behavior of Biomedical Materials*, *71*, 441–447. doi:10.1016/j.jmbbm.2017.03.011 PMID:28499606

Mosavar, A., Ziaei, A., & Kadkhodaei, M. (2015). The effect of implant thread design on stress distribution in anisotropic bone with different osseointegration conditions: A finite element analysis. *International Journal of Oral & Maxillofacial Implants*, *30*(6), 1317–1326. doi:10.11607/jomi.4091 PMID:26478976

O'Sullivan, D., Sennerby, L., & Meredith, N. (2004). Influence of implant taper on the primary and secondary stability of osseointegrated titanium implants. *Clinical Oral Implants Research*, *15*(4), 474–480. doi:10.1111/j.1600-0501.2004.01041.x PMID:15248883

Roy, S., Das, M., Chakraborty, P., Biswas, J. K., Chatterjee, S., & Khutia, N., & RoyChowdhury, A. (2017). Optimal selection of dental implant for different bone conditions based on the mechanical response. *Acta of Bioengineering and Biomechanics*, *19*(2). PMID:28869633

Roy, S., Dey, S., Khutia, N., Chowdhury, A. R., & Datta, S. (2018). Design of patient specific dental implant using FE analysis and computational intelligence techniques. *Applied Soft Computing*, *65*, 272–279. doi:10.1016/j.asoc.2018.01.025

Roy, S., Panda, D., Khutia, N., & Chowdhury, A. R. (2014). Pore geometry optimization of titanium (Ti6Al4V) alloy, for its application in the fabrication of customized hip implants. *International Journal of Biomaterials*. PMID:25400663

Shafi, A. A., Kadir, M. R. A., Sulaiman, E., Kasim, N. H. A., & Kassim, N. L. A. (2013). The effect of dental implant materials and thread profiles—A finite element and statistical study. *Journal of Medical Imaging and Health Informatics*, *3*(4), 509–513. doi:10.1166/jmihi.2013.1199

Sugiura, T., Yamamoto, K., Horita, S., Murakami, K., & Kirita, T. (2018). Micromotion analysis of different implant configuration, bone density, and crestal cortical bone thickness in immediately loaded mandibular full-arch implant restorations: A nonlinear finite element study. *Clinical Implant Dentistry and Related Research*, *20*(1), 43–49. doi:10.1111/cid.12573 PMID:29214714

Chapter 11
Modelling Stress Distribution in a Flexible Beam Using Bond Graph Approach

Jay Prakash Tripathi
Thapar Institute of Engineering and Technology, India

ABSTRACT

The bond graph (BG) approach of modelling provides a unified approach for modelling the systems having components belonging to multi-energy domains. Moreover, as evident by its name, it is a graphical approach. The graphical nature provides a tool for conceptual visualization of the model. It also provides some algorithmic tools because of its formal structure and syntax, thereby enabling model consistency checks such as checking algebraic loops, etc. There are a large number of texts published in recent years that may be refereed for background material on the BG methodology. Though used in many applications, its use in modelling stress distribution in the system is limited. Finite element (FE) modelling has found wide applicability for the same. This chapter is aimed at providing background knowledge, a comparison of BG approach with the FE approach, and a review of research progress of past two decades in this direction.

INTRODUCTION

Interdisciplinary research is common, with its broad or complex problems needing to be solved by a single discipline. Conceptual unification must shrink the many interdisciplinary techniques into one common method. One example of conceptual unification, the bond graph (BG) technique, includes the modeling, simulation, control, and synthesis of dynamics of physical systems with different energy domains. Paynter (1961) invented the BG technique at the Massachusetts Institute of Technology. Apart from conceptual unification, the BG also provides the following benefits:

DOI: 10.4018/978-1-7998-1690-4.ch011

- **Modular Approach of Modeling:** Each component can be modeled into individual blocks for easy conversion, correction, and reuse.
- **Dynamic Model from Kinematic Constraints:** Using the BG algorithm, a fully dynamic model and equations of motion are directly obtained from kinematic constraints and energy balance.
- **Graphical Approach of Modeling andAnalysis:** BG modeling is done in a well-defined, systematic order. This provides a structure and causality within the model for qualitative information.

The BG approach provides a unified approach for modeling systems with components belonging to multienergy domains. Moreover, as evident by its name, it is a graphical approach. Its graphical nature provides a tool for conceptual visualization of the model. It also provides algorithmic tools due to its formal structure and syntax, enabling a model consistency check (for example, checking algebraic loops).

There is a large number of texts published in recent years that may be refereed for background material on the BG methodology. Though used in many applications, there is limited information on its use in modeling stress distribution. However, finite element (FE) modeling has found wide applicability.

The objective of this chapter is to achieve two goals. First, it reviews recent advancements in the field of modeling stress distribution using the BG technique. Second, it is concise, attempting to incorporate all information required to understand the basics of modeling stress distribution using the BG technique.

BG VS. FINITE ELEMENT METHOD (FEM)

The FEM is either computationally expensive or inaccurate. Its heavy computational load arises because system boundaries are included after modeling the grid over the entire body. Otherwise, significant inaccuracy may arise.

BASICS OF BG MODELING

Power is a common variable exchanged between domains. It provides a basis for unification.Variables of energy domains may be generalized based on two factors of power: (1) effort (e); and (2) generalized flow (f). The effort and flow variables in mechanical domains are given in Table 1.

The unified approach of the BG may prove to be a good tool in modeling, simulation, control, or synthesis of dynamics of a system with the variables in Table 1.

Table 1. Effort (e) and flow (f) variables in mechanical domains

Effort (e)	**Flow** (f)
Force(F)	Velocity(v)
Torque(τ) or moment	Angular velocity (ω)
Pressure (P)	Volume flow rate (Q)
Temperature (T)	Entropy change rate ($\dot{S}$)
Stress (σ)	Strain rate ($\dot{\varepsilon}$)

Figure 1. Power bonds and causality

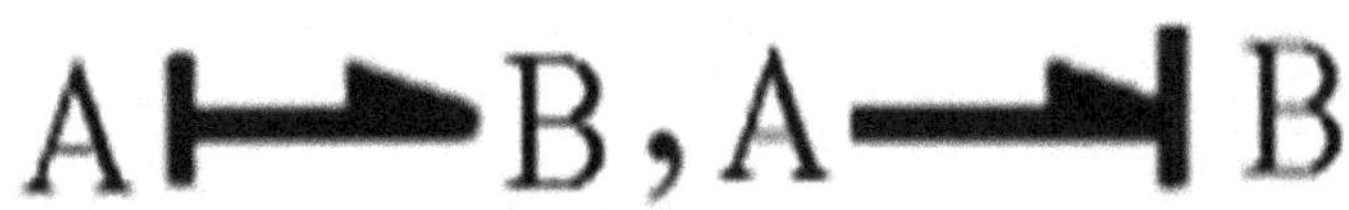

Tracking the exchange of power is required to make the BG model. Power is exchanged through power bonds (see Figure 1).

The two *power bonds* in Figure 1 model the power direction situation and causality situation. The *half arrow* denotes the source and receiver of power. A small transverse stroke at the end of the bond represents *causality*, denoting the element causing the flow variable by receiving effort information. The end with the *transverse stroke* receives the effort information and causes or generates the information of flow. On the left side, the BG element *A* supplies power to *B*. Causality of the bond allows *A* to receive effort information from *B*. In return, it gives flow information to element *B*.

The mechanism behind causing effort from the information of flow and vice versa depends on the constitutive equations of the BG element. Three elements convert effort into flow and flow into effort: (1) inertances(*I*); (2) compliances (capacitance) (*C*); and (3) dissipaters (*R*), respectively.

The direction of power for the elements is the same because they all are power recieptor. The causality of the *C* element receives flow information and gives information of effort. Reverse implies the *I* element. The causality of the*R* element is not fixed (unlike inertia or compliant elements). The *I* and *C* elements integrate effort or flow variables. They also develop a relationship between the flow and the effort on a bond. The resistive element directly relates the effort to flow, or viceversa, without integration. It sometimes represents resistivity or conductivity. The causality of *R* (the rightmost position of Figure2) represents a conductive causality. The other *R* element is resistive causality. When modeling a system,it is important to have a clear idea of a strong bond, power direction, and causality.

Figure 3 shows the constitutive equations for different elements in regards to respective power direction and causality.

In Figure3, the variables e and f represent effort and flow. K and m are stiffness and mass of the compliance and inertia elements, respectively. The terms φ and ψ are single-valued functions of flow or effort, respectively. The terms t and t_0 are time limits;ξ is the integration variable.

Other elements of the BG are the transformer element (TF), flow source (SF), and effort source (SE). See Figure 4.

The termμ in the constraint equations denotes the modulus of the transformer or gyrator.

Apart from these BG elements, two junction elements are represented by 0 and 1. These are used for algebraic summation of power. Moreover, the 1 and 0 junction elements equate flow and effort in their associated power bonds.The 1 and 0 junctions should have one *strong bond*. For 1 junction, *strong bond*

Figure 2. Power and causality orientation for the BG elements

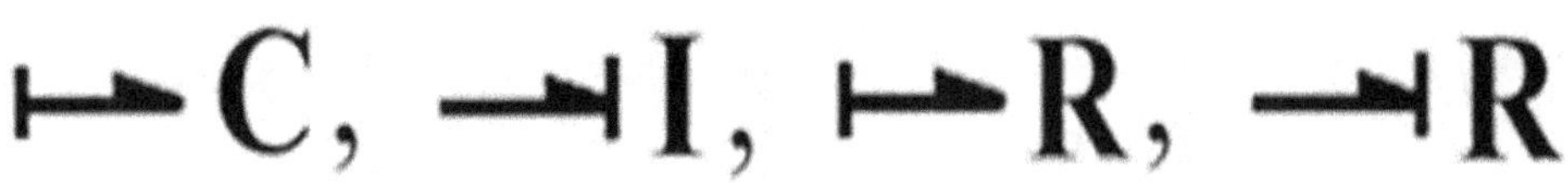

Figure 3. Constraint imposed by I, C, and R elements

C $$e = K\int_{t_0}^{t} f(\xi)d\xi$$

I $$f = \frac{1}{m}\int_{t_0}^{t} e(\xi)d\xi$$

R $$e = \phi(f)$$

R $$f = \psi(e)$$

Figure 4. Constraint imposed by TF, SF, and SE elements

1 **TF** 2 $$f_2 = \mu f_1$$

1 **TF** 2 $$f_1 = \mu f_2$$

1 **GY** 2 $$e_2 = \mu f_1$$

1 **GY** 2 $$e_1 = \mu f_2$$

SF $$f = SF$$

SE $$e = SE$$

has flow causality toward the junction. The 0 junction has effort causality toward the junction. In Figure 5, bond 1 is a *strong bond* for 1 junction and bond 3 is the *strong bond* for 0 junction.

The constitutive equations for 1 junction are:

$$f_1 = f_2 = f_3 = f_4, \tag{1}$$

Applying power conservation provides:

$$f_1e_1 - f_2e_2 - f_3e_3 - f_4e_4 = 0, \tag{2}$$

Equations (1) and (2) provide:

Figure 5. Constraint imposed by TF, SF, and SE elements

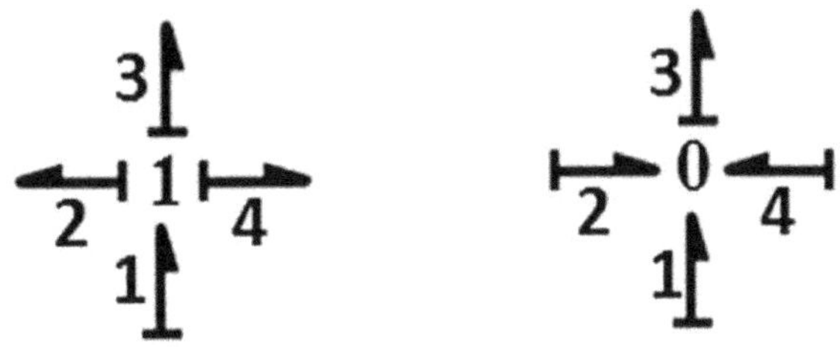

$$e_1 - e_2 - e_3 - e_4 = 0, \tag{3}$$

Similarly, the constitutive equations for 0 junction are:

$$e_1 = e_2 = e_3 = e_4, \tag{4}$$

$$f_1 + f_2 - f_3 + f_4 = 0 \tag{5}$$

The BG approach is commonly used for modeling a multidisciplinary system. Therefore, it is important to reference Merzouki, Samantaray, Pathak, and Bouamama (2012) and Mukherjee and Karmakar (2000) so readers can locate simple examples related to the systems of electrical, thermal, hydraulic, pneumatic, and energy domains.These references will help beginners get familiar with the BG technique.

EXAMPLE OF EULER-BERNOULLI BEAM MODEL

This section reviews an example of a simple deformable-body and a model of the stress and strain induced through the BG. For example, a beam portion (or element) is shown in Figure 6.The whole beam is made up of a finite number of the same elements. This element is considered to be i^{th} element. Let us assume that the geometric axis and the neutral axis are the same and are taken along the z-axis. Also assume that the beam is subjected to pure bending. The deformation in the neutral axis occurs along the y-axis; rotation occurs along the z-axis (rotation termed θ). The length of the beam portion is Δx, deflection is y, F is shear force, and M is the bending moment.

Figure 6. Small beam segment with representation of forces and moments acting on it

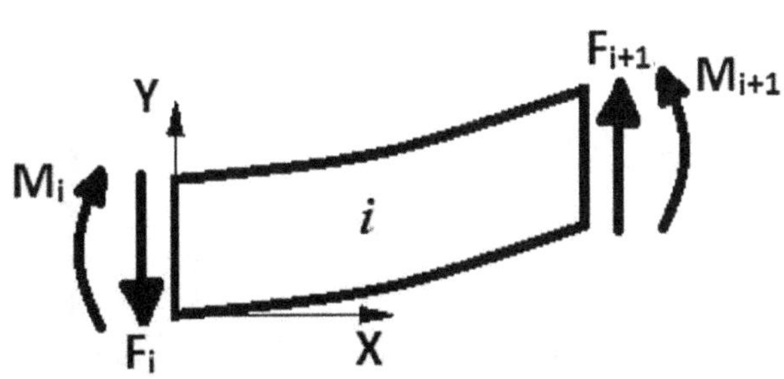

$$Fi_{+1} - Fi = \rho A\Delta x\ddot{y} \tag{6}$$

$$\Delta\dot{\theta} = \Delta\dot{y} \times \frac{2}{\Delta x_i + \Delta x_{i+1}} \tag{7}$$

$$\mathrm{M} = \mathrm{EI}\frac{\Delta\theta}{\Delta x} \tag{8}$$

$$\theta = \frac{\Delta y}{\Delta x} \tag{9}$$

$$Fi = -\frac{\Delta M}{\Delta x} \tag{10}$$

Equations (6-10) provide:

$$\rho A\ddot{y} + EI\frac{\partial^4}{\partial x^4}y = 0 \tag{11}$$

In equations (6-11), the term ρ is the density of the beam element, A is the cross-sectional area, and EI is flexural rigidity.

Let us create the BG model corresponding to equations (6-11). The BG model in Figure 7 begins with equation (6).

In Figure 7, the effort and flow expressions (flow equations [1-3]) are represented in Figure 7(a) across the 1 junction without any conversion between effort and flow. The causality represents an expression for the conversion between effort and flow. In Figure7(b), the causality of I denotes that it is receiving effort and giving flow information toward 1 junction. Since 1 junction represents a common flow junction, the flow is given by the I element (equated in bonds 1, 2, and 3). Effort in bond 1 is F_{i+1}, bond 2 is F_i, and bond 3 is F_{i+1} - F_i. In this way, effort and flow variables in each bond are determined. The BG model for equation (7) is given in Figure 8.

Figure 7. (a) Power bonds in the BG model corresponding to equation (6). (b) Causality in the BG model corresponding to equation (6).

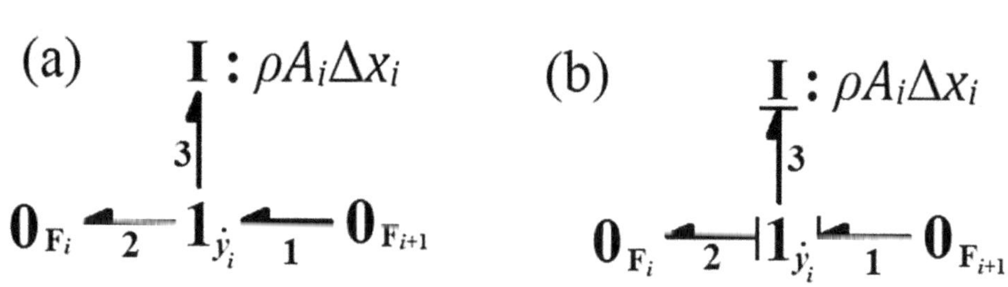

Figure 8. BG model corresponding to equation (7)

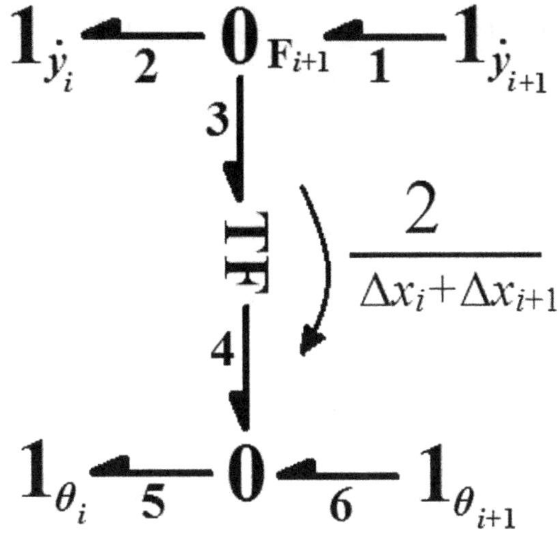

The flow in bond 1 is $\dot{y}_{i+1}$. The flow in bond 2 is $\dot{y}_i$. Therefore, using a constitutive equation of 0 junctions, we obtain flow in bond 3 as $\dot{y}_{i+1} - \dot{y}_i$ or $\Delta\dot{y}$. The *TF* element multiplies the flow in bond 3 by $\frac{2}{\Delta x_i + \Delta x_{i+1}}$. This makes the flow at bond 4 equal to $\Delta\dot{y} \times \frac{2}{\Delta x_i + \Delta x_{i+1}}$, which is equal to $\Delta\dot{\theta}$. Here, $\Delta\dot{\theta} = \theta_{i+1} - \theta_i$.

Equation (9) is used to convert the flow $\Delta\dot{\theta}$ into effort (i.e., bending moment). See Table 1. The conversion of flow to effort is done using the *C* element.Therefore, a *C* element is added at 0 junction in Figure 9.

Combining Figure 7(b) and Figure 9 provides Figure 10.

Similarly, the BG model of the other beam elements can be obtained. Combining the BG model of two segments of the beam provides the BG model in Figure 10. Similarly, the BG model of the complete beam can be obtained (see Figure 11).

To practice more complex problems of BG modeling, the reader may refer to Merzouki et al. (2012). Recent work is reviewed in the next section to understand the research problem being studied by researchers.

A BRIEF REVIEW OF BG MODELING OF DEFORMABLE BODIES

This section reviews approaches to model stress distribution in a system using BG. Only the literature published in the past two decades has been considered.

Researchers are puzzled by the serious problem of bending in deformable bodies. As a result, several types of beam models have been developed for applications like robotic linkages, flexible panels of spacecraft, wind turbines, etc. Researchers' main goals in model development include ease of usage, reuse, compactness, and simulation efficiency. The BG model of deformable multibodies has been reported in Hwang (2005), Raymond and Granda (2003), and Schiavo, Virani, and Ferretti (2006).

Figure 9. BG model corresponding to equations (7-8) with causality

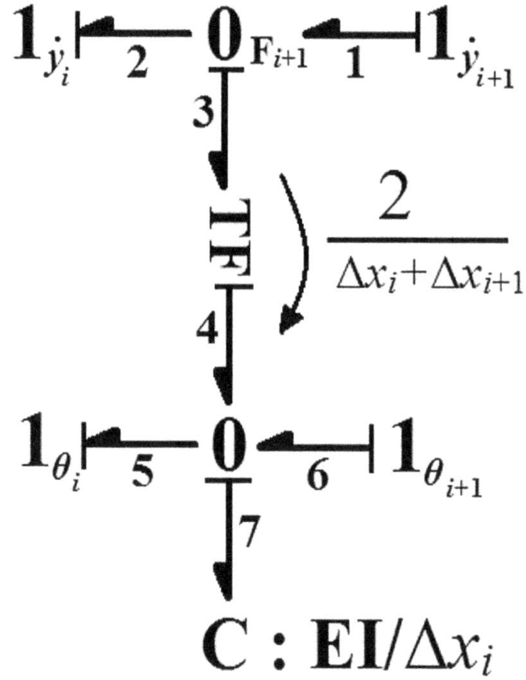

Figure 10. BG model of the ith beam element

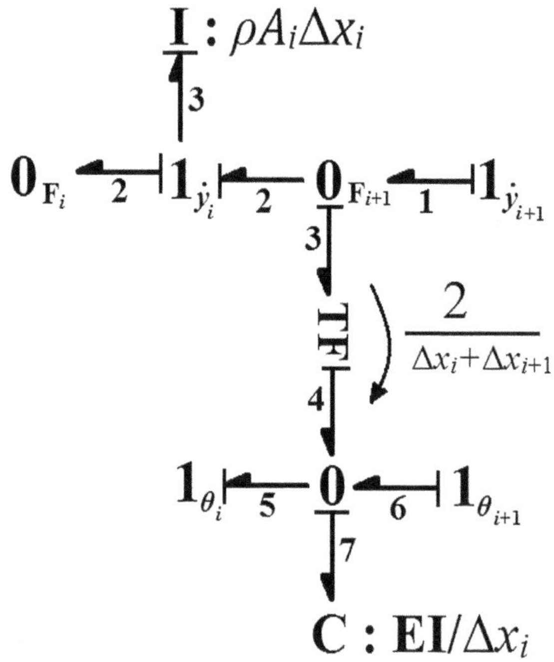

Figure 11. BG model of ith and $(i\text{-}1)^{the}$ beam combination

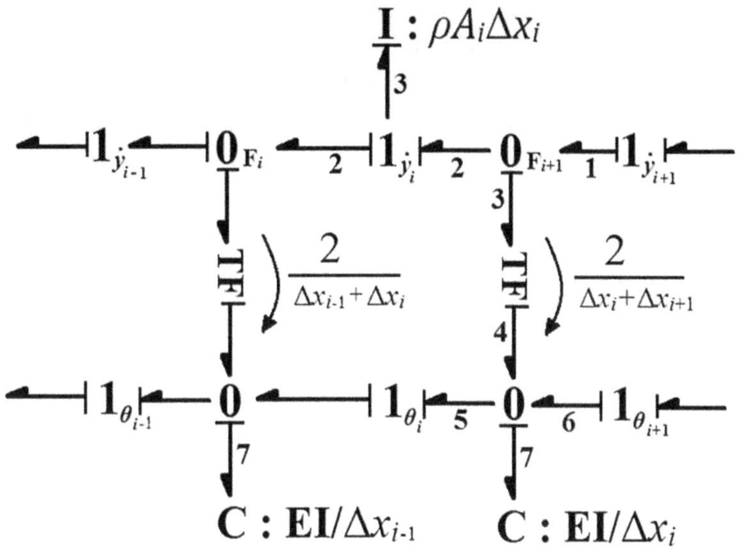

There are three approaches to modeling a deformable multibody system: (1) floating frame of reference method; (2) absolute nodal coordinate formulation (ANC); and (3) corotational formulation (CR). The first approach poses a lesser computational load compared to other approaches (Damić, 2006). Using modal superposition, the order of elastic deformation equations is reduced. Extending the multibody floating frame approach develops a novel inertia-capacitance (IC) beam in Xing, Pedersen, and Moan (2011). In it, elastic deformations are coupled with rigid body motion in a nonlinear fashion. The corotational approach is given in Cohodar (2005). It is used to model flexible system dynamics in the BG framework. A comparison between the corotational approach and the well-known ANC approach is given in Cohodar, Borutzky, and Damic (2009). The corotational formulation is applied in the BG framework to model flexible multibodies by Damic and Cohodar(2005). Similar implementation of the finite element absolute nodal coordinate formulation and finite element CR in the BG framework is presented and compared in Cohodar, Borutzky, and Damic (2009).The authors conducted this comparison on the slider-crank mechanism and planar flexible pendulum. The comparison indicated that the CR needs less computation.

The BG equivalence of the FEM, developed by Choi and Bryant (2002), developed a new model of a gearbox using a scalar BG and FEM by incorporating the planar dynamics of the bending of the shaft. A three-dimensional (3D) model of deformation and stress was developed in Sainthuile et al. (2012). It included stress distribution along the different dimensions of the transducer. The model computes the piezoelectric transducer response with respect to deformation. A similar application is flexible solar panels of an international space station (ISS) module (Granda, Nguyen, & Hundal, 2011). The panel is a complex multibody assembly. Some links may be considered rigid. The BG model is developed by modal analysis, incorporating a flexible panel like the Bernoulli-Euler beam. The rotating beam problem is a similar problem investigated in many research articles.

In the case of rotating-beam problems, the stress stiffening effect is a popular problem. The stress stiffening problem is nonlinear, arising due to the presence of centrifugal forces. It increases the stiffness of the beam in bending significantly such that it must be coupled with axial forces. Similarly, there is much literature wherein the derivation and analysis of differential equations of similar problems are presented (Apiwattanalunggarn, Shaw, Pierre, & Jiang, 2003; Yang, Jiang, & Chen, 2004). In addition, literature combines the finite element method with the BG technique (Derkaoui, Bideaux, & Scavarda, 2005).

CONCLUSION

The BG models the dynamics of various mechanical systems. In most cases related to the BG model, a flexible body with bending is described as a lumped parameter model. This chapter provides a review of the recent trend of incorporating appropriate dynamics of the flexible body in the BG model. To understand this complex research, it is necessary to start with a simple problem. Keeping this in mind, the Euler-Bernoulli beam model has been demonstrated in the BG framework.

ACKNOWLEDGMENT

The author acknowledges support given by the Department of Mechanical Engineering at Thapar Institute of Engineering and Technology Patiala.

REFERENCES

Apiwattanalunggarn, P., Shaw, S. W., Pierre, C., & Jiang, D. (2003). Finite-element-based nonlinear modal reduction of a rotating beam with large-amplitude motion. *Modal Analysis*, *9*(3-4), 235–263.

Choi, J., & Bryant, M. (2002). Combining lumped parameter bond graphs with finite element shafts in a gearbox model. *Computer Modeling in Engineering & Sciences*, *3*(4), 431–446.

Cohodar, M. (2005). *Modelling and simulation of robot system with flexible links using bond graphs* (PhD thesis). University of Sarajevo.

Cohodar, M., Borutzky, W., & Damic, V. (2009). Comparison of different formulations of 2D beam elements based on bond graph technique. *Simulation Modelling Practice and Theory*, *17*(1), 107–124. doi:10.1016/j.simpat.2008.02.014

Damić, V. (2006). Modelling flexible body systems: A bond graph component model approach. *Mathematical and Computer Modelling of Dynamical Systems*, *12*(2-3), 175–187. doi:10.1080/13873950500068757

Damic, V., & Cohodar, M. (2005). A bond graph approach to modelling of spatial flexible multibody system based on co-rotational formulation. *Proceedings of the 2005 International Conference on Bond Graph Modelling and Simulation.*

Derkaoui, A., Bideaux, E., & Scavarda, S. (2005). Finite element local structure assembly and shape functions influence on bond graph modelling. *International Conference on Bond Graph Modelling and Simulation (ICBGM'05).*

Granda, J., Nguyen, L., & Hundal, S. (2011). Modeling the Completed Space Station a Three Dimensional Rigid-Flexible Dynamic Model to Predict Modes of Vibration and Stress Analysis. In Infotech@ Aerospace 2011 (p. 1535). doi:10.2514/6.2011-1535

Hwang, Y. L. (2005). A new approach for dynamic analysis of flexible manipulator systems. *International Journal of Non-linear Mechanics*, *40*(6), 925–938. doi:10.1016/j.ijnonlinmec.2004.12.001

Merzouki, R., Samantaray, A. K., Pathak, P. M., & Bouamama, B. O. (2012). *Intelligent mechatronic systems: Modelling, control and diagnosis*. Springer Science & Business Media.

Mukherjee, A., & Karmakar, R. (2000). *Modelling and simulation of engineering systems through bondgraphs*. Alpha Science International Ltd.

Paynter, H. (1961). Analysis and design of engineering systems. *Class Notes for MIT Course*, *2*, 751.

Raymond, C., &Granda, J. (2003). Using bond graphs for articulated, flexible multi-bodies, sensors. *Actuators, and Controllers with Application to the International Space Station. ICBG03*.

Sainthuile, T., Grondel, S., Delebarre, C., Godts, S., & Paget, C. (2012). Energy harvesting process modelling of an aeronautical structural health monitoring system using a bond-graph approach. *International Journal of Aerospace Sciences*, *1*(5), 2.

Schiavo, F., Vigano, L., & Ferretti, G. (2006). Object-oriented modelling of flexible beams. *Multibody System Dynamics*, *15*(3), 263–286. doi:10.100711044-006-9012-8

Xing, Y., Pedersen, E., & Moan, T. (2011). An inertia-capacitance beam substructure formulation based on the bond graph method with application to rotating beams. *Journal of Sound and Vibration*, *330*(21), 5114–5130. doi:10.1016/j.jsv.2011.05.025

Yang, J., Jiang, L., & Chen, D. C. (2004). Dynamic modelling and control of a rotating Euler–Bernoulli beam. *Journal of Sound and Vibration*, *274*(3-5), 863–875. doi:10.1016/S0022-460X(03)00611-4

Chapter 12
A Theoretical Study of Thermal Stress for Engineering Applications

Mrinal Bhowmik
Indian Institute of Technology, Guwahati, India

Payal Banerjee
National Institute of Technology, Rourkela, India

Manoj Kumar Bhowmik
Tripura Institute of Technology, Agartala, India

ABSTRACT

The stress generated due to the temperature difference is called thermal stress. Generally, the temperature gradients, thermal shocks, and thermal expansion or contraction are most effective contributors to thermal stress. The improper temperature profile of a body results in the formation of cracks, fractures, or plastic deformations at single or multiple spots depending upon two factors (i.e., the magnitude of temperature distribution or other variables of heating and material properties). So, this chapter analyses the causes of thermal stress and their measurement techniques. However, as most of the engineering problems, the thermal stress is due to the thermal expansion or sudden temperature changes happening in the body. Therefore, a brief analysis of temperature measurement devices with their proper data capturing methodology is also discussed.

INTRODUCTION

Generally, materials expand or contract when heated or cooled. Any change in temperature produces a thermal strain (Balaji & Yadava, 2013). This strain is positive for the increase in temperature. Thermal strains are usually reversible in such a way that the object retains its original shape when the temperature returns to its original value. Most materials exhibit reversibility of thermal strain. However, some materials are exceptions.

DOI: 10.4018/978-1-7998-1690-4.ch012

A special type of stress is developed if an external force prevents the free expansion of the material. Similary, fluid forces have a significant influence on the development of stress and strain at different operating conditions on the material surface (Annamalai & Karuppasamy, 2016; Karthik, Kumar, & Kumaraswamidhas, 2015; Kumar & Kumaraswamidhas, 2018). This has an auxiliary effect on certain phenomenon like thermal expansion and contraction on subjected materials. The magnitude of this stress is equal to stress produced in a bar if it is allowed to expand freely in length direction and when enough force is applied to return the bar to original length. This stress, termed "thermal stress," is caused by thermal contraction or expansion. Thermal stress may also be destructive (for example, expansion of gasoline can rupture a tank). Thermal stress can explain many day-to-day phenomena.

For exmaple, metals are widely used for knee and hip implants. As metal does not bond with bone, most implants need to be replaced over time. There is current research in pursuit of better metal coatings to sustain the metal-to-bone bond. It is challenging to find a material whose thermal expansion coefficient is close to that of the metal. If the thermal expansion coefficient of metal and coating material are too different, cracks may occur during the manufacturing process due to thermal stress at the coating-metal interface.

Another case of thermal stress is associated with oral health. Dental fillings and tooth enamel experience different expansions. Thus, cracks can occur in the filling, causing pain when eating something hot or cold (i.e., tea, coffee, ice cream). Composite fillings (porcelain) have replaced metallic fillings (gold, silver, etc.) due to their relatively smaller expansion coefficient, which is similar to that of the teeth.

Thermal expansion of brittle materials should also be controlled. For example, glass and ceramics are both brittle. When subjected to uneven temperatures, nonuniform thermal expansion results in thermal stress and fractures. Ceramics, for instance, are used in a variety of materials. Therefore, their expansion should be matched to the application. Products like Corningware and spark plugs found success due to their control over thermal expansion. Ceramic bodies are fired to create crystalline varieties that affect the overall thermal expansion of the material in the desired direction. In the case of porcelain wares and porcelain equipment, the chemical composition and firing schedule can be changed to control the thermal expansion of glazes to attach to the porcelain (or another body type). Therefore, their thermal expansion should be controlled to "fit" the body and avoid shivering. Temperature gradients, thermal strain (i.e., expansion or contraction), and thermal shock parameters contribute to thermal stress.

TEMPERATURE GRADIENT

The rapid heating or cooling of a material causes a temperature difference between the surface and its center. During the heating process, the surface of the material becomes relatively hotter, expanding more than the center. This difference in temperature causes thermal stress. The center of the material eventually reaches the same temperature as the surface.

THERMAL EXPANSION OR CONTRACTION

The contraction and expansion of material depends on the thermal expansion coefficient of the material. If the material moves freely, the material expands or contracts without generating stress. However, thermal stresses are generated if the material connects to a rigid body at one end.

Thermal expansion is the change in size or volume of a given mass with respect to the change in temperature of the mass. The same thing happens in fluids, both gas and liquid forms. A common example of this is the expansion of alcohol in a thermometer. Hot air goes up due to its increase in volume. This makes the density of the hot air to be less than the density of the surrounding air. It leads to an upward (buoyant) force in nature, causing the hot air to rise.

Solids also experience thermal expansion. For example, railway tracks and bridges have expansion joints to expand or contract with temperature changes. Water and steam pipes often have a U-bend to allow for thermal expansion.

Reasons for Thermal Expansion

Unlike gases, the atoms or molecules in solids are closely packed together. An increment in temperature leads to an increase in the kinetic energy of the individual atoms. This increased kinetic energy causes propagation of vibration (small and rapid pulses) across the object. This pushes neighboring atoms or molecules away from each other, which results in a slightly larger body size due to the distance between neighbors. Under ordinary conditions, the size of a solid body (in most materials) increases by a fraction in each dimension (no preferred direction) with an increment of temperature (Bansal, 2016).

Mathematically, the change in length ΔL is proportional to length L. The dependence of thermal expansion on temperature, material, and length is expressed by equation (1).

$$\Delta L = \alpha L \Delta T \qquad (1)$$

where ΔL is the change in length L, ΔT is the change in temperature. α, the coefficient of linear expansion of the material, varies slightly with temperature.

Linear expansion means a change in length (one dimension) against change in volume (volumetric expansion). The measurement of change in length due to thermal stress is related to temperature change by a "linear expansion coefficient." It is the fractional change in length per degree of temperature change. Neglecting the effect of pressure, we may write:

$$\alpha_L = \frac{1}{L}\frac{dL}{dt} \qquad (2)$$

where L is a particular length of the object and $\frac{dL}{dt}$ is the rate of change of the linear dimension per unit change in temperature.

The change in the linear dimension can be calculated as (see Figure 1):

$$\frac{\Delta L}{L} = \alpha_L \Delta T \qquad (3)$$

Figure 1. Change in length of a rod due to thermal expansion

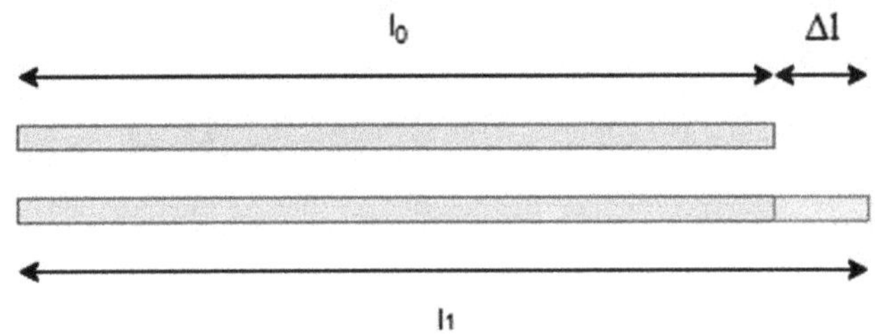

Equation (3) is valid when the linear-expansion coefficient does not vary drastically with the change in temperature ΔT and the fractional change in length $\Delta L/L$ is very small. The equation must be integrated if any of these conditions are not fulfilled. The list of the coefficient of thermal expansion (CTE) of some common materials is represented in Table 1.

Factors Affecting Thermal Expansion

Generally, thermal expansion decreases with an increase in intermolecular bond energy. This bond energy also affects the melting point of solids. Materials with a high melting point experience lower thermal expansion. Solid materials, unlike fluids, tend to keep their shape when undergoing thermal expansion. In general, fluids expand more than solids.

Thermal expansion also depends on the arrangement of the constituent particles of matter. For example, glass has a higher thermal expansion as compared to crystals. Rearrangements in an amorphous material (at the glass transition temperature) lead to characteristic discontinuities of coefficient of thermal expansion and specific heat. These discontinuities allow for the detection of the glass transition temperature where a supercooled liquid transforms to a glass.

Table 1. Linear coefficient of thermal expansion (α at 20 °C)

Materials	α (10^{-5} K^{-1})	Materials	α(10^{-5} K^{-1})
Aluminium	2.31	Mercury	6.1
Brass	1.9	Molybdenum	0.48
Carbon steel	1.08	Nickel	1.3
Concrete	1.2	Platinum	0.9
Copper	1.7	PVC	5.2
Diamond	0.1	Silicon	0.256
Glass	0.85	Silver	1.8
Gold	1.4	Titanium	0.86
Iron	1.18	Tungsten	0.45
Lead	2.9	Magnesium	2.6

Figure 2. Schematic of a cylindrical material fixed to a rigid body at one end

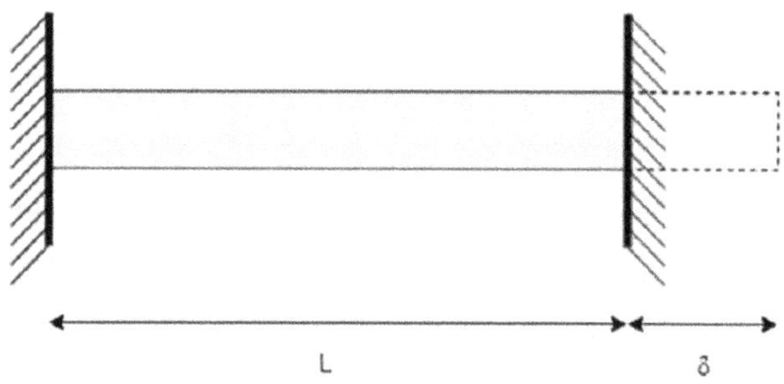

The size of many common materials can be changed with absorption or desorption of water (or other solvents). Many organic materials change their size due to this rather than thermal expansion. Common plastics, when exposed to water, can (in the long term) be expanded by many percents.

THERMAL SHOCK

The thermal shock phenomenon can be attributed to brittle materials. This is a condition of rapid temperature change over a large temperature gradient in a very short time. The change in temperature causes stresses (tension) on the surface, resulting in crack formation and propagation. For example, when quenched in cold water, glass temperature falls rapidly. Stresses are induced in the glass body, causing fractures/cracks or shattering. Ceramics are also susceptible to thermal shock.

Thermal Stress Formulation

Thermal stresses can be created if a material is fixed to a rigid body at one end (see Figure 2). This stress can be estimated by multiplying the change in temperature, material's thermal expansion coefficient, and material's young's modulus as in equation (4). E is young's modulus, α is thermal expansion coefficient, T_0 is original temperature, and T_f is the final temperature (Hetnarski, Eslami, & Gladwell, 2009).

For a homogeneous rod mounted between unyielding supports, the thermal stress in the rod is computed as:

$$E = \frac{\sigma}{\varepsilon} \tag{4}$$

$$\sigma = \frac{P}{A} \tag{5}$$

$$E = \frac{P}{AE} \tag{6}$$

$$\varepsilon = \frac{\delta_l}{L} \tag{7}$$

$$E = \frac{PL}{A\delta_l} \tag{8}$$

Deformation due to temperature changes:

$$\delta_T = \alpha L \Delta T \tag{9}$$

Deformation due to equivalent axial stress:

$$\delta_l = \frac{PL}{AE} = \frac{\sigma L}{E} \tag{10}$$

$$as, \delta_l = \delta_T \text{ then}, \alpha L \Delta T = \frac{\sigma L}{E} \tag{11}$$

$$\sigma = \alpha E \Delta T \tag{12}$$

When $T_f > T_0$, heating takes place. Constraints exert a compressive force on the material. When $T_f < T_0$, cooling takes place. The stress is in the expansion.

Welding is an example of thermal expansion, contraction, and temperature gradients with both heating and cooling of metal. The unit of the parameter α (CTE) is a reciprocal temperature (K^{-1}), such as μm/m · K or 10^{-6} K.

Temperature, expansion coefficient, and young modulus are calculated to find thermal stress. The next section will discuss the measurement of thermal expansion coefficient, young modulus, and temperature.

MEASUREMENT OF THERMAL EXPANSION COEFFICIENT

Two physical quantities, displacement and temperature, must be measured to determine the thermal expansion coefficient of a sample undergoing a thermal cycle. Dilatometry, interferometry, and thermo-mechanical analysis are three techniques used to measure CTE. Optical imaging may be used at high-temperature conditions. X-ray diffraction can be used to study changes in the lattice parameter. It cannot be used for bulk thermal expansion (Johns, 2013).

Dilatometry

Mechanical dilatometry is used on a wide scale. In this, a specimen is heated in a furnace and, with the help of the push rod, displacement of the end of the specimen is transmitted to a sensor. The precision of this test is inferior to interferometry. The test is commonly related to materials with CTE above 5×10^{-6}/K (2.8×10^{-6}/°F) in the temperature range of -180°C to 900°C (-290°F to 1650°F). Push bars might be of the vitreous silica type, high-immaculateness alumina type, or isotropic graphite type. Alumina frameworks can stretch the temperature range to 1600°C (2900°F) and graphite systems to 2500°C (4500°F). ASTM test method cove the determination of linear thermal expansion of inflexible solid materials by using a vitreous silica push bar or tube dilatometers.

Interferometry

Optical interference techniques measure the displacement of the specimen ends in terms of the number of wavelengths of monochromatic light. Precision is more noteworthy than thermo-mechanical and dilatometry.

Thermo-Mechanical Analysis

The thermo-mechanical analyzer takes measurements of a specimen holder and a probe. It transmits changes in length to a transducer, which interprets movements of the probe into an electrical signal. The setup includes a heater for uniform heating, a temperature-detecting component, callipers, and a means for accounting results. The ASTM test is the standard test technique for linear thermal expansion of solid materials by the thermo-mechanical investigation. In this method, the minimum value for CTE is 5×10^{-6}/K (2.8×10^{-6}/°F). However, it might be utilized at lower or negative extension levels with diminished exactness and accuracy. The appropriate temperature range is -120°C to 600°C (-185 to 1110°F). However, the temperature range can expand depending on instrumentation and calibration materials.

MEASUREMENT OF YOUNG'S MODULUS

The measurement of Young's modulus can be done with a different technique. The value of Young's modulus is differs with materials (Venkateshan, 2015).

Determination of Young's Modulus by Flexure Method for a Rectangular Bar

Measurement of the elasticity of a rectangular bar can be done with the help of the following model. This model has a solid bar, two stands with a knife-edged top and a vertical pointer, a traveling microscope, standard weights, hanger, meter scale, and slide callipers (see Figure 3). First, we measure the length of the bar and put it symmetrically on the knife-edges. We then place the hanger on the midpoint of the bar. We hang the loads one at a time to a certain weight. Then, we remove the loads in the same manner. During the increasing and decreasing of the loads, we use the traveling microscope to record the depression in the bend. The average of the depression in the bar while increasing and decreasing the loads at each stage is recorded. The graph between the depression and loads will give a straight line through the origin. With the use of equation (7) and slope of this line, Young's modulus can be calculated.

$$\mathbf{Y} = MgL^3/4bd^3x \qquad (13)$$

Figure 3. Schematic diagram of Young's modulus experiment

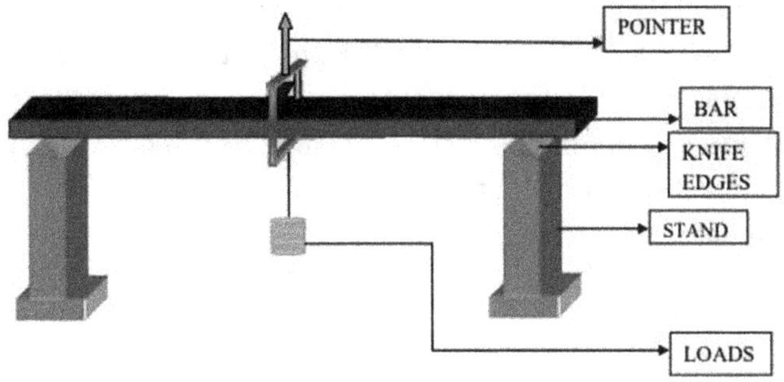

where **Y** = Young's modulus, L = length of the bar between the two knife-edges, b = breadth of the bar, d = thickness of the bar, g = acceleration due to gravity at the laboratory, M = load suspended at the middle point of the bar, and x = depression at the middle point of the bar.

Determination of Young's Modulus of Spherical Metallic Bodies

Measurement of the elasticity of metallic spheres occurs through its free fall from a certain height. This determines the compression time (τ) and contact area (A) of the sphere with the impacting surface (a plate with aluminium sheet covering) as shown in Figure 4. Compression time can be found by using a digital timing circuit and the compression area by measuring the area of the impression formed by the falling impact of the sphere over the aluminium sheet.

After obtaining the values of τ and **A**, with the help of following equation, elasticity (**E**) can be calculated.

$$\mathbf{A}^{1/2}\,\tau^2 = 110\,\rho\,R^3/E \tag{14}$$

where ρ = density of sphere, R = radius of the sphere, and E = elasticity of the sphere.

Figure 4. Diagram of a compressed elastic sphere

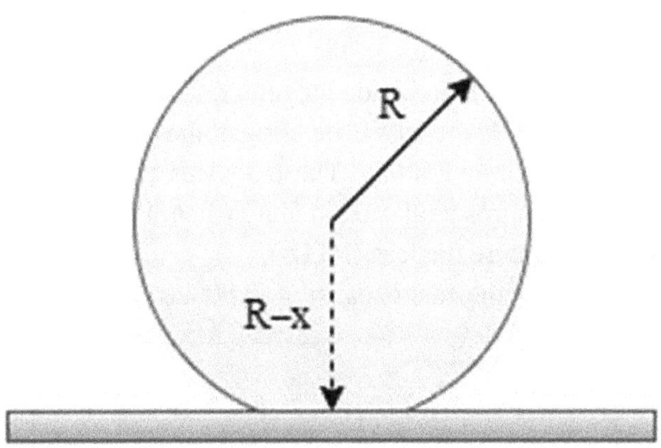

MEASUREMENT OF TEMPERATURE

There are two main methods to measure temperature. The physical contact method is thermal equilibrium (i.e., thermocouple, thermometer, etc.). The noncontact method measures the thermal radiant power of the infrared or optical radiation. According to McGee (1988), this is received from a known or calculated area on its surface (the pyrometer, inferred, etc.).

Temperature measurement references:

$$\text{Degrees Fahrenheit} = \frac{9}{5}{}^{\circ}C + 32\,; \tag{15}$$

$$\text{Degrees Celsius} = \frac{5}{9}{}^{\circ}F - 32\,; \tag{16}$$

$$\text{Degrees Kelvin} = {}^{\circ}C{+}273 \tag{17}$$

$$\frac{C}{5} = \frac{F-32}{9} = \frac{R}{4} = \frac{K-273}{5} \tag{18}$$

The relationship between the scales is C = temperature in degree Centigrade, F = temperature in degree Fahrenheit, R = temperature in degree Rankin, and K = temperature in degree Kelvin. The general temperature reference points are mentioned in Table 2.

Classification of Temperature Measurements

The classification of temperature measurements is shown in Figure 5 (Holman, 1966).

Changes in Physical Properties

Bimetallic Temperature Measurement Devices

A bimetallic strip converts the temperature change into mechanical displacement (see Figure 6). Bimetallic devices work based on the difference in the rate of thermal expansion. Strips of two metals (different thermal expansion coefficient) are bonded together. Once the device is subjected to heat, one side will expand more than the other. This results in bending of the strip. The mechanical linkage pointer translates this bending into a temperature reading. These devices are stable in operation, portable, and do not require a power supply. It requires negligible maintenance. However, they are not as accurate as thermocouples or resistance temperature devices (RTDs).

Take for example, nickel with low thermal expansion (0.0000017 mm/°C) and SS, Brass, Cu, Al for high expansion.

Table 2. General temperature reference points

	Celsius	Fahrenheit	Kelvin
Absolute zero	-273	-460	0
Liquid Nitrogen (boiling)	-196	-321	77
Liquid Helium (boiling)	-268.8	-452.1	4.2
Water (freezing)	0	32	273
Water (boiling)	100	212	373

Figure 5. Classification of temperature measurements

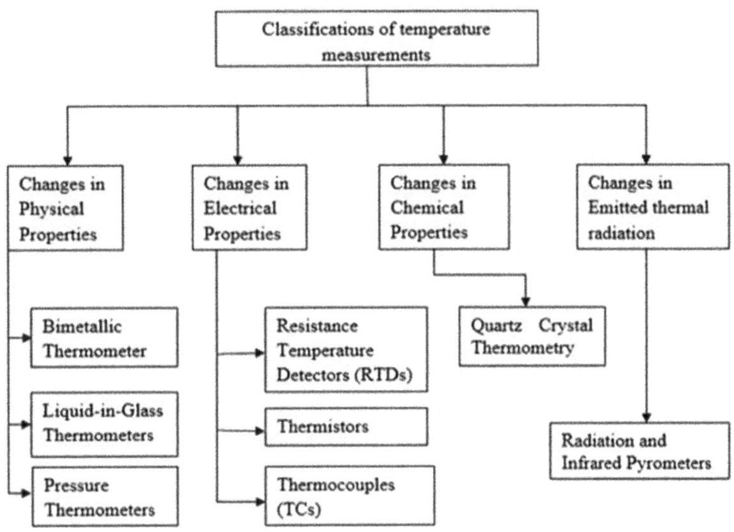

Figure 6. Bimetallic strips

Final curvature of a bimetallic beam can be represented as:

$$r = \frac{t\left\{3\left(1+m\right)^2 + \left(1+m\right)\left[m^2 + \left(1/mn\right)\right]\right\}}{6\left(\alpha_2 - \alpha_1\right)\left(T - T_0\right)\left(1+m^2\right)} \tag{19}$$

where:

- t = combined thickness of the bonded strip (in m)
- m = ratio of thickness of low and high expansion materials
- n = ratio of moduli of elasticity of low to high expansion materials
- α_1, α_2 = higher and lower coefficient of expansion per °C
- T = Temperature, °C
- T_o = Initial bonding temperature, °C

The pictorial view of bimetallic temperature measurement devices (indicators/pointers) are shown in Figure 7.

In general, the range of use is -65°C to 430°C and accuracy vary with range ± 0.5 to 12°C. The bimetallic thermometer is widely used in thermal cut-off relay.

Fluid-Expansion Temperature Measurement Devices

Temperature estimation depends on the difference between the expansion of glass and liquid. This difference occurs due to the transfer of heat from earth's surface to the bulb. Heat is transferred to the bulb from the stem. Less stem conduction is preferable. Special marking for depth is to be immersed. The differential thermal expansion between the liquid and glass is the principle of the system. Several liquids are used like the fluid-expansion temperature measurement devices (Noda, 2018).

Mercury-filled thermometers have the following: Range -37°C to around 400 °C with Accuracy ±0.3°C. Alcohol-filled thermometers have the following: Range: -75°C to 120°C with Accuracy ±0.6°C. Thermometers are broadly utilized in innovation and industry as monitors in meteorology, prescriptions, and logical research. They are easy to use, low in cost, and have lower/upper limits determined by freezing/boiling points of liquid. However, it is difficult for remote operation. A sample of mercury in a glass thermometer is shown in Figure 8.

A mercury glass thermometer cannot be used below -37.8°C or above 538°C. Desirable properties include:

- Linear expansion
- Liquid with large coefficient of expansion (for high sensitivity)
- Liquid suitable for a reasonable temperature range without change of phase

Figure 7. Pictorial view of bimetallic temperature measurement devices (indicators/pointers)

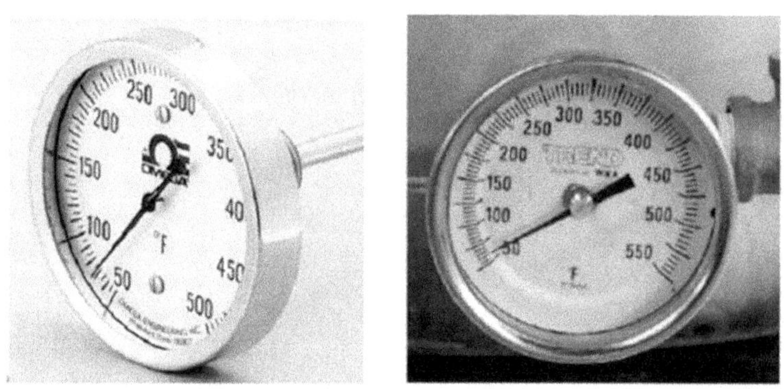

- Liquid clearly visible when drawn into a fine capillary
- Liquid should not stick to the glass
- High thermal conductivity (for better response)

A fluid-expansion sensor does not require electric power, does not create explosion risks, and is in a stable condition even after continuous cycling. They do not generate data that can be recorded or transmitted easily. These are not used for dynamic measurements. Stem correction must be considered when the calibration ambient temperature differs from the actual usage due to the wrong usage of the thermometer.

Pressure Thermometer

A bulb filled with fluid is attached to a pressure measuring device (i.e., manometer, bourdon tube, bellows) through a capillary tube (see Figure 9). Fluid expands when heated; pressure also increases. Temperature can measure -150 to 750°F with xylene. The height "h" is calibrated to represent the measure of temperature. High response is obtained with a smaller bulb and a short capillary with an electric pressure transducer. Generally, the accuracy of such a pressure thermometer is a ±0.5 - 1.5°C or ±0.5% range.

Advantages and disadvantages of pressure thermometers are:

- Lower in cost
- Stable operation
- Simple (widely used for industrial applications)

Figure 8. Mercury in glass thermometer

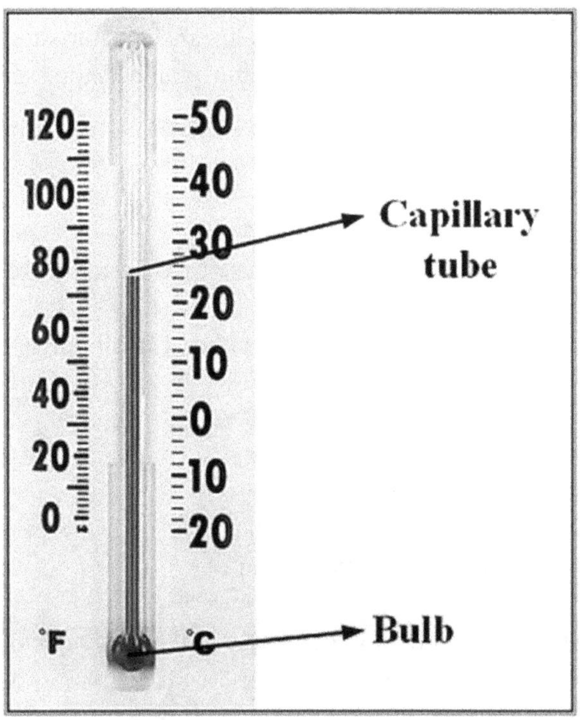

Figure 9. Pressure thermometer (A = sensor bulb in hot substance)

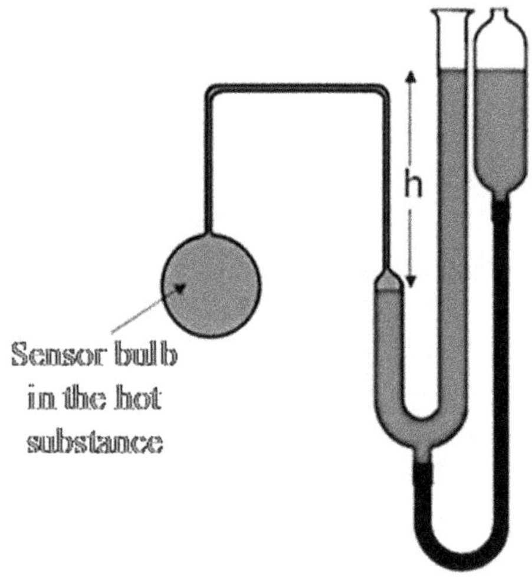

- Remote readings
- Response depends on bulb volume and capillary tube dimensions

Changes in Electrical Properties

RTD

A resistance thermometer or RTD measures the resistance of a pure electrical wire for determining temperature. This electrical wire is called a temperature sensor. RTD is widely used in industries for measuring temperature with high accuracy. RTD shows linear characteristics over a wide range of temperature. The variation of metal resistance with temperature can be expressed as follows:

$$R_t = R_0\left[1 + \left(t - t_0\right) + \beta\left(t - t_0\right)^2 + \cdots\cdots\cdots\right] \tag{20}$$

where R_0 is the resistance at t_0°C and R_t is the resistance at t°C. α and β are constants that depend on the metals.

Equation 11 is valid for a wide range of temperatures. For a small range, it can be expressed as:

$$R_t = R_0[1 + (t - t_0)] \tag{21}$$

The advantages of RTD are:

- Wide operating range -260 to 1000°C with platinum
- Low resistance
- High sensitivity -200 to 450°C with Ni • (compared to thermocouples)
- High repeatability and stability circuit

- Different range may be higher accuracy ±0.1 to 1.5°C possible with variable R
- Low drift -0.0025°C/year (industrial models drift < 0.1 °C/year)

The disadvantages of RTD are:

- Potentially significant lead wire resistance
- Sensitive to shock and vibration
- Slow response time
- Fragile
- Internal/self-heating (I2R)

Copper, nickel, and platinum are widely used in RTDs. The resistance-temperature characteristics of the three metals are shown in Figure 10. Platinum, copper, and nickel have a temperature range of 650°C, 120°C, and 300°C, respectively. Purity is verified by measuring R_{100}/R_0 because materials for making RTD should be pure. If metal is impure, it deviates from the conventional resistance-temperature graph. The value of α and β will change depending on the metal.

Thermistors

A thermistor is a type of resistor whose resistance is dependent on temperature. The resistance is inversely proportional to temperature. It drops nonlinearly with the rise in temperature. Ceramics and some semiconductors have a negative temperature coefficient. A typical schematic view of the thermistor is shown in Figure 11.

$$R(T) = R_{ref} \exp\left(\beta\left(1/T - 1/T_{ref}\right)\right) \tag{22}$$

Figure 10. Resistance-temperature characteristics

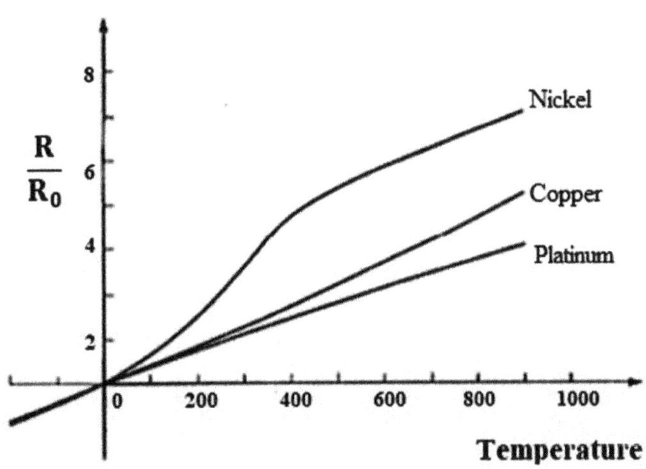

Figure 11. Schematic view of the thermisto

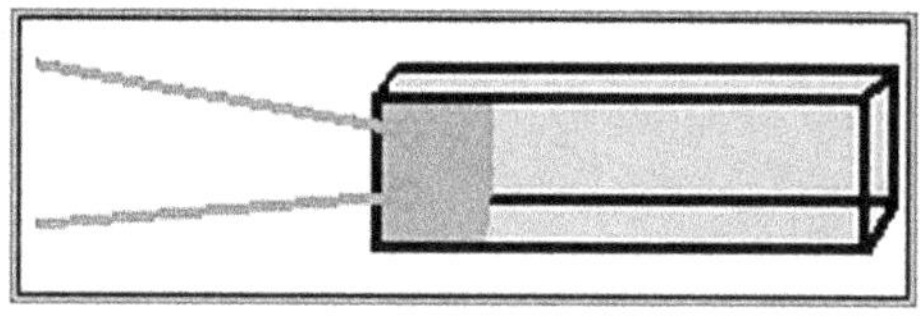

$$\frac{1}{T}=\frac{1}{T_{ref}}+\frac{1}{\beta}\left[In\left(R\right)-In(R_{ref}\right] \tag{23}$$

$$T=\frac{T_{Ref}\beta}{\beta+T_{ref}\left[In\left(R\right)-In\left(R_{ref}\right)\right]};\beta=3,000\,to\,5,000K \tag{24}$$

Advantages of thermistor include:

- Accuracy {~±0.02°C (±0.36°F)} better than RTDs and thermocouples
- Sensitivity significantly better than RTDs and thermocouples
- High sensitivity with insignificant errors in lead wire and self-heating
- Long-term stability and repeatability
- Small size compared to thermocouples
- Shorter response time than RTDs (but same as thermocouples)

Disadvantages of thermistor include:

- Working temperature ranges from -100°C to 150°C (-148°F to 302°F)
- Resistance-temperature relationship nonlinear (unlike RTDs with a linear relationship)
- Sensitivity ±6mV/°C Silicon with x% of Boron Germanium doped arsenic, gallium, arsenic for cryogenic temperature

Thermocouple

This is the most commonly used passive method for temperature measurement. It works on the principle of the seebeck effect. The thermocouple is a sensor with two wires of different metals. An electromotive force (EMF) develops in the circuit when two dissimilar metals are joined to form a junction while remaining at different temperatures (see Figure 12). One junction is often kept at a reference point (temperature). The other is kept at the environment the temperature is to be measured.

Change in temperature at junctions induces a change in the EMF (E). As the temperature rises, the output (EMF) of the thermocouple rises. This may not be linear. Induced EMF depends on the material and temperature of the junctions.

Then:

$$E = \int_{T_1}^{T_2} \varepsilon_A \frac{dT}{dx} dx + \int_{T_2}^{T_1} \varepsilon_B \frac{dT}{dx} dx = \int_{T_1}^{T_2} \left(\varepsilon_A - \varepsilon_B\right) dT \tag{25}$$

Then, $E=(\varepsilon_A - \varepsilon_B)(T_1 - T_2)$ (26)

$$E = \varepsilon_{AB}(T_1 - T_2) \tag{27}$$

Where ε_{AB} = proportionality constant/seeback coefficient, which depends on different materials. The seebeck coefficient can be defined as the open-circuit voltage, which is produced between two points on a conductor when a uniform temperature difference of 1K exists between those two points. Thermocouples are inexpensive, small in size, passive, and offer high accuracy when properly used.

Key Points

- Thermoelectric current generates the magnetic field
- Voltage generated when two types of wires join to form a junction
- Voltage measures temperature
- Voltage in range of microvol.
- Current flows very little

Material EMF versus temperature plot is shown in Figure 13 where:

- Constantan: 55% Cu with 45% Ni
- Chromel: 90% Ni with 10% chromium
- Alumel: 95% Ni, 2% AL, 2% Mn, and 1% silicon

Thermojunctions are formed by welding, soldering, or pressing two materials together. The different types of thermocouples and their operating ranges are shown in Table 3.

Thermocouple sensitivity is shown in Figure 14. A thermocouple is considered the first-order instrument in which the time constant is a significant parameter.

Energy, regardless of its type, is stored in a thermocouple. Therefore, at any point, energy stored should be equal to the summation of conduction, convection, and radiation.

Figure 12. Thermocouple principle

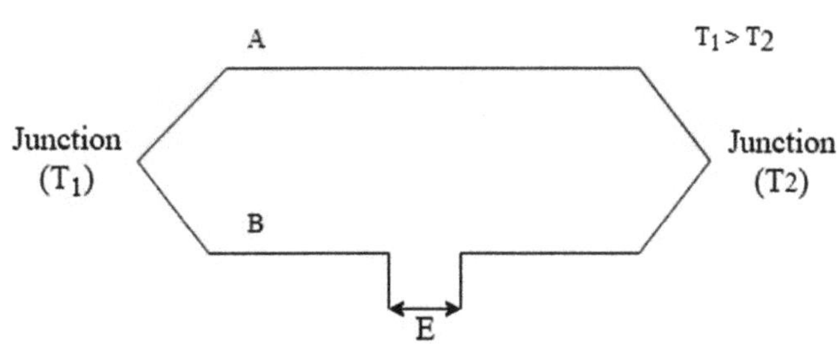

Figure 13. Material EFM vs. temperature

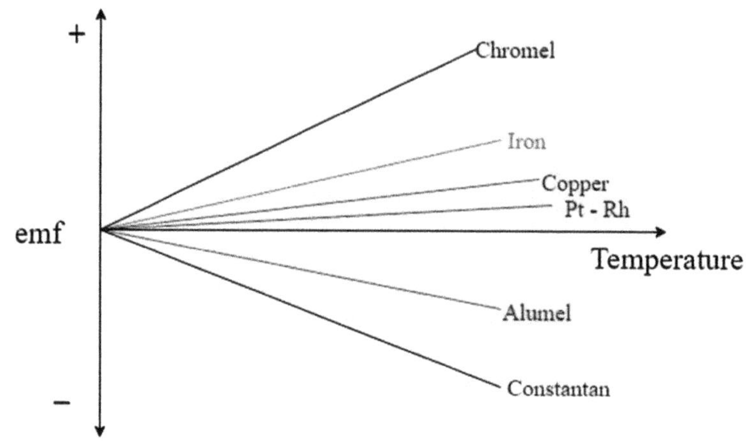

Table 3. Types of thermocouples and their operating ranges

Type	Material	Temp. Range	Voltage Swing Over the Range (mv)
K	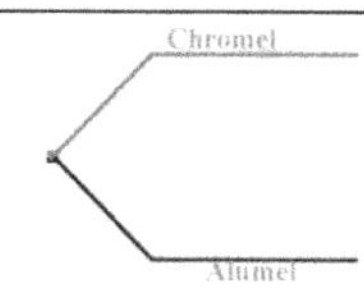 Chromel / Alumel	-200 to 1300 °C	56.0
T	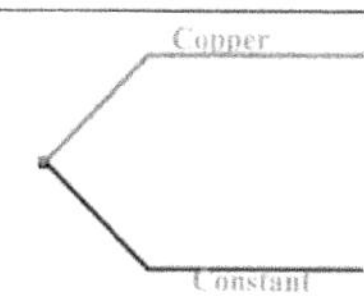 Copper / Constant	-200 to 350 °C	26.0
J	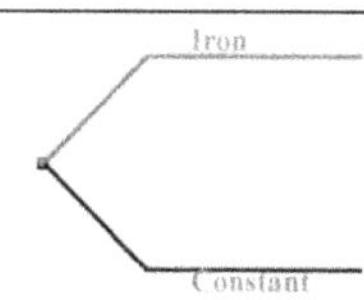 Iron / Constant	-150 to 1000 °C	50.0
S	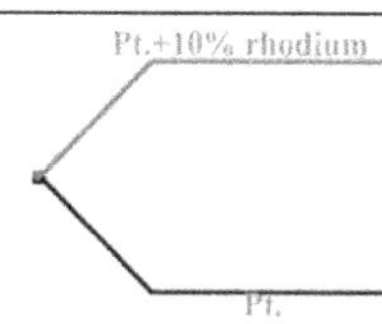 Pt.+10% rhodium / Pt.	0 to 1538 °C	16.0
R	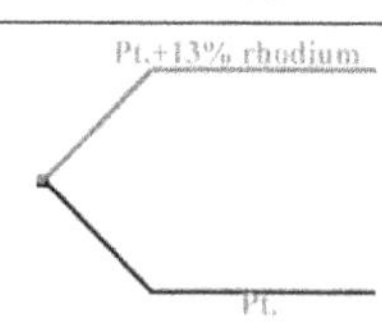 Pt.+13% rhodium / Pt.	0 to 1593 °C	18.7
E	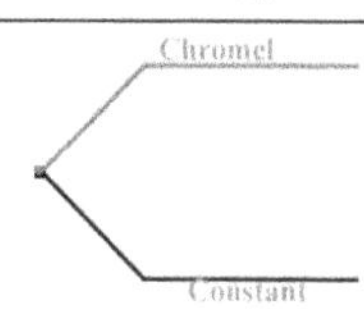 Chromel / Constant	0 to 982 °C	75.0

Figure 14. Thermocouple sensitivity

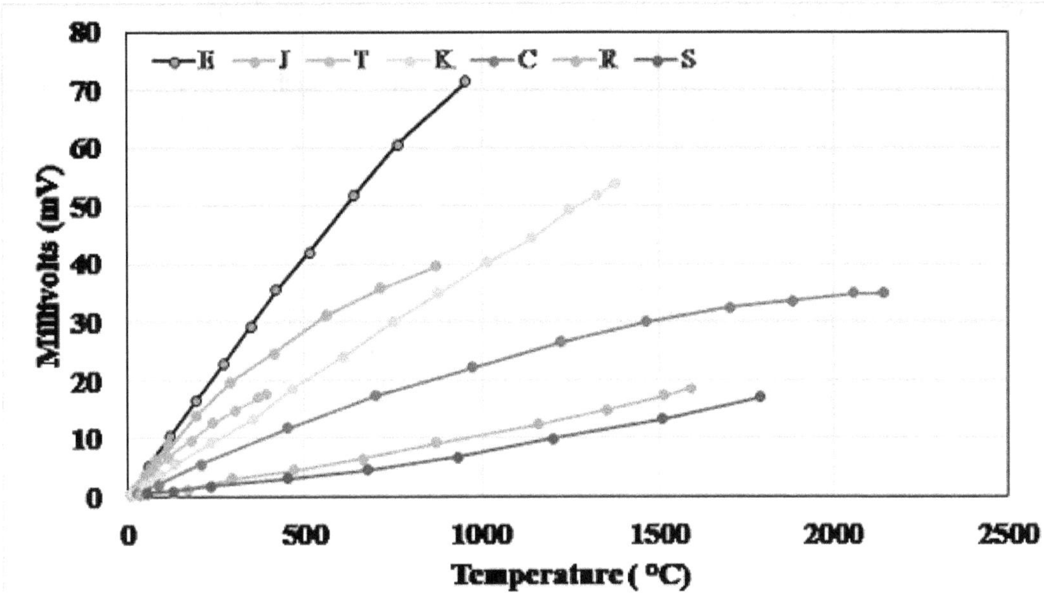

From, energy balance:

$$mc\frac{dT}{dt} = Q_{cod} + Q_{con} + Q_{rad} \tag{28}$$

If conduction and radiation heat transfer are neglected, then:

$$mc\frac{dT}{dt} = Q_{con}$$
$$E'mc\frac{dT}{dt} = hA\left(T - T_{\alpha}\right)$$
$$\frac{mc}{hA}\int_{T_i}^{T}\frac{dT}{\left(T - T_{\alpha}\right)} = \int_{t=0}^{t} dt$$
$$t = \frac{mc}{hA}\ln\left|\frac{T - T_{\alpha}}{T_i - T_{\alpha}}\right|; t = response\, time\, of\, the\, thermocouple$$

Finally:

$$\frac{T - T_{\alpha}}{T_i - T_{\alpha}} = e^{-\frac{hA}{mcE'}t} \tag{29}$$

Where mc/hA is the time constant of the thermocouple. The time constant of a first-order system can be defined as the time needed by the sensor to respond to 63.2% of its total output from a step change in temperature.

The response of the thermocouple refers to the time it will take to reach its full output. The response of a first-order system is mathematically described by a first-order differential equation.

$$c(t) = k\left(1 - e^{-t/\tau}\right) \tag{30}$$

Where c: output, k = gain, t = time, and τ = time constant of an instrument.

Figure 15 expresses the color codes for thermocouples in different countries. The comparison of thermocouple, thermistor, and RTD are portrayed in Table 4 and Figure 16.

Figure 15. Thermocouple extension cable color codes

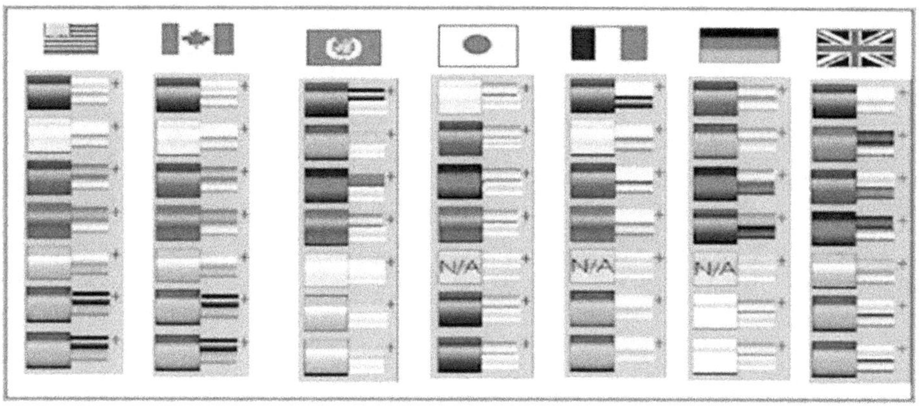

Figure 16. Comparison between thermocouple RTD and thermistor

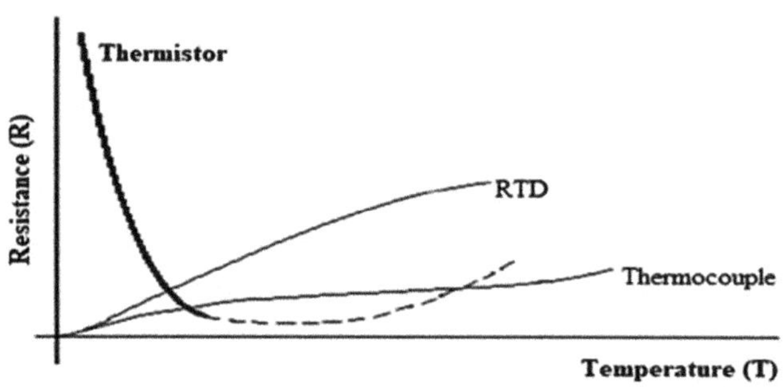

Table 4. Comparison of thermocouple, thermistor, and RTD

RTD	Thermistor	Thermocouple
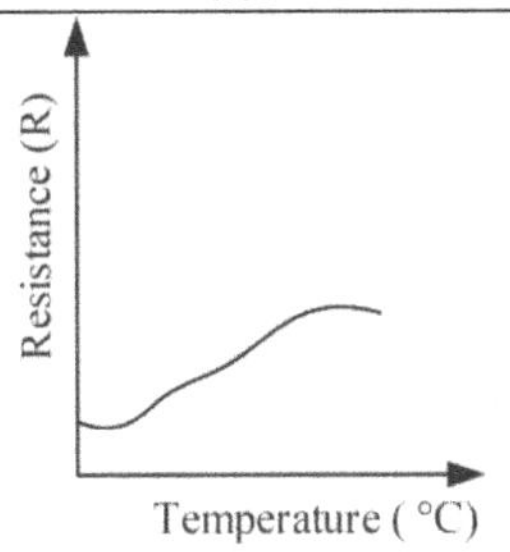	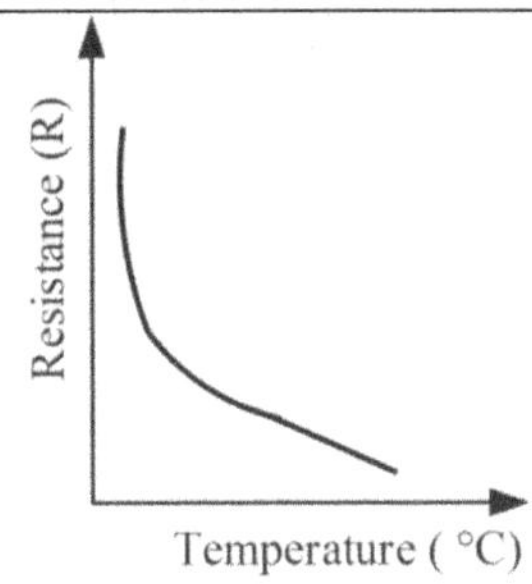	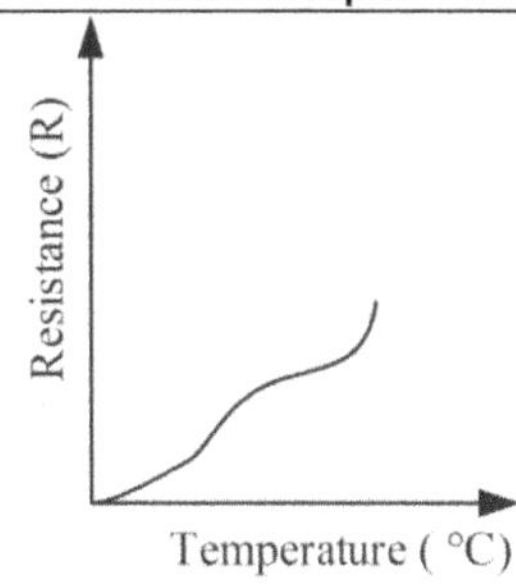
Advantages ➢ Most accurate ➢ Most stable ➢ More linear than thermocouple	Advantages ➢ Fast ➢ High output ➢ Two-wire ohms	Advantages ➢ Simple ➢ Inexpensive ➢ Wide variety ➢ Self-powered ➢ Rugged ➢ Wide temperature range
Disadvantages • Self-heating • Expensive • Low absolute resistance • Small ΔR • Current source required	Disadvantages • Self-heating • Fragile • Nonlinear • Limited temperature range • Current source required	Disadvantages • Low voltage • Least stable • Nonlinear • Least sensitive • Reference required

Changes in Chemical Properties

Liquid-Crystal Thermography

Liquid crystals have the mechanical properties of a liquid. Yet they encompass the optical properties of a single crystal. The color of the crystal, which is affected by change in temperature, allows for the measurement of temperature. By varying the chemical composition, the range and resolution of liquid crystal thermometers are varied. It ranges from 0 to several hundred resolution of liquid crystal sensors ± 2°C (based on range). Disposable liquid crystal thermometers are developed for domestic and medical applications. These includecholesterol liquid crystals and temperature paints. Polyvinyl alcohol coating prevents the cracking of crystal. A typical setup of liquid-crystal thermography is shown in Figure 17.

The heat transfer coefficient is calculated using:

$$h = \frac{q_w}{\left(T_w - T_\alpha\right)}$$

With the transient technique, the change in color of the liquid crystal can be observed to deduce temperature rise on the test model. Liquid crystals are applied as a thin sheet over the heated wall. Liquid crystal properties, by virtue, react chemically and change color (red, green, and blue). A charge-coupled

Figure 17. Setup of liquid-crystal thermography

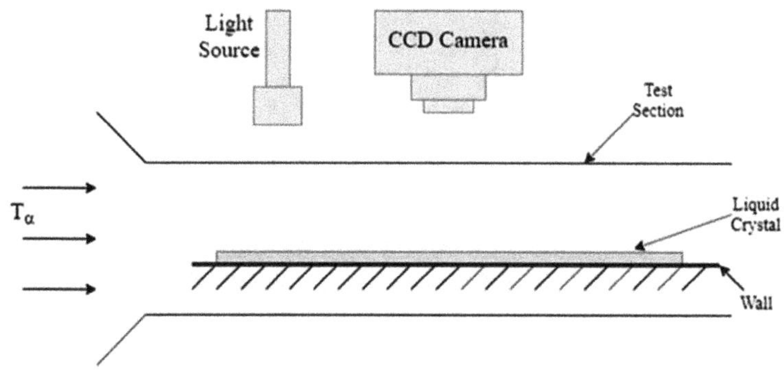

device (CCD) camera is placed over the heated wall to capture an image of the liquid crystal sheet. An image digitizer gives the R, G, B values.

Different Colour Model

RGB (Red Green Blue) Model

$$r = \frac{R}{R+G+B} \tag{30}$$

$$g = \frac{G}{R+G+B} \tag{31}$$

$$b = \frac{B}{R+G+B} \tag{32}$$

HSI (Hue Saturation Intensity) Model

$$H = 90^\circ - \tan^{-1}\left(\frac{F_1}{\sqrt{3}}\right) + F_2; \quad \begin{matrix} \text{if } F_2 = 0, \ G \geq B \\ \text{if } F_2 = 180, \ G < B \end{matrix} \tag{33}$$

where,

$$F_1 = \frac{2R - G - B}{G - B} \tag{34}$$

$$S = 1 - \frac{E' \min\left(R, G, B\right)}{I} \tag{35}$$

$$I = \frac{R + G + B}{3} \tag{36}$$

where H, S, I are hue, saturation, intensity, respectively.

With in-camera measurements, the individual values of R, G, B vary from 0 to 255 (8-bit resolution). In addition, the color temperature response of the surface coated with a liquid sheet is calibrated.

Changes in Emitted Thermal Radiation

Pyrometer

This remote-sensing thermometer measures surface temperature. According to electromagnetic radiation, every hot body emits visible radiation. This varies in color as the temperature increases. This principle is elaborated on in Figure 18.

Optical pyrometer consists of:

- An eyepiece (optical lens on the right) and a reference lamp power-driven by battery
- Rheostat for changing the current (brightness intensity)
- Absorption screen between the optical lens and reference bulb to increase the temperature range
- Red filter between the reference bulb and eyepiece to narrow the wavelength band

Radiation emitted from the source is captured by an optical objective lens. Thermal radiations are focused by the lens to the reference bulb. The observer watches the process, correcting it so the reference lamp filament has a sharp focus. It is superimposed on the temperature source image. Rheostat values and the current in the reference lamp are changed to alter its intensity. The change in current can be observed as follows:

Figure 18. Optical pyrometer

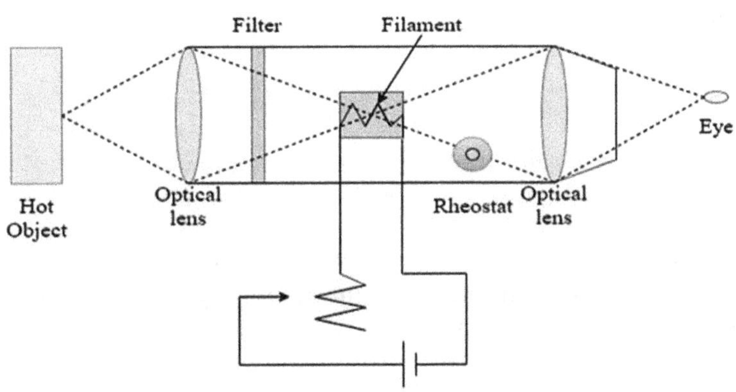

- If the filament is dark, it is cooler than the temperature source.
- If the filament is bright, it is hotter than the temperature source.
- If the filament disappears, there is equal brightness between temperature source and filament. At this condition, the current flowing through the reference lamp is measured. When calibrated, it gives the value of the temperature of the radiated light.

The most common features of an optical pyrometer include:

- Most commercially used instruments for measuring temperature ranges of 700°C to 3000°C
- Very stable instrument with easy calibration
- All types of furnaces

SUMMARY

This chapter discusses temperature gradients, thermal shocks, and thermal expansion (or contraction effects). Like most of the engineering problems, the thermal stress is due to the thermal expansion or sudden temperature changes in the body. Therefore, a constructive analysis is detailed regarding temperature measurement devices and their proper data capturing methodologies.

REFERENCES

Annamalai, K. L., & Karuppasamy, S. K. (2016). Experimental investigation on stability of an elastically mounted circular tube under cross flow in normal triangular arrangement. *Journal of Vibroengineering*, *18*(3), 1824–1838. doi:10.21595/jve.2016.16708

Balaji, P. S., & Yadava, V. (2013). Three dimensional thermal finite element simulation of electro-discharge diamond surface grinding. *Simulation Modelling Practice and Theory*, *35*, 97–117. doi:10.1016/j.simpat.2013.03.007

Bansal, R. K. (2016). *A textbook of strength of materials - Mechanics of solids*. Laxmi Publications.

Hetnarski, R. B., Eslami, M. R., & Gladwell, G. M. L. (2009). *Thermal stresses: Advanced theory and applications* (Vol. 41). New York, NY: Springer.

Holman, J. P. (1966). *Experimental methods for engineers*. McGraw Hill.

Johns, D. J. (2013). *Thermal stress analyses* (Vol. 2223). Elsevier.

Karthik, K., Kumar, S., & Kumaraswamidhas, L. A. (2015). Experimental investigation on flow-induced vibration excitation in an elastically mounted square cylinder. *Journal of Vibroengineering*, *17*(1), 468–477.

Karthik, K., & Kumaraswamidhas, L. A. (2015). Experimental Investigation on Flow-Induced Vibration Excitation in an Elastically Mounted Circular Cylinder in Arrays. *Fluid Dynamics Research*, *47*(1), 015508. doi:10.1088/0169-5983/47/1/015508

Kumar, K. S., & Kumaraswamidhas, L. A. (2018). Investigation on Stability of an Elastically Mounted Circular Tube under Cross Flow in Inline Square Arrangement. *Iranian Journal of Science and Technology, Transactions of Mechanical Engineering,* 1-13.

McGee, T. D. (1988). *Principles and methods of temperature measurement*. John Wiley & Sons.

Noda, N. (2018). *Thermal stresses*. Routledge. doi:10.1201/9780203735831

Venkateshan, S. P. (2015). *Mechanical measurements*. John Wiley & Sons.

KEY TERMS AND DEFINITIONS

Electromotive Force (EMF): It is the source of energy which enables electrons to move around an electric circuit. It is determined by the potential difference between the two electrodes.

Repeatability: It is defined as the closeness of agreement between independent test results, obtained with the same method, on the same test material, in the same laboratory.

APPENDIX

A1. Steel railroad reels 15 m long are placed with a clearance of 4 mm at a temperature of 25°C (see Figure 19). At what temperature will the rails just touch? What stress would be induced in the rails at that temperature if there was no initial clearance? Consider $\alpha = 22.1$ gm/(m °C) and E = 210 GPa.

Figure 19. Steel railroad

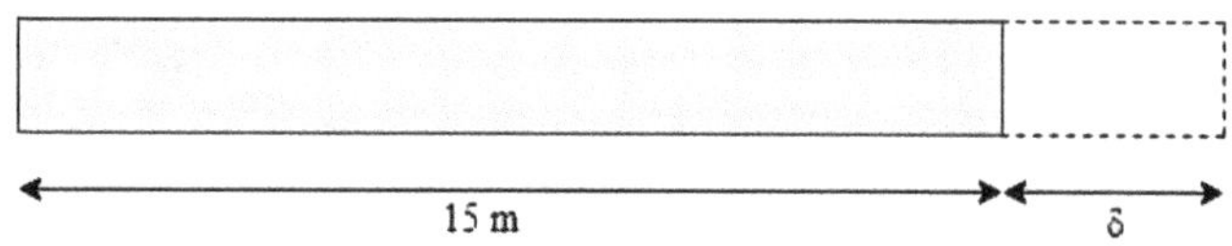

Answer:

$\delta_T = 4$ mm: $\delta_T = \alpha L \Delta T$, $\delta_T = \alpha L(T_f - T_i)$,

where

$\delta_{T=}$ 4; $\alpha = 22.1$ gm/(m°C) and $T_i = 25$°C

then

$T_f = 37.06$°C (Ans.)

And the stress:

$\delta = \delta_{T}$, $\sigma L/E = \alpha L \Delta T$, $\sigma = \alpha E(T_f - T_i)$

where

$\alpha = 22.1$ gm/(m°C)

and

$T_i = 25$°C; $E = 210\times10^6$ Pa

Then

$\sigma = 118.9\times10^6$Pa (Ans.)

A2. Find the thermal stress for section 1 and section 2 of Figure 20. Consider that $\alpha = 13\times10^{-6}$/°C; E = 210 GPa.; $\Delta T = 75$°C.

Figure 20. Composite bar

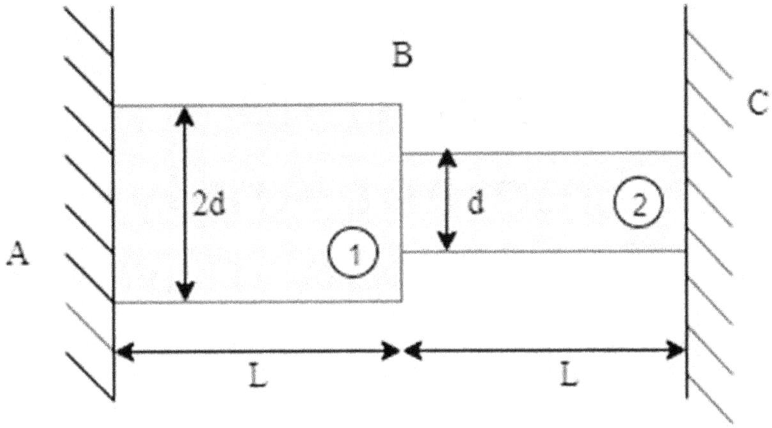

Answer:

Total thermal deformation = total axial deformation

$$\alpha_1 T_1 L_1 + \alpha_2 T_2 L_2 = \frac{\alpha_1 L_1}{E_1} + \frac{\alpha_2 L_2}{E_2},$$

$\alpha_1 + \alpha_2 = 2\alpha TE$

where $\alpha = 13\times10^{-6}/°C$; E=210 GPa;

$\alpha_1 + \alpha_2 = 409.5\times10^3$ kPa. (Ans)

As $R_A = R_B = R_C \sigma_1 A_1 = \sigma_2 A_2$

$$\sigma_1 \frac{\pi}{4}(2d)^2 = \sigma_2 \frac{\pi}{4}(d)^2$$
$$4\sigma_1 = \sigma_2$$

Then $5\sigma_1 = 409.5$; $\sigma_1 = 81.9$ MPa and $\sigma_2 = 327.6$ MPa (Ans)

Chapter 13
Finite Element Analysis of Chip Formation in Micro-Milling Operation

Leo Kumar S. P.
PSG College of Technology, India

Avinash D.
PSG College of Technology, India

ABSTRACT

Finite element analysis (FEA) is a numerical technique in which product behavior under various loading conditions is predicted for ease of manufacturing. Due to its flexibility, its receiving research attention across domain discipline. This chapter aims to provide numerical investigation on chip formation in micro-end milling of Ti-6Al-4V alloy. It is widely used for medical applications. The chip formation process is simulated by a 3D model of flat end mill cutter with an edge radius of 5 μm. Tungsten carbide is used as a tool material. ABACUS-based FEA package is used to simulate the chip formation in micro-milling operation. Appropriate input parameters are chosen from the published literature and industrial standards. 3-D orthogonal machining model is developed under symmetric proposition and assumptions in order to reveal the chip formation mechanism. It is inferred that the developed finite element model clearly shows stress development in the cutting region at the initial stage is higher. It reduces further due to tool wear along the cutting zone.

INTRODUCTION

The modelingand simulation of micromachining are used as a supplementary method to analyzea physical experiment. Finite element simulation predicts the optimum process variables that are complicated to obtain by experimental investigations (Arrazola et al., 2013). Micro-milling is a flexible manufacturing method for the production of a functional 3D micro-product. The progress of the micro-milling process eventually depends on themicro cutter, since micro-milling is a tool-based process. Micro-end milling

DOI: 10.4018/978-1-7998-1690-4.ch013

has numerous advantages such as, for instance, low-cost investment and process flexibility. It is the most suitable method for manufacturing of micro parts and devices with complex shapes (David, 2004).

The scale of micro-milling process is very small when compared with macro milling. It is more complicated to systematically conduct and investigate micro-milling by experimentation. Hence, investigation on micro-milling is done by modeling and simulation using the finite element method. Finite Element Modling(FEM) provides results in terms of temperature, displacement, stress, and strain (Souza et al., 2008). To enhance manufacturing quality, to increase tool performance, and to discover optimum cutting parameters, many research centers and industries are using FEM and simulation that help for better understanding of the metal cutting process. The numerical analysis of chip formation in micromachining needs research attention, since it severely affects the part functionality and high investment involved in fabrication with macro machining (Leo Kumar et al., 2014). The numerical method is able to distinct the domains in order to solve differential equations; and it is a frequently used technique to simulate the outcome of an actual process without the necessity ofexecuting a sequence of experimental trials (Vaz et al., 2007).

ANALYZING TECHNIQUE

Finite Element Analysis (FEA) is capable ofresolving complex non-linear problems. The first simulation of the machining process was done in the 1980s, and it was performed in 2D for orthogonal cutting using plain strain models (Shrot et al., 2012). To perform a simulation of the machining process, the most commonly used software programs are ANSYS, ABAQUS, and DEFORM3D.FEA is a vital tool appropriate for solving nonlinear and linear problems in a static and dynamic environment. Many pieces of literaturehave dealt with diverse models and simulation for specific machining problems (Vaziri et al., 2010). Mesh area of workpiece have diverse geometry and boundary conditions for testing, but most of the models aim at investigating the effects of chip morphology and cutting parameters (Arrazola et al., 2010). Every model has the same types of elements whichare usually quadrilateral in nature, with thermo-mechanical properties (CPE4RT,CP4RT,etc). A 2Dstrain analysis can be done in ABAQUS for the orthogonal cutting process by using a quadrilateral with the thermo-mechanical assumption (Munoz-Sanchez et al., 2011).

In FEA, complexities like varying shapes and loading conditions lead to an approximate solution. Due to flexibility and diversity, much attention has been given to such complexities in the engineering domain (Thanongsak et al., 2013). FEA was originated as a way to perform stress analyses, and it was initially began as the addition of a matrix method for structural analysis. Currently, this method is utilized for analysis in solid mechanics, fluid flow, heat transfer, and many other fields of interest. Many civil engineers use FEA extensively for analysing beams, plates, folded plates, space frames, shells, and rock mechanics-related problems. Analysis of both static and dynamic problems can be done by using FEA. Many FEA packages are currently available, and a few packages are NASTRAN, ANSYS,ABACUS, STAAD-PRO, and NISA (Rosa et al., 2007).

MODELING

In machining simulation, three methods are generally incorporated: Lagrangian, Eulerian, and Arbitrary Lagrangian-Eulerian (ALE) approach. The ALE approach integrates the characteristics of Lagrangian and Eulerian formulations and adopts a clear incorporation of solutions for rapid convergence. A major advantage of this technique is the freedom it provides in dynamically defining mesh configuration. Performing machining simulation using this ALE approach does not require separation criterion during formulation (Escamilla et al., 2010; Zhang et al., 2011). In ABACUS, meshing determines the type of algorithm used. Generally, two different methods are used to assign element shapes for 2D and 3D problems. Figure 1 shows the meshview of the tool and the workpiece.

The ALE adaptive mesh area is engaged in the workpiece where deformation occurs. The adaptive mesh area defines portions of a finite element model, where mesh movement is independent of material deformation. It helps in preserving the properties of the nodes (Nasr et al., 2008).During machining, mesh sweep is related to the mesh domainwhich escalates the concentration of adaptive meshing. In this step, ALE remeshing of nodes is rearranged based on the recent positions of the adjacent nodes/elements. At each sweep, nodes are oftenly used to decrease element distortion. Large sweep values will augment the accuracy and computational time of the model (Vaziri et al., 2011).

MATERIAL MODEL

Model development in ABACUS is carried out by using the ALE approach. The number of elements used is approximately 6,000. As the initial boundary condition, all nodes at the bottom of the workpiece are constrained for displacement. The motion of the tool nodes is allowed only in the cutting direction. Figure 2 shows applied boundary conditions and tool geometry. To simulate the machining process, a

Figure 1. Meshview of the tool and the workpiece

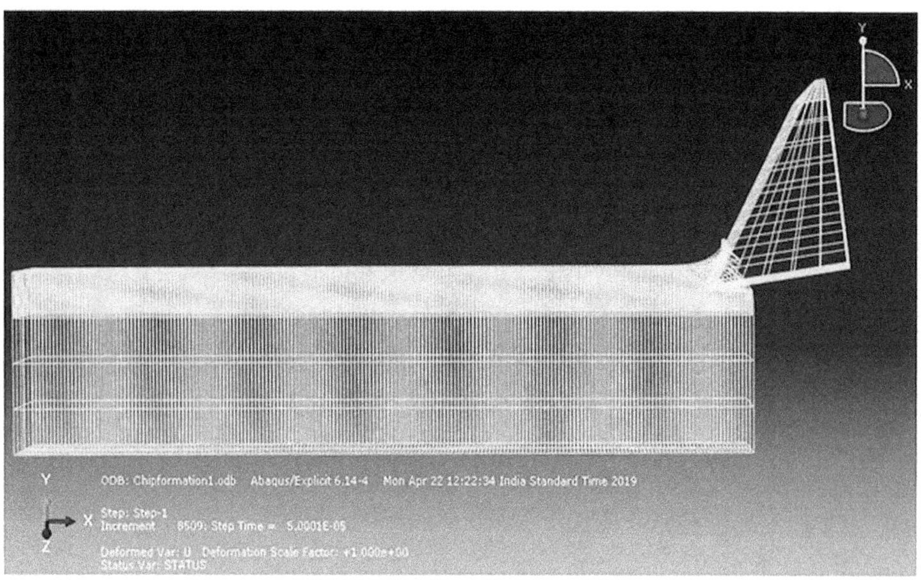

workpiece 100x 20 mm in dimensional size is modeled. Tungsten Carbide is used as the tool material and assumed that it possess a rigid cutting edge radius of 5 μm. Rake angle vary from -10° to 20°.The clearance angle is fixed at10°and the depth of the cut is maintained at 0.5 mm. The properties of the workpiece and the tool material are shown in Table 1.

In the Johnson-Cook Equation (1), flow stress is described as being a product of equivalent strain, temperature-dependent terms, strain rate and several other parameters in order to mimic the behavior of the Ti-6Al-4V alloy. The yield strength of the workpiece at room temperature is represented as A and B. It indicates the pre-exponential factor relating to yield strength at the beginning of plastic deformation. *C* is a factor of strain rate. ε represents equivalent plastic deformation. $\dot{\varepsilon}$ is reference strain rate. ε_o' is the strain-rate of the material; n is the work-hardening exponent;and m is the thermal-softening exponent.T's are variables related toroom temperature and melting temperature. Table 2 shows the parameters used for Ti-6Al-4V in the Johnson-Cook equation.

$$\sigma = \left[A + B\left(\varepsilon\right)^{n}\right] * \left[1 + C * \ln\left[\varepsilon' / \varepsilon_o'\right] * \left(1 - \frac{T - T_{room}}{T_{melt} - T_{room}}\right)^{m}\right. \quad (1)$$

Table 1. Properties of Tool And Workpiece Material

Sl.No.	Properties	Unit	The Workpiece (Ti-6Al-4V)	Tool (Tungsten Carbide)
1.	Density	Kg/m^3	4,430	11,900
2.	Elastic Modulus	GPa	106	534
3.	Poisson Ratio	Nil	0.3	0.22
4.	Ultimate Tensile Strength	MPa	950	3000
5.	Thermal Conductivity	$W.m^{-1}.K^{-1}$	17	15

Figure 2. Boundary conditions and tool geometry

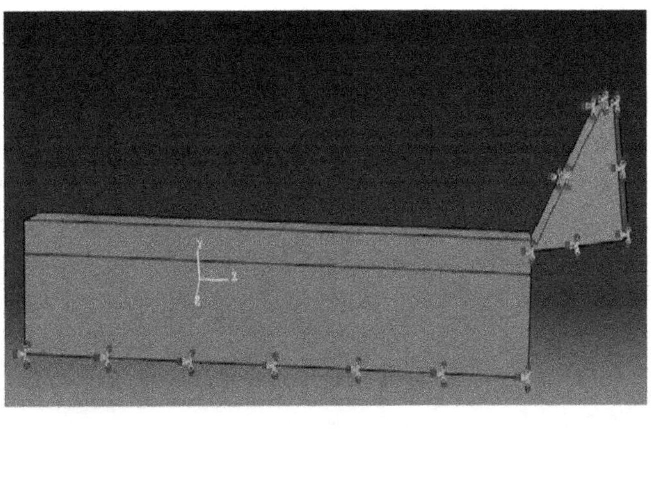

Table 2. Parameters ofthe Johnson-Cook equation

Material Model	A [MPa]	B [MPa]	n	m	C
Values	782.7	498.4	0.28	1.0	0.028

The workpiece and the tool contact determine the surface quality and the stress developed during machining. According to characteristics based on interfacial friction, a tool-to-chip and work-to-tool contact must be modeled using standard methods in order to provide accurate details of stress developed due to friction (Jin et al., 2012). The Coulomb friction model,shown in Equation (2), is used for modeling.The expression for this friction model is as follows:

$$\gamma = \sigma * \mu \tag{2}$$

Here,γ is frictional shear stress, μ is the friction coefficient, and σ represents normal stress exerted on the surface. This is a simple model for the machining process.Hence, it is not suitable for machining conditions where high stress is developed. In this case, the shear friction model,shown in Equation (3), is more accurate for machining in cases of high-stress variation (Schulze et al., 2010).

$$\gamma = m * k \tag{3}$$

Here, k is average flow stress.

When the distribution of normal stress on the rake face of the tool is fully defined, and μ is known, the frictional stress can be determined by assuming the sliding region and sticking region on the secondary deformation zone. At the sticking region, critical frictional stress is reached due to the development of high normal stress on the cutting tool (Calamaz et al.,2008). The sliding contact region satisfies the Coulomb friction law with a constant coefficient of friction and affirms that, in a sliding region, the frictional stress is lower than yield shear stress..

Damage Model

The initialization of breakdown and of material separation aredefined by the nodal distance of the elements and by local damage criterion. Difficulties in the FEA of machining process include need to identify proper separation criterion that is related to a material breakage. Among many damage model, "Johnson Cook" damage model is preferred for micromachining applications(John et al., 2013). The ductile damage model defines the points where the initiation of stiffness degradation occurs over the workpiece. Ductile damage is generally related to the occurrence of huge plastic deformation near crystalline defects (Gao et al., 2013). The Johnson-Cook model assumes that a fracture begins when accumulated plastic strain attains a crucial value.Equation (4) represents the model as follows:

$$\int_{0}^{\varepsilon_f} f\left(stressstate\right) d\varepsilon = D \tag{4}$$

Here, f is the damage function and ε_f is an equivalent strain-to-fracture. D represents the material constant. By extending the model with $f=1/\varepsilon_f$, equivalent strain-to-fracture can be expressed, as shown in Equation (5).

$$\varepsilon_f = [D_1 + D_2 \exp\left(D_3\sigma^*\right].\left[1 + D_4 \ln \varepsilon'^*\right].\left[1 + D_5 T^*\right] \quad (5)$$

Here, σ^* is a stress triaxiality parameter, and ε'^* is the representation of the dimensionless plastic strain. The fracture constants are D_1, D_2, D_3, D_4 and D_5.This new set of parameters of the Johnson-Cook damage model is used for the ALE solution in the machining simulation. Necessary values are obtained by post-processing calculations based on the available data in ALE simulation output files (Shams et al.,2012; Molinari et al., 2011). In this chapter, the least-squares (nonlinear) optimization method is used to discover five fracture constants of the Johnson-Cook damage model. Table 3 shows the parameters used in the model.

FINITE ELEMENT SIMULATION OF CHIP FORMATION

Machining simulation is performed using the finite element method. Damage circulation relating to material deprivation during initialization of the chip formation process is presented in Figure 3. Under machining conditions, periodic cracks and segmented chip morphology are identified. The separation of material is initiated at a side-free surface, closer to the tooltip, and it continues to propagate within the material (Bhavikatti et al., 2005). It is noticed that fracture phenomenon is mostly contained at the tool-to-chip interface and also in the shear band. Figure 3 shows the initial stage of chip formation. From the figure, it is identified that higher stress has been developed at the intersection point of the tool and the chip. The rack face of the tool is subjected to high stress during chip formation (Sima et al., 2010).

The morphology of the segment is significantly affected by the rake angle. It is identified that loading conditions lead to failure, which differs based on rake angle changes. For the nodal points near the tooltip, the state of strain is severely compressive. By moving away from the tooltip, shear occurrence becomes more significant (Vavourakis et al., 2013). The crucial shear area does not withstand shear alone. The development of stress also plays a vital role in chip formation. Characterisation of mechanical loading arehighly connected to cutting geometry and segmented shape (Kug et al., 1999). The stochastic nature of segment shapes is explicitely predicted by computation, as is shown in Figure 4.

Usage of the finite element model improved the perceptive of the notched chip generation process. Specific attention is given to the mechanism of chip formation. The results obtained lead to the division of the chip generation process into three successive steps. The first phase in chip formation is germination. In this phase,the linear development of stress is observed. It is due to the presence of constant compressive loading at tooltip (List et al., 2012).Elastic energy is generated and stored within the region

Table 3. Parameters for the Johnson-Cook Damage Model

Damage Model	D_1	D_2	D_3	D_4	D_5
Value	-0.09	0.25	-0.5	0.014	3.87

Figure 3. The initialization of chip formation

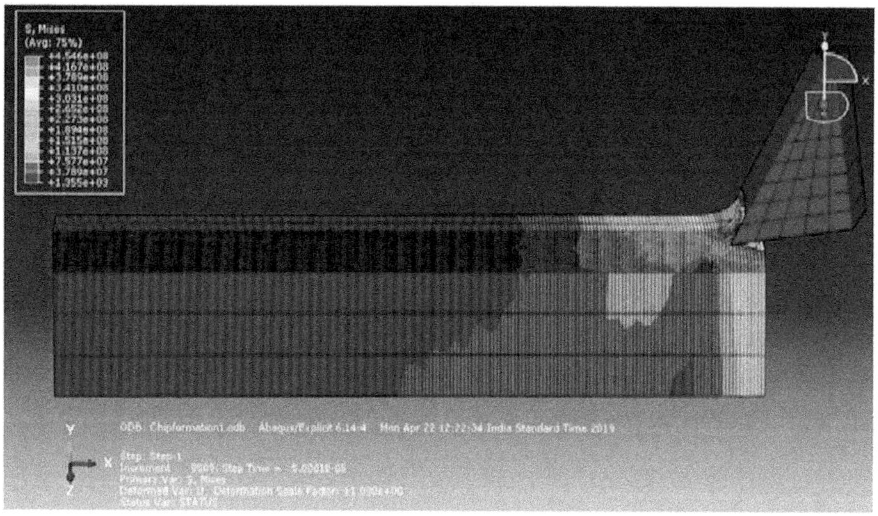

Figure 4. Segment shapes predicted by computation

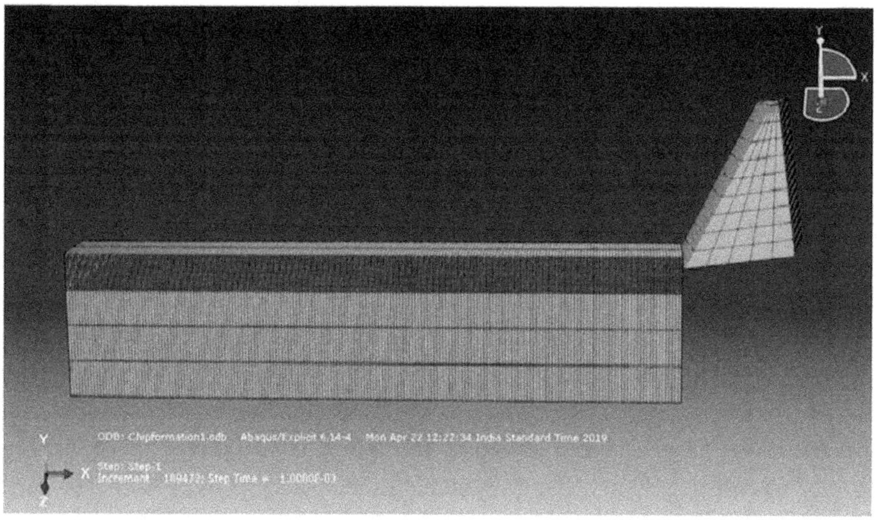

where no thermal debauchery is identified.Similarly, plasticity is developed entirely on the primary shear zone (Umbrello, 2008).This loading induces one out-of-plane deformation of the chip that is clearly visible in Figure 5. From the figure, a strong hydrostatic compressive zone is observed near the tooltip. Identification of the end of the germination stage is defined by the rebellion of a microcrack. Hence, the damage parameter attains unity in the first element which is finally deleted in this stage.

The second phase in chip formation is the growth of crack propagation. It describes the evolution of the crack along the principal shear plane. It is regarded as involving a rapid increase both instrain rate and intemperature. Due to hardening, Von Misses strain remains approximately constant in this phase. Triaxiality ratio clearly shows that shear is the main driving mechanism in this phase. These phenomena

Figure 5. Plane deformation of the chip

lead to the activate accumulation of strain in the shear band. Crack propagation along the same direction is identified, as shown in Figure 6. These cracks are initiated at the tooltip and develop inside the shear zone towards the chip free surface. At the end of this phase, as cracks reach the upper surface and no further strain can be generated in the shear band.

The final phase of chip formation is extraction. Von Mises stress is decreased as the chip formed moves upward and the loaded zone is left behind. The formation of the discontinuity in the chip is identified in this stage, as shown in Figure 7. Damage factors were found to have been increasing at the tool-to-chip interface.This leads to a minor increase of the hydrostatic compressive area. The friction on the next section leads to a faintly positive triaxiality ratio which remains constant as strain tends toward zero. From Figure 7, the breakage formed in the chip,and the stress developed, is clearly identified. But it is not an absolute solution forunderstandingthe machining process. Parameters that are directly connected

Figure 6. Crack propagation along the tool tip

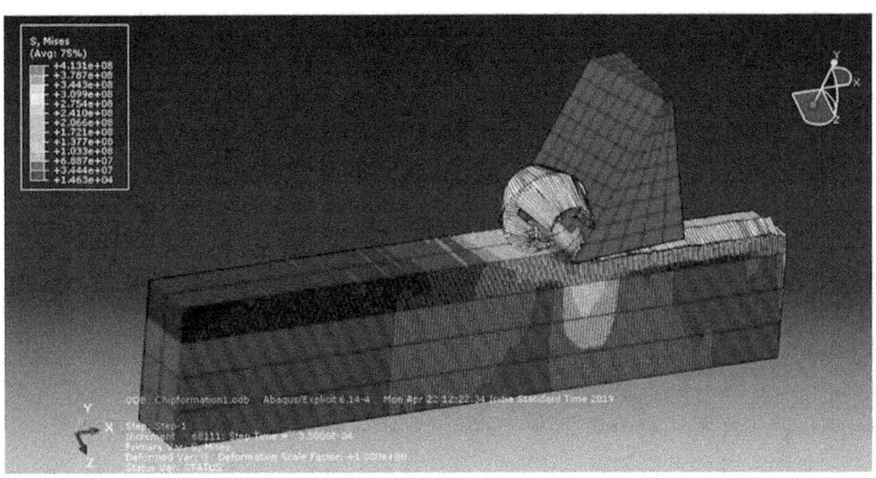

Figure 7. Discontinuous chip formations

to crack transmission are chip separation criterion. Changing the tensile failure rate is licit. It is temperature dependent, even though high strain rates are generated in machining. The temperature-dependent tensile failure rate of Ti-6Al-4V is unavailable in the literature, and measuringit is highly challenging.

SUMMARY

In this chapter, finite element simulation of the micro-milling process inTi-6Al-4Valloy was carried out, and a chip formation study was exemplified. The main aim is to perform the numerical modeling of physical phenomenon which induce segmented chip formation in micromachining. The Johnson-Cook model is calibrated by using a contrary recognition method. The model properly predicts the experimental data and gives better results.

- Chip formation results obtained by shear phenomena revealed that the chosen maximum shear damage factor clearly predictsductile fracture behavior during chip formation.
- The rack angle dependency, andaccurate results from the model, both lead to abetter depiction ofdamage phenomenon.
- A three-dimension orthogonal machining model is developed under symmetric proposition and assumptions in order to reveal the chip formation mechanism.
- The finite element model developed here clearly shows that stress developed in the cutting region at the initial stage is higher. The higher stress value obtained is 4.56 GPa.
- Based on the micro-milling simulation result, it is inferred that the chip formation process occurs from the development of the crack propagation and the adiabatic shear band. It starts at the tooltip and develops within the shear area towards the free surface.
- Proper accord is obtained in FEA results in terms of chip geometry. Chip generation problem is formore stochastic than has previously beenconsideredin the published literature. More research attention is needed in this regard.

REFERENCES

Arrazola, P. J., & Ozel, T. (2010). Investigations on the effects of friction modeling in finite element simulation of machining. *International Journal of Mechanical Sciences*, *52*(1), 31–42. doi:10.1016/j.ijmecsci.2009.10.001

Arrazola, P. J., Ozel, T., Umbrello, D., Davies, M., & Jawahir, I. S. (2013). Recent advances in modeling of metal machining processes. *Journal of Manufacturing Technology*, *62*, 695–718.

Bhavikatti, S. S. (2005). *Finite element analysis*. New Delhi: New Age International.

Calamaz, M., Coupard, D., & Girot, F. (2008). A new material model for 2D numerical simulation of serrated chip formation when machining titanium alloy Ti–6Al–4V. *International Journal of Machine Tools & Manufacture*, *48*(3-4), 275–288. doi:10.1016/j.ijmachtools.2007.10.014

David, V. H. (2004). *Fundamentals of finite element analysis*. NewYork: McGraw-Hill.

Escamilla, I., Zapata, O., Gonzalez, B., Gamez, N., & Guerrero, M. (2010). 3D finite element simulation of the milling process of a Ti-6AL-4V alloy. *Simulia Customer Conference*.

Gao, C., & Zhang, L. (2013). Effect of cutting conditions on the serrated chip formation in high-speed cutting. *Machining Science and Technology: An International Journal*, *17*(1), 26–40. doi:10.1080/10910344.2012.747887

Jin, X., & Altintas, Y. (2012). Prediction of micro-milling forces with finite element method. *Journal of Materials Processing Technology*, *212*(3), 542–552. doi:10.1016/j.jmatprotec.2011.05.020

John, M. R. S., Shrivastava, K., Banerjee, N., Madhukar, P. D., & Vinayagam, B. K. (2013). Finite element method-based machining simulation for analyzing surface roughnessduring turning operation with HSS and carbide insert tool. *Arabian Journal for Science and Engineering*, *38*(6), 1615–1623. doi:10.100713369-013-0541-1

Kug, W. K., Woo, Y. L., & Hyo-chol, S. (1999). A finite element analysis for the characteristics of temperature and stress in micro-machining considering the size effect. *International Journal of Machine Tools & Manufacture*, *39*(9), 1507–1524. doi:10.1016/S0890-6955(98)00071-6

Leo Kumar, S. P., Jerald, J., Kumanan, S., & Prabakaran, R. (2014). Review on current research aspects in tool based micromachining processes. *Materials and Manufacturing Processes*, *2*(11-12), 1291–1337. doi:10.1080/10426914.2014.952037

List, G., Sutter, G., & Bouthiche, A. (2012). Cutting temperature prediction in high speedmachining by numerical modeling of chip formation and its dependence with craterwear. *International Journal of Machine Tools & Manufacture*, *54*, 1–9. doi:10.1016/j.ijmachtools.2011.11.009

Molinari, A., Cheriguene, R., & Miguelez, H. (2011). Numerical and analytical modeling oforthogonal cutting: The link between local variables and global contact characteristics. *International Journal of Mechanical Sciences*, *53*(3), 183–206. doi:10.1016/j.ijmecsci.2010.12.007

Munoz-Sanchez, A., Canteli, J. A., Cantero,J. L., &Miguelez, M. H. (2011). Numericalanalysis of the tool wear effect in the machining induced residual stress. *Journal of Simulation Modeling Practice and Theory, 19*, 872–886.

Nasr, M. N. A., & Elbestawi, M. A. (2008). A modified time-efficient FE approach for predicting machining induced residual stress. *Journal of Finite Element in Analysis and Design*, *44*(4), 149–161. doi:10.1016/j.finel.2007.11.005

Rosa, P. A. R., Martins, P. A. F., & Atkins, A. G. (2007). Revisiting the fundamentals of metalcutting by means of finite elements and ductile fracture mechanics. *International Journal of Machine Tools & Manufacture*, *47*(3–4), 607–617. doi:10.1016/j.ijmachtools.2006.05.003

Schulze, V., Autenrieth, H., Deuchert, M., & Weule, H. (2010). Investigation of surface near residual stress after micro-cutting by finite element simulation. *CIRP Annals–Manufacturing Technology, 59*, 117–120.

Shams, A., & Mashayekhi, M. (2012). Improvement of orthogonal cutting simulation with non-local damage model. *International Journal of Mechanical Sciences*, *61*(1), 88–96. doi:10.1016/j.ijmecsci.2012.05.008

Shrot, A., & Baker, M. (2012). Determination of Johnson-Cook parameters from machining simulations. *Journal of Computational Materials Science*, *52*(1), 298–304. doi:10.1016/j.commatsci.2011.07.035

Sima, M., & Ozel, T. (2010). Modified material constitutive models for serrated chipformation simulations and experimental validation in machining of titanium alloy Ti–6Al–4V. *International Journal of Machine Tools & Manufacture*, *50*(11), 943–960. doi:10.1016/j.ijmachtools.2010.08.004

Souza, N. E. A. (2008). *Computational methods for plasticity: theory and applications.* Willey Publications.

Thanongsak, T., & Ozel, T. (2013). Experimental and finite element simulation based investigation on micro-milling Ti-6Al-4V titanium alloy: Effects of cBN coating on tool wear. *Journal of Materials Processing Technology*, *213*, 535–542.

Umbrello, D. (2008). Finite element simulation of conventional and high speed machiningof Ti6Al4V alloy. *Journal of Materials Processing Technology*, *196*(1-3), 79–87. doi:10.1016/j.jmatprotec.2007.05.007

Vavourakis, V., Loukidis, D., Charmips, D. C., & Panastasiou, P. (2013). Assessment of remeshing and remapping strategies for large deformation elastoplastic finite element analysis. *Journal of Computers and Structures*, *114-115*, 133–146. doi:10.1016/j.compstruc.2012.09.010

Vaz, M. Jr, Owen, D. J. R., Kalhori, V., Lundblad, M., & Lindgren, L. E. (2007). Modeling and simulation of machining processes. *Archives of Computational Methods in Engineering*, *14*(2), 173–204. doi:10.100711831-007-9005-7

Vaziri, M. R., Salimi, M., & Mashayekhi, M. (2010). A new calibration method for ductile fracture models as chip separation criteria in machining. *Journal of Simulation Modeling Practice and Theory*, *18*(9), 1286–1296. doi:10.1016/j.simpat.2010.05.003

Vaziri, M. R., Salimi, M., & Mashayekhi, M. (2011). Evaluation of chip formation simulation models for material separation in the presence of damage models. *Journal of Simulation Modeling Practice and Theory*, *19*(2), 718–733. doi:10.1016/j.simpat.2010.09.006

Zhang, Y. C., Mabrouki, T., Nelias, D., & Gong, Y. D. (2011). Chip formation in orthogonal cutting considering interface limiting shear stress and damage evolution based on fracture energy approach. *Journal of Finite Elements in Analysis and Design*, *47*(7), 850–863. doi:10.1016/j.finel.2011.02.016

Chapter 14
Prominence in Understanding the Position of Drill Tool Using Acoustic Emission Signals During Drilling of CFRP/Ti6Al4V Stacks

A. Prabukarthi
https://orcid.org/0000-0002-0467-5262
PSG College of Technology, India

M. Senthilkumar
https://orcid.org/0000-0002-3720-0941
PSG College of Technology, India

V. Krishnaraj
PSG College of Technology, India

ABSTRACT

CFRP/Ti6Al4V stacks are widely used in aerospace and automobile industries as structural components. The parts are made to near net shape and are assembled together. Aerospace standards demand rigid tolerance for the holes. While drilling stacks, during the exit of drill from CFRP and entry into Ti6Al4V, there is a change in the overall behavior of the drilling process due to changes in the mechanical properties of the two materials. Hence, stacks should be drilled under their optimal machining conditions in order to achieve better hole quality. The machining parameters and tool geometry are different for CFRP and Ti6Al4V. This requires knowing the thickness of the CFRP and Ti6Al4V layers beforehand so that at the time of drill tool transition from CFRP to Ti6Al4V the machining parameters can be altered. But in aircraft bodies the cross-section varies along the profile and the thickness of the individual layers at different locations. The current study proposes the use of acoustic emission (AE) signals to monitor the drill position while drilling of CFRP/Ti6Al4V stacks.

DOI: 10.4018/978-1-7998-1690-4.ch014

INTRODUCTION

A recent trend in the aerospace industry is an increase in the use of composites and aluminum and/or titanium alloys because of the outstanding mechanical properties that can be provided at critical- load load-carrying locations of the aircraft. The central wing box is a critical component made out of CFRP/Ti6Al4V stacks, located at the top of the fuselage; it forms the attachment point for both wings and all the engines. Similarly, CFRP/Ti6Al4V stacks are also used along the tail sections. Composites on the B787 account for 50% of the aircraft's structural weight. Aluminum comprises only 12% of the mentioned weight; and titanium makes up a greater percentage than aluminum, namely 15%. Steel comprises 10%, and other metals share 5%(Alberdi,Artaza, Suárez, Rivero, & Girot, 2016).

Condition monitoring during drilling is aimed at understanding the state of the tool, by acquiring the different process parameters by direct or indirect methods. These data are then used to bring changes in either the machining parameters or to replace a worn tool. The timely replacement of tools can help in avoiding the rejection of parts with poor hole quality. The monitoring is done online, which means that the tool is inspected during the actual drilling process. The advantage of online inspection is that it is possible to influence the drilling process during the hole-making process—thereby correcting a potential hole quality problem before the hole is completed. Therefore, the timely replacement of worn tools can be achieved; and the possibility of getting a hole with poor quality is eliminated. This leads to an economy in the drilling process. By contrast, direct methods involve measurement of tool wear using optical techniques. The principle disadvantage of the direct method is that the tool can be inspected only after the machining process. In indirect methods, process variables such as changes in hole quality, cutting force, vibrations, and spindle current or power can be monitored using acoustic emission signals with the help of different sensorssuch as thoselisted below:

- Accelerometers for measuring vibration
- Acoustic emission sensors for measuring acoustic signals during metal cutting
- Dynamometers for measurement of cutting force and torque
- Sensors to measure current or power of feed drives and main spindle

Among these, acoustic emission sensors are widely used for monitoring metal-cutting applications (Arul, 2007; Byrne, 1995). The principle advantages of these sensors are that they are not affected by machine noise and that they are sensitive to the changes taking place in the material during the metal-cutting process. These are transient elastic waves generated by the rapid release of energy from a localized source within a material when subjected to a state of stress. This energy release is associated with the abrupt redistribution of internal stresses; and as a result of this, a stress wave propagates through the material. The above definition indicates that processes that are capable of changing the internal structure of a material—processes such as dislocation motion, directional diffusion, creep, and grain boundary sliding which results in plastic deformation and fracture—are sources of Acoustic Emissions (AE). In metal cutting, the sources of AE signals are

- Plastic deformation in the primary and secondary shear zones
- Chip friction
- Chip collision and breakage
- Tool wear and breakage

In recent years, the methods of the cold expansion process of the holes, the residual stress around the extended holes, the behavior of the initiation and propagation of the fatigue cracks and the fatigue lives after the cold expansion are extensively investigated through many experiments and finite element simulation of cold expansion process. By analyzing the current characteristics and the defects of the cold expansion technology of the hole, combined with the real needs in the design and manufacture of next-generation aircraft, the development trends and the new research directions for achieving precision and efficiency in a high-fatigue manufacturing process are presented. As the compressive residual stress field produced by cold expansion is responsible for improving the stress concentration at faster holes, the accurate measurement of these fields has been at the center of many experimental programs. In the early studies, researchers had performed a careful analysis of the residual stress around an expansion hole, starting with the deformation of the material and its is spring back theory of elastoplasticity. It is significant to note that due to the limitation of measurement techniques, stress was not directly measurable in the initial research. Due to the progress of the measuring means, different test methods are used to study the residual stress after cold expansion. Depending on the measurement area, it can be divided into two types: namely, point measurement and face measurement. Point measurement methods—for example, the voltage measurement method, the 31-hole drilling method, the X-ray diffraction method, laser beam interferometry, etc.—have been widely used to measure the residual stress at the hole.

However,it should be emphasized that the residual voltage cannot accurately be obtained at a point in the region near the hole, because the experimental measurements are averaged over a surface of 1 mm × 1 mm. To overcome this problem, face measurement methods such as moire interferometry and photoelastic coating method are used. Although the mechanism and accuracy of these methods of measurement are different, the typical distribution of residual voltage measured by different methods around an extended cold hole is practically similar. The typical distribution of the tangential residual stresses around a cold extended circular hole is shown in Figure 1(Fu, Ge, Su, Xu,& Li,2015).

Figure 1. Representative tangential residual stress distribution around stretched hole

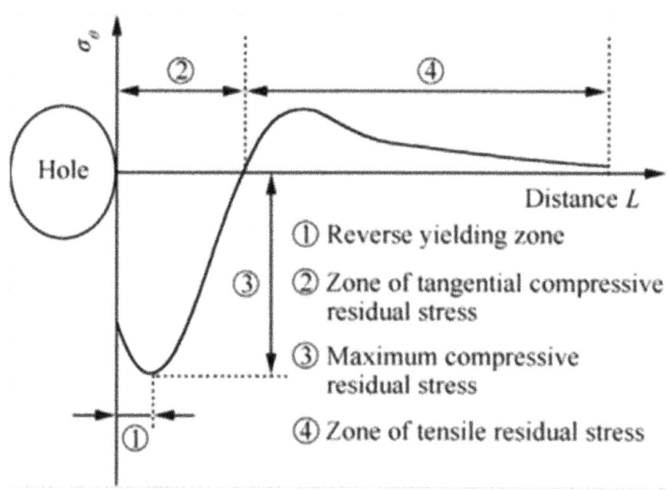

BACKGROUND

Arul, Vijayaraghavan, and Malhotra (2007) made a time-domain analysis of AE signals relating AE-RMS with tool flank wear. Analysis of AE signals revealed that AE-RMS value decreased with flank wear. A power spectrum was calculated and found that the AE signals peaked at around 70 kHz. At the 30th hole as the increase in the tool wear can be understood based on increase in the AE power which helps in fixing the threshold limit for the AE signal value in order to monitor tool life online.

Brinksmeier and Janssen (2002) performed experiments on drilling of multi-layer composite materials consisting of Carbon-Fiber-Reinforced Plastics (CFRP), plus titanium and aluminum alloys. Different carbide drill designs with improved geometries and coatings were investigated and compared by characterizing the cutting forces along with tool wear, hole quality, and chip formation. Investigations have shown that dry machining of titanium workpiece layers leads to increased tool wear, chip formation problems, and surface damage in the aluminum and CFRP-layers. Also, some investigations were made with regard to analyzing surface defects. It was understood that the variation of the hole diameter is attributable to the high mechanical and thermal loads generated when machining titanium. The use of adapted step drills improves diameter tolerances, surface quality, and tool wear. When drilling multilayer materials, MQL with internal supply should be used to improve the hole's quality and increase the tool's life.

Kim, Beal, and Kwon(2016) experimentally studied the machinability of Ti/CFRP composites using the drilling process and found that titanium chips were long and continuous at low feeds and became shorter and stiffer as the feeds increased. Polymer composite chips were continuous at low feeds and became dust type chips as the feed increased. Hole damage was similar at the entrance and at the exit of the drilled holes. Ti burrs and delamination between Ti and Polymer Matrix Composite (PMC) occurred at the exit as well as the entrance. High feed rates (greater than one-ply thickness) resulted in deep fiber pull-out, with a resulting detrimental effect on hole quality.

Poutord, Rossi, Poulachon, M'Saoubi, and Abrivard (2013) studied tool wear in the drilling of CFRP and Ti6Al4V individually and as stacks. Results revealed that the thrust force was higher in drilling Ti6Al4V than CFRP. It was concluded that the majority of the tool wear (abrasive wear) comes from drilling CFRP and cutting edge chipping occurs during drilling Ti6Al4V.

Ramulu, Branson, and Kim (2001) conducted a study on the drilling of composite arid titanium stacks. They found that high-temperature-induced material damage occurs near and around the hole region, at the interface of Gr/Bi-Ti. As a result, fewer holes were produced when high spindle speeds and a slow feed were used. It is found that carbide drills outperformed all other tools in terms of tool life, minimal surface damage, and heat-induced damage on both workpiece materials.

Neugebauer, Ben-Hanan, Ihlenfeldt, Wabner, and Stoll (2012) applied AE signals for identifying tool position in CFRP/Al stacks using a standard drill and a stepped drill. For a standard twist drill, the time difference in detecting the drill transition was smaller for the Al entry and exit but was high for the CFRP exit. The stepped drill offered better detection of drill transition. Identifying the drill exit out of Al was not possible because of the noisy AE signals generated by the chips remaining in the flute of the drill.

Shyha et al. (2010) experimented with the drilling of Titanium/CFRP/Aluminum stacks and obtained the best tool life/performance (310 drilled holes) with the more conventional uncoated carbide drills at a lower cutting speed and feed rate. Typically, thrust forces increased from 300 N for the first hole to -2200 N for the last hole drilled; while torque values were generally below 600 Nm for worn tools.

Velayudham, Krishnamurthy, and Soundarapandian (2005) performed a frequency domain analysis using wavelet packet transformation upon AE signal during drilling of glass polymeric composites. The AE signals were decomposed into 4 levels (16 packets). The first packet was chosen with a range of 0-153.25 kHz. As the tool became more worn, the amplitude of the wavelet coefficient increased. Crest factor was calculated for the first packet and was used as the monitoring index for tool wear.

Xiaoli and Zhejun (1998) used a decision making systems like fuzzy logic for tool condition monitoring was done by extracting feature packets out of decomposed AE signals. The AE signal was decomposed into 16 packets and the RMS for each packet was used for describing the different tool wear states. Packets sensitive to tool wear were chosen as the monitoring index

Zitoune, Krishnaraj, and Collombet (2010) studied the drilling of a composite material and an aluminum stack. The experiment was carried out with a CFRP/Al2024 stack without coolant, with plain carbide (K20) drills (4, 6, 8 mm diameter). CFRP circularity was 6 μm at a low feed rate, when feed increases reach 25 μm. The surface roughness value is around 2-4 μm. The surface roughness and circularity of Al is better compared to CFRP.

From the above literature, it can be understood that in order to achieve better hole quality when drilling stacks, it is necessary to drill each material using its optimized process parameter and tool geometry. This necessitates the online monitoring of tool position during multi-material stack drilling,which seeds the importance of identifying the time of the drill entry into, and exit from, CFRP and Ti6Al4V layers using Acoustic Emission (AE) signals and compares it with the actual time of the drill entry into, and exit from, each material, so as to dynamically change the machining parameters for each layer of CFRP and Ti6Al4V by means of using an adaptive drilling system.

Experimental Analysis

Carbon fiber reinforced polymer (CFRP) and Ti6Al4V (Titanium alloy) were stacked together and fastened. The CFRP layer has eight plies, with the orientation of the plies being symmetric $(0°/90°/0°/90°)_{2s}$ and the overall plate being 2.4mm (each ply is 0.3 mm thick) thick. The titanium alloy used is Ti6Al4V grade 5. The plate is 3mm thick with a cross-section of 160×60 mm.

The experiments were carried out using Makino S-33 vertical machining center. A solid carbide twist drill of 5mm diameter coated with TiAlN coating was used for drilling. 3 drill geometries of different point angles (130°, 135°, and 140°) were used.The specifications of the tool geometry are presented in Table 1.The machining parameters were chosen based on the results obtained from the past research work carried out in the area of the drilling of the CFRP/Ti6Al4V stack (Krishnaraj et al.,2012a, 2012b;Prabukarthiet al.,2016;Senthilkumar et al.,2017,2018).

Kistler Piezotronsensors (model no.8152B) were used to understand the machining characteristics while drilling stacks. The sensing element is made of piezoelectric ceramic and is mounted on a thin steel diaphragm. The coupling surface of the diaphragm is welded to the housing and is made slightly protrudingto measure the AE signals. This sensor features very high sensitivity to the surface wave (the Raleigh wave), and to longitudinal waves,over a broad frequency range. The specification of the sensor is given in Table 2.

The sensor can be either magnetically clamped or fastened to the workpiece. For experimental purposes, the sensor was fastened to the workpiece in the manner presented in Figure 2. Since CFRP is non-magnetic, the magnetic clamp could not be used. Silicone grease was used as the acoustic coupling between the workpiece and the sensor.

Table 1. Tool geometry specification

Parameters	Value
Helix angle	35°
Point clearance angle	7°
Core diameter	2.1 mm
Margin	0.25 mm
Body clearance diameter	4.6 mm
Web thickness	0.8 mm

Table 2. AE sensor specification

Technical Data	Value
Sensitivity dBref 1V/(m/s)	57
Frequency range +/- 10dB kHz	50 to400
Ground isolationMΩ	>1
Operating temperature range °C	-40 to 60
Sensing element type	Ceramic
Output impedance Ω	<10
Voltage full-scale V	+/- 2

Figure 2. Mechanical fastening of the AE sensor onto the workpiece
(Kistler Instrument Corporation, 2007, acceleration, Piezotron® sensor, acoustic emission sensor, accessed August 1, 2019, from http://www.intertechnology.com/Kistler/pdfs/Accelerometer_Model_8152B.pdf)

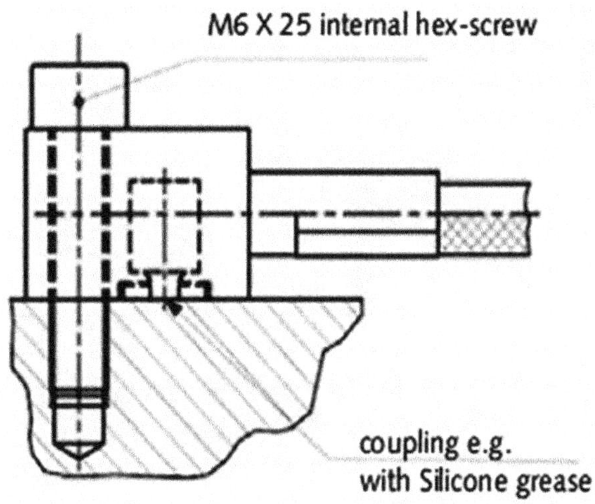

The Kistler 5125B Piezotron coupler is used to amplify and filter the AE signals from the sensor. The outputs of the coupler are AE RMS and AE filtered signals. The RMS value is calculated by the built-in RMS converter circuit. The time constant of the RMS circuit is 1.2 ms. Signals are bandpassfiltered between 50 kHz and 1000 kHz by passing the signals through a high-pass filter and a low-pass filter in series. The high-pass filter, low-pass filter, and RMS circuits are available as plug-in modules. The specification of the coupler is presented in Table 3.

A strain-gauge-type drilling dynamometer (model SI-674) was used for measuring axial thrust force and torque, which is presented in Figure3. The output from the dynamometer was fed to an amplifier (model SI-223D) which produces an output voltage in digital form that is proportional to the applied load.

Applied thrust force or torque causes an imbalance of the Wheatstone bridge, thus producing an output voltage proportional to the force. A data acquisition card (NI PCI-6132 DAQ) was used for data acquisition. The DAQ card is used for signal amplification and also for analog-to-digital conversion. The specification of the DAQ card is presented in Table 4.

Table 3. Specification of AE Piezotron coupler

Technical Data	Value
High-pass filter kHz	50
Low-pass filter kHz	1000
Frequency rangekHz	10 to 1000
Integration time constant Ms	1.2
Output impedance Ω	10
Supply voltage VDC	15 to 36
Supply current mA	<70

Table 4. Specifications of DAQ

Product Family	Multifunction DAQ
No.of channels	4
Maximum voltage range	-10 to 10 V
Sensitivity	432μV
Sampling rate	2.5 MS/s per channel
Sampling mode	simultaneous
Over-voltage protection	+/- 36 V

Figure 3. Experimental setup

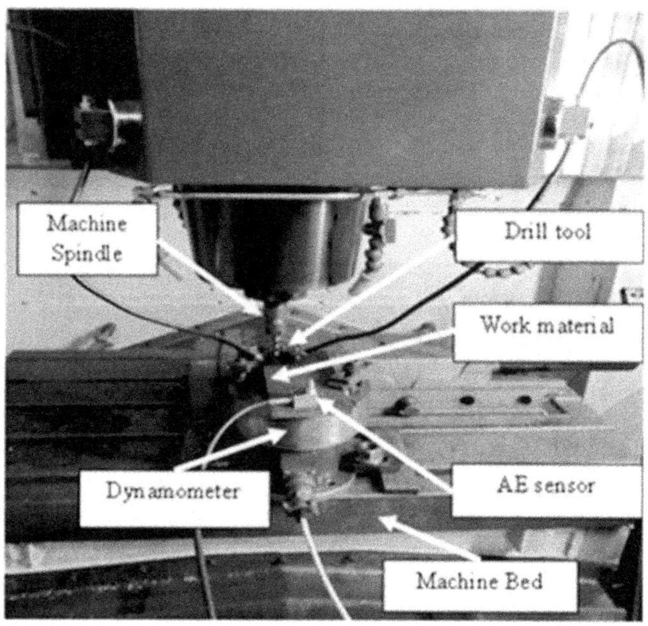

AE RMS and AE filtered signals were acquired during experimentation. Figure 4 shows AE RMS and AE filtered signals for a speed of 895 RPM, a feed of 0.1 mm/rev, and a point angle of (130°).

AE Signals and Evolutionof Thrust Forcein CFRP andTi6Al4V

During drilling, when the drill touches the workpiece, cutting action is initiated by the chisel edge of the drill tool, and the AE signal starts increasing. AE signal in CFRP is generated due to

- Contact friction (rubbing which occurs between the tool chisel edge and the uncut material)
- Drilling (in generating microchips)
- Peripheral friction (rubbing which occurs between the drill body and the whole wall)

Figure 4. AE RMS and AE filtered signal acquired during drilling

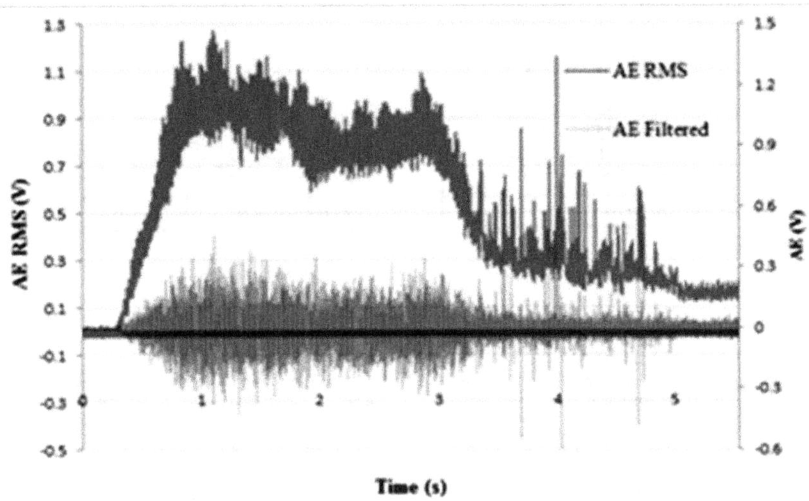

The AE RMS and AE filtered signals, captured during drilling experiments, are studied. The complete process of drilling CFRP using a twist drill can be divided into 4 major stages:

1. In the initial stage, the chisel edge comes into contact with the workpiece, and the workpiece is extruded to form an indentation.Spindle rotation causes frictional rubbing. No cutting action occurs during this stage
2. Further traverse of the drill bit engages the main cutting edge (the cutting lips) in order to generate cutting forces. During this stage, actual drilling action begins, and alarge volume of material is removed.
3. As the drill enters further, it cuts the alternate layers until it reaches the last few plies before the drill completely exits out. At this point, due to the decrease in the thickness of the workpiece, the resisting stiffness of the laminate is reduced, and the last few plies elastically bend due to the applied thrust force. This is the major reason for generating thrust force, which results in interlaminar stress. When this stress becomes larger than the inter-laminar bond strength, a crack is initiated and propagates, resulting in damage around the hole. This is pushed out by delamination. This stage is called the piercing stage.
4. The last stage involves rotation of the drill bit within the hole to induce peripheral friction. No cutting action occurs. The four stages of signal processing during drilling of CFRP are presented in Figure5.

The continuous AE signals are generated due to contact and peripheral friction. Transient AE signals are due to the cutting of plies, and to the micro-chipping of the tool. As the drill moves out of CFRP and enters into Ti6Al4V—i.e., as the chisel edge enters Ti6Al4V—the AE signals drop; and their amplitude remains low till the end of the drilling process. This drop is due to the air gap between the two layers. In Ti6Al4V, AE has both continuous and transient constituent-based signals. Continuous signals are due to

- Plastic deformation of the Ti6Al4V grains

Figure 5. Stages in drilling CFRP
(Neugebauer et al., 2012)

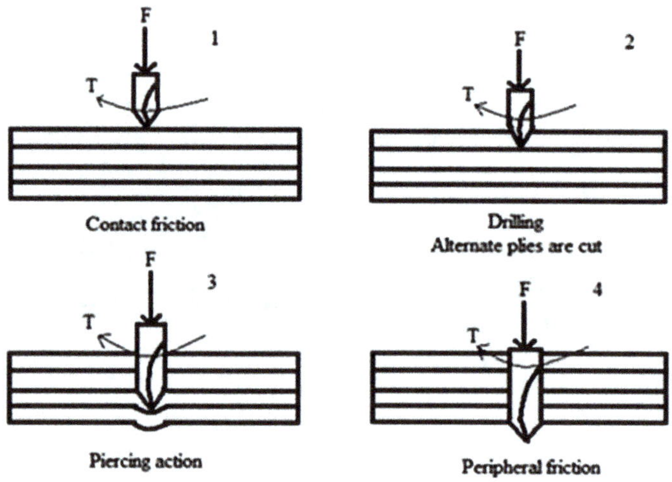

- Friction between the tool margin and the hole wall
- Friction between the chisel edge and the uncut material

Transient AE signals are due to

- Chip breakage
- Collision of chips with the walls of the hole
- Tool breakage
- Cutting edge chipping

The various stages in the drilling of Ti6Al4V and the evolution of AE are discussed below

- When the drill touches the workpiece, the chisel edge extrudes the work material. No cutting action occurs during this stage. The friction between the chisel edge and the workpiece results in continuous AE signals. This was verified by protruding the chisel edge of the drill into the workpiece and by running the drill at the same position without any feed. The friction between the drill and the workpiece resulted in continuous AE signals without any transients.
- As the drill enters further into the material, cutting action is initiated by the main cutting edge. Plastic deformation in the primary and secondary shear zone result in AE signals, that travel through the material and reach the sensor. Friction between the drilling margin and the hole walls also generates AE signals. Plastic deformation and friction are a continuous source of AE signals. During this stage, chips are formed. Chip breakage, collision of chips with the walls of the hole, and microchipping of the tool result in transient AE signals. The effect of peripheral friction on continuous AE signals was verified by removing the chips from the hole and running the drill inside the pre-drilled hole. The AE signals recorded did not show any transient peaks. Later, some chips were carefully inserted into the flute region of the drill, and the drill was run inside the pre-drilled hole. Continuous and transient peaks were recorded.

- At the exit, the thickness of the workpiece under the drill is very less. This small thickness of the material does not have the necessary stiffness to withstand the compressive thrust force induced by the drill. Hence, the drill pierces through these final layers rather than cutting through them, resulting in burrs. The continuous and transient AE signals in drilling Ti6Al4Vare shown in Figure6.

Based on the AE signals studies, it is possible to visibly identify the differences in AE signals while drilling CFRP and Ti6Al4V. AE signals in the CFRP and Ti6Al4V layers have been separated and are shown in Figure 7.

The evolution of thrust force in drilling CFRP and Ti6Al4Vstacks as the drill progresses into the workpiece can be divided into 6 stages. This evolution is depicted in Figure 8.

1. The chisel edge of the tool comes in contact first. There is a sharp rise in the thrust force (zone a) since the chisel cuts with a negative rake angle.

Figure 6. Continuous and transient AE signals during drilling in Ti6Al4V

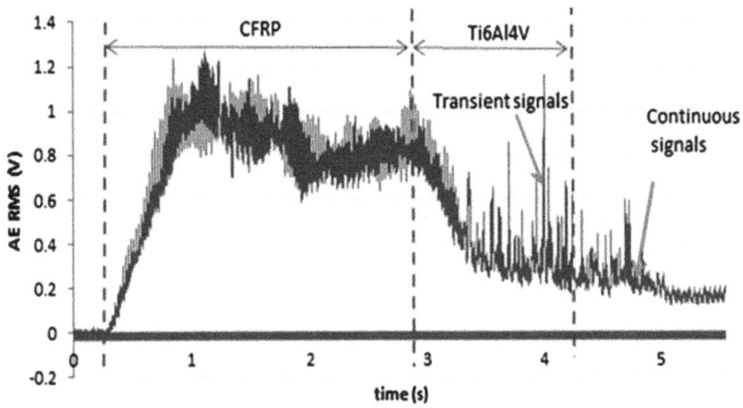

Figure 7. AE signals in CFRP and Ti6Al4V

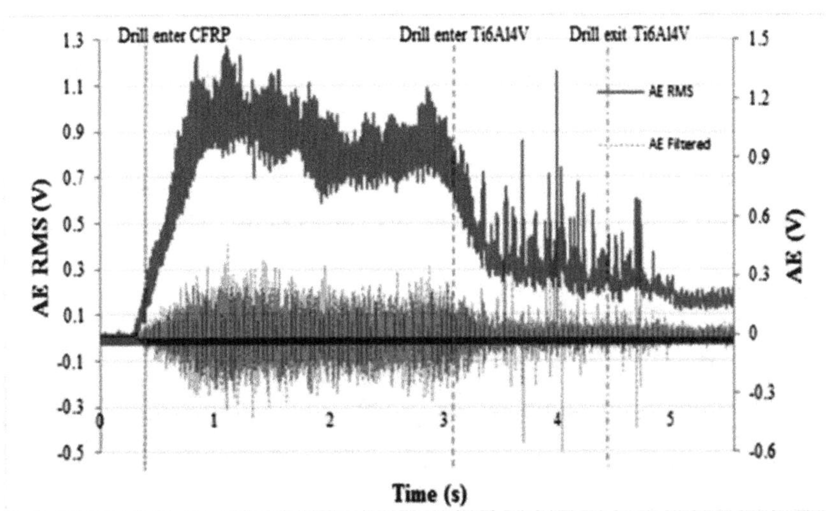

Figure 8. Evolution of thrust force as the drill progresses while drilling Ti6Al4V

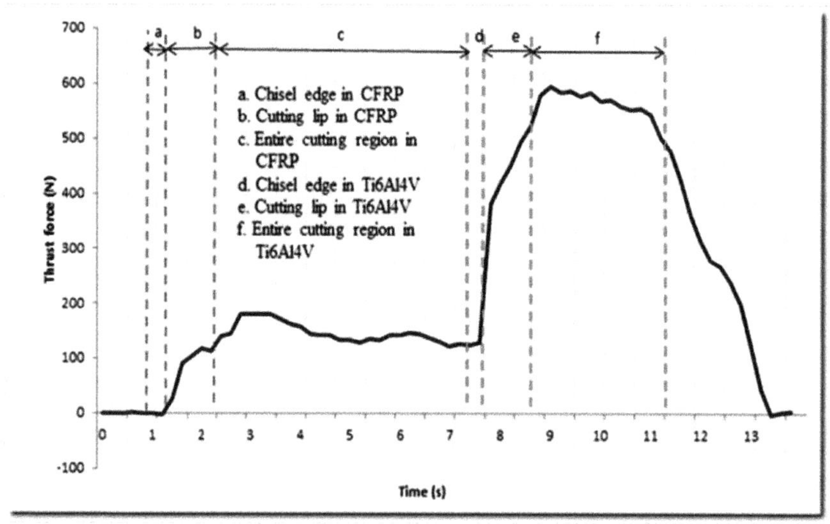

2. As the tool drills further, the lip region of the tool cuts the workpiece and the rate at which the thrust force increases are comparatively less (zone b), since the cutting lip has a positive rake angle.
3. After the entire cutting region is inside the workpiece, the thrust force remains almost a constant (zone c).
4. When the drill enters Ti6Al4V, the thrust force shows a sharp increase, since the Ti6Al4V layer is harder (zones d,e).
5. It can be seen that the thrust force decreases during the last part of zone f. This is because, during the exit of the drill from the Ti6Al4V layer, the cutting region of the tool slowly comes out of the workpiece. Hence, the effective cutting area is reduced, resulting in decreased thrust force.

Tool Position Monitoring

From the nature of the AE signals, it can be observed that there is a difference in its characteristics in Ti6Al4V and CFRP. Hence, any process monitoring setup must look for developing a monitoring index by taking advantage of this difference. A Time-Domain analysis of the signal was done. Several statistical parameters like standard deviation, kurtosis, skewness, and crest factor were applied to the signal. Among these, crest factor showed the most considerable difference between drilling in CFRP and Ti6Al4V.

Crest factor is defined as the ratio of maximum signal amplitude to the Root Mean Square (RMS) of the signal. Crest factor was calculated using Equation 1 and Equation 2:

$$Crest\ factor = \frac{V_{peak}}{V_{RMS}} \tag{1}$$

$$RMS = \sqrt{\frac{\sum_{i=1}^{N} X_i^2}{N}} \tag{2}$$

Here, V_{peak} is peak value, V_{RMS} is RMS value, N is the number of samples in a set, and x_i is the value of the discrete AE signal.

The definition of the crest factor means that when a signal has high transient peaks, its crest factor will be high. From the AE signals acquired during CFRP/Ti6Al4V drilling, it can be observed that there are high transient peaks in Ti6Al4V. Hence, crest factor value will be high for Ti6Al4V, as compared to CFRP.Crest factor for all holes is calculated using MATLAB. In calculating the RMS of a signal, it is important to choose the correct size for N. The value of N must be chosen such that the set contains a transient peak signal in it. Careful analysis of the signal reveals that a transient peak arrived, on average, every 0.05 seconds. Based on this, the value of N is chosen as 5,000 samples. The variation in the crest factor with variation in N is given in Figure 9; and it can be seen that the value of N that can better distinguish between the CFRP layer and the Ti6Al4V layer is N=5000. For N=10,000, there is a slow increase in the crest factor from CFRP to Ti6Al4V. This will lead to a delay in predicting the time of entry and exit. For N=1,000, the value of N is so small that even a small transient signal in CFRP will result in high crest factor values. Hence, it would be difficult to distinguish between CFRP and Ti6Al4V. Crest factor and AEfiltered signal for one complete hole are shown in Figure 10.

Figure 9. Crest factors for (a). N=10000, (b). N=5000, and (c). N=1000

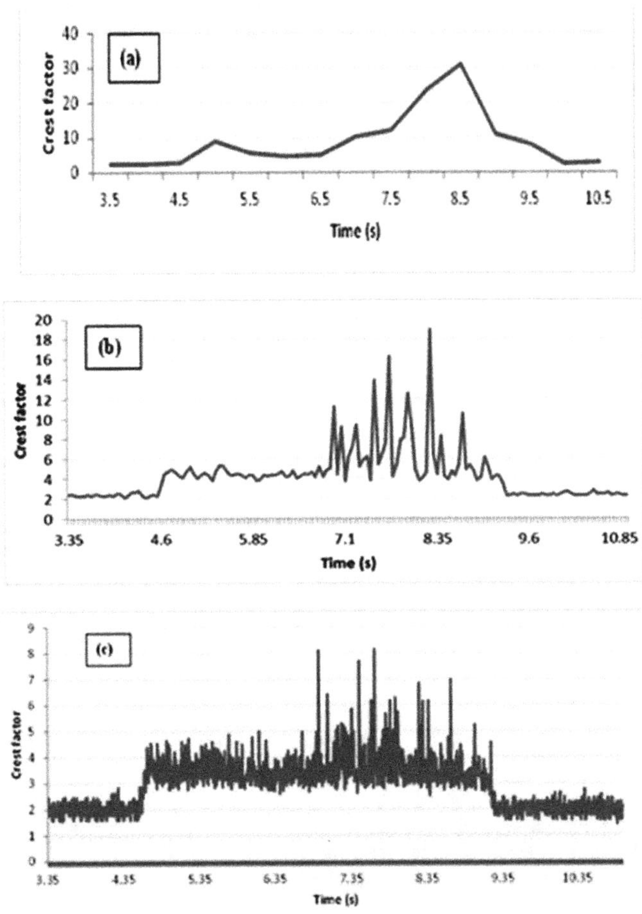

After analyzing the crest factor, threshold limits are set. The threshold limit for CFRP and Ti6Al4V is shown in Figure 11. The time at which the drill enters into CFRP is taken as being the time at which the crest factor value crosses the minimum value of the threshold limit for CFRP. The time of the drill entry into Ti6Al4V is taken as being the time at which the crest factor value crosses the threshold for Ti6Al4V.

Using the crest factor, the time of the drill entry into, and exit from, each material is detected. When the drill touches CFRP, there is a rise in the crest factor. When the drill enters Ti6Al4V, there is a sharp rise in the crest factor. Figure 12 shows the time of the drill entry and exit as detected by the crest factor; and the actual time of the drill entry and exit.

To analyze how efficient the crest factor is in identifying the drill position, a comparison is made between the actual time of the drillentry and exit and the predicted time, which is presented in Table 5. The actual time of the drill entry is obtained from the thrust force signals. The thrust force signals are

Figure 10. Evolution of the crest factor and AE for one complete hole

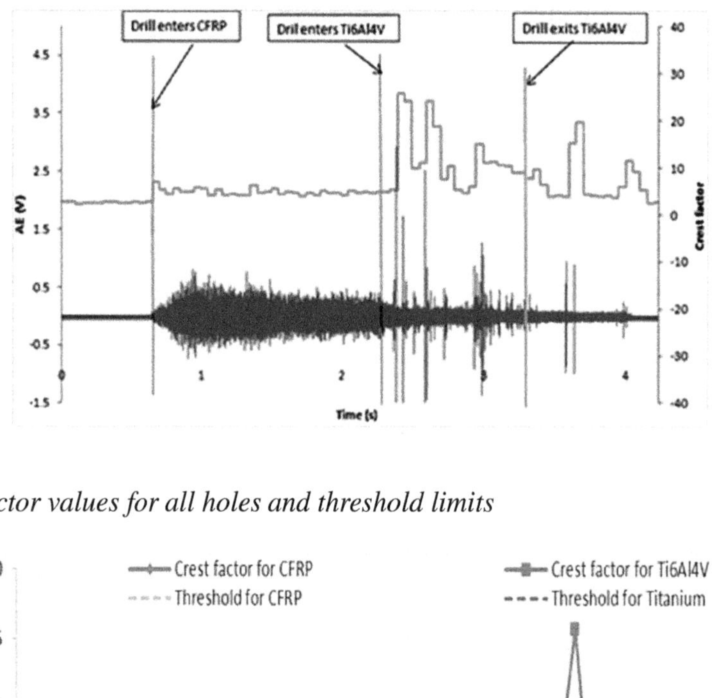

Figure 11. Crest factor values for all holes and threshold limits

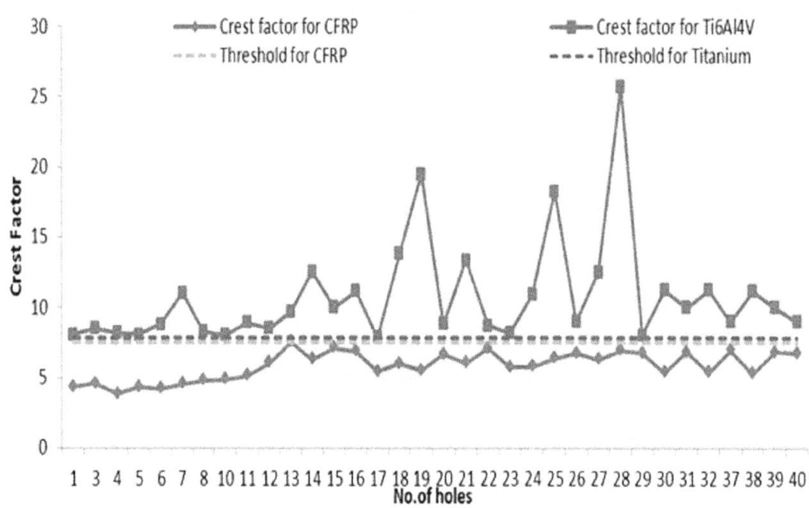

Figure 12. Crest factor value for one complete hole

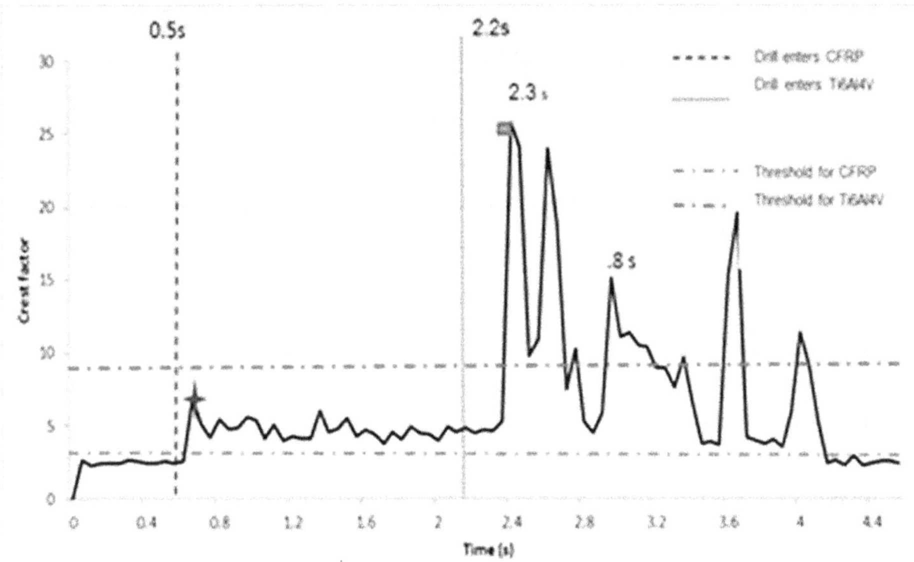

Table 5. Comparison of the time of the drilentry and exit

Exp.No.	Feed Rate (mm/ min)	CFRP			Ti6Al4V		
		t_{entry} (From Thrust Force) s	t_{entry} (From AE) s	ΔT (s)	t_{entry} (From Thrust Force) s	t_{entry} (From AE) s	Δt (s)
1	44.75	3.02	3.05	0.03	9.5	10.35	0.85
2		0.71	0.8	0.09	7.41	7.95	0.54
3		1.56	1.6	0.04	8.05	8.4	0.35
4		1.52	1.55	0.03	8.01	8.39	0.38
5	89.5	3.42	3.45	0.03	6.67	6.9	0.23
6		0.7	0.75	0.05	3.96	4.35	0.39
7	101.06	11.01	11.05	0.04	13.92	14.25	0.33
8		0.11	0.2	0.09	2.96	3.2	0.24
9		1.32	1.35	0.03	4.16	4.4	0.24
10		0.63	0.65	0.02	3.48	3.7	0.22
11	134.75	4.62	4.65	0.03	6.81	6.95	0.14
12	90	0.85	0.9	0.05	4.05	4.55	0.5
13		0.61	0.65	0.04	3.81	4.23	0.42
14	135	4.61	4.65	0.04	6.7	6.95	0.25
15	180	0.64	0.65	0.01	2.25	2.4	0.15
16		4.92	4.95	0.03	6.57	6.7	0.13

very sensitive to drilling in different materials. Hence, thrust force signals are taken as the standard for the drill entry and exit. The difference in time is estimated by the difference between the drill actual entry and the drillentry as detected by crest factor time. This difference is presented in Equation 3.

$$\Delta t = a - b \tag{3}$$

Here, Δt represents changes in time, a is the actual time of the drill entry, and b is the time of the drill entry as detected by crest factor.

CONCLUDING REMARKS

In the current work, AE signals have been used to identify a drill position in CFRP/Ti6Al4V stacks. The monitoring index is developed to identify the time of the drill entry into, and exit from, each material. The following results can be concluded from this work:

- Crest factor is used as the monitoring index for drill transition. It shows a good correlation with AE and thrust force signal—and there is a significant difference in its value between drilling in CFRP / Ti6Al4V.
- Experimental results are analyzed to detect the time of the drill entry into the two layers. The time difference between the actual time of transition and the time of transition as detected by the crest factor is compared.
- The time difference Δtis small in the case of the drill entry into CFRP and is high for the drill entry into Ti6Al4V. This is because the crest factor crosses the threshold only when the Ti6Al4V chips start to collide with the walls of the tool, leading to high peaks in the signal. Chips start forming only when the cutting edge of the tool is completely inside the workpiece. This leads to some degree of time delay in detecting the drill entry into Ti6Al4V.
- Δt is low for CFRP entry and high for Ti6Al4V entry. It is also observed that Δtis low for high feed rates and that it is high for low feed rates.
- When the drill touches CFRP, AE is generated due to friction between the drill chisel edge and the workpiece surface. These generate transient elastic stress waves that reach the sensor as AE. The same phenomena occur when the drill touches Ti6Al4V; but since the two workpieces are mechanically fastened, the AE signals in Ti6Al4V do not reach the AE sensor.
- At a high feed rate, the chisel edge and the cutting lips enter the workpiece at once; and chips are formed instantaneously. Hence, transient AE signals appear much faster for high feed rates than for low feed rates.

REFERENCES

Alberdi, A., Artaza, T., Suárez, A., Rivero, A., & Girot, F. (2016). An experimental study on abrasive waterjet cutting of CFRP/Ti6Al4V stacks for drilling operations. *International Journal of Advanced Manufacturing Technology*, *86*(1–4), 691–704. doi:10.100700170-015-8192-x

Arul, S., Vijayaraghavan, L., & Malhotra, S. K. (2007). Online monitoring of acoustic emission for quality control in drilling of polymeric composites. *Journal of Materials Processing Technology*, *185*(1–3), 184–190. doi:10.1016/j.jmatprotec.2006.03.114

Brinksmeier, E., & Janssen, R. (2002). Drilling of multi-layer composite materials consisting of carbon fiber reinforced plastics (CFRP), titanium and aluminum alloys. *CIRP Annals-Manufacturing Technology*, *51*(1), 87–90. doi:10.1016/S0007-8506(07)61472-3

Byrne, G., Dornfeld, D., Inasaki, I., Ketteler, G., König, W., & Teti, R. (1995). Tool condition monitoring (TCM)— The status of research and industrial application. *CIRP Annals-Manufacturing Technology*, *44*(2), 541–567. doi:10.1016/S0007-8506(07)60503-4

Fu, Y., Ge, E., Su, H., Xu, J., & Li, R. (2015). Cold expansion technology of connection holes in aircraft structures: A review and prospect. *Chinese Journal of Aeronautics*, *28*(4), 961–973. doi:10.1016/j.cja.2015.05.006

Iwata, K., & Moriwaki, T. (1978). Cutting state identification and in-process tool wear sensing by acoustic emission analysis. *Bulletin of the Japan Society of Precision Engineering*, *12*(4), 213–214.

Kim, D., Beal, A., & Kwon, P. (2016). Effect of tool wear on hole quality in drilling of carbon fiber reinforced plastic–titanium alloy stacks using tungsten carbide and polycrystalline diamond tools. *Journal of Manufacturing Science and Engineering*, *138*(3), 031006. doi:10.1115/1.4031052

Kistler Instrument Corporation. (2007). *Acceleration. Piezotron sensor. Acoustic emission sensor*. Accessed August 1, 2019, from http://www.intertechnology.com/Kistler/pdfs/Accelerometer_Model_8152B.pdf

Krishnaraj, V., Prabukarthi, A., Ramanathan, A., Elanghovan, N., Kumar, M. S., Zitoune, R., & Davim, J. P. (2012a). Optimization of machining parameters at high-speed drilling of carbon fiber reinforced plastic (CFRP) laminates. *Composites. Part B, Engineering*, *43*(4), 179–1799. doi:10.1016/j.compositesb.2012.01.007

Krishnaraj, V., Prabukarthi, A., Santhosh, M., Senthilkumar, M., & Zitoune, R. (2012b, June). Optimization of machining parameters in CFRP/Ti stacks drilling. In *ASME 2012 International Manufacturing Science and Engineering Conference collocated with the 40th North American Manufacturing Research Conference and in participation with the International Conference on Tribology Materials and Processing* (pp. 269–275). American Society of Mechanical Engineers. 10.1115/MSEC2012-7216

Neugebauer, R., Ben-Hanan, U., Ihlenfeldt, S., Wabner, M., & Stoll, A. (2012). Acoustic emission as a tool for identifying drill position in fiber-reinforced plastic and aluminum stacks. *International Journal of Machine Tools & Manufacture*, *57*, 20–26. doi:10.1016/j.ijmachtools.2012.01.013

Poutord, A., Rossi, F., Poulachon, G., M'Saoubi, R., & Abrivard, G. (2013). Local approach of wear in drilling Ti6Al4V/CFRP for stack modelling. *Procedia CIRP*, *8*(1), 316–321.

Prabukarthi, A., Senthilkumar, M., & Krishnaraj, V. (2016). Study on drilling of CFRP/Ti6Al4V stack with modified twist drills using acoustic emission technique. *Steel and Composite Structures*, *21*(3), 573–588. doi:10.12989cs.2016.21.3.573

Ramulu, M., Branson, T., & Kim, D. (2001). A study on the drilling of composite and titanium stacks. *Composite Structures*, *54*(1), 67–77. doi:10.1016/S0263-8223(01)00071-X

Ravindra, H. V., Srinivasa, Y. G., & Krishnamurthy, R. (1997). Acoustic emission for tool condition monitoring in metal cutting. *Wear*, *212*(1), 78–84. doi:10.1016/S0043-1648(97)00137-3

Ravishankar, S. R., & Murthy, C. R. L. (2000). Application of acoustic emission in drilling of composite laminates. *NDT & E International*, *33*(6), 429–435. doi:10.1016/S0963-8695(00)00014-1

Senthilkumar, M., Prabukarthi, A., & Krishnaraj, V. (2017). Multi-response optimisation in drilling of CFRP/Ti6Al4V stacks: A degree of similarity approach. *International Journal of Materials Engineering Innovation*, *8*(3–4), 222–249. doi:10.1504/IJMATEI.2017.090235

Senthilkumar, M., Prabukarthi, A., & Krishnaraj, V. (2018). Machining of CFRP/Ti6Al4V stacks under minimal quantity lubricating condition. *Journal of Mechanical Science and Technology*, *32*(8), 3787–3796. doi:10.100712206-018-0731-6

Shyha, I., Soo, S. L., Aspinwall, D. K., Bradley, S., Dawson, S., & Pretorius, C. J. (2010). Drilling of titanium/CFRP/aluminum stacks. *Key Engineering Materials*, *447*, 624–633. doi:10.4028/www.scientific.net/KEM.447-448.624

Velayudham, A., Krishnamurthy, R., & Soundarapandian, T. (2005). Acoustic emission based drill condition monitoring during drilling of glass/phenolic polymeric composite using wavelet packet transform. *Materials Science and Engineering A*, *412*(1–2), 141–145. doi:10.1016/j.msea.2005.08.036

Xiaoli, L., & Zhejun, Y. (1998). Tool wear monitoring with wavelet packet transform—Fuzzy clustering method. *Wear*, *219*(2), 145–154. doi:10.1016/S0043-1648(98)00165-3

Zitoune, R., Krishnaraj, V., & Collombet, F. (2010). Study of drilling of composite material and aluminum stack. *Composite Structures*, *92*(5), 1246–1255. doi:10.1016/j.compstruct.2009.10.010

Chapter 15
Simulation of Mn2-x Fe1+x Al Intermetallic Alloys Microstructural Formation and Stress-Strain Development in Steel Casting

Malaidurai Maduraipandian
https://orcid.org/0000-0002-1882-154X
Indian Institute of Technology, Danbad, India

ABSTRACT

In this simulation, the permeation of the n-phase precipitation to the Mn2 Fe Al crystallization is induced by the steel casting solidification process by JMatPro. Using the model, the morphological evolution of the Fe and Mn in different percentages was obtained, in which the heated data obtained by simulating casting and extreme heat treatment processes were adopted. This chapter describes a model of the computer model for calculating the phase transition and properties of materials required to predict the deviation during the heat treatment of steel. The current model has the advantage of using a variety of shape memory alloys including medium to high aluminium-based Heusler alloys. Even for an arbitrary cooling profile, a wide range of physical, thermodynamic, and mechanical properties can be calculated as a function of time/temperature/cooling with different proportions. TTT (time-temperature transfer) curves are exported to FE-/FD-based packages to reduce the data distortion of materials. The test results are displayed as a stress-strain diagram.

INTRODUCTION

The microstructure of Mn_2FeAl-shaped memory composites is closely related to their Fe to Mn ratio of the crystal structures of FeAl and Mn_2FeAl intermetallic composites are thought to be useful for high temperature structural materials (Hiroki, Yamamoto, Takeda, Suzuki, & Sakuma, 2010; Otsuka, 1998;

DOI: 10.4018/978-1-7998-1690-4.ch015

Otsuka & Ren, 2005). These intermediate structures exhibit good corrosion and antioxidant resistance, good mechanical properties at low density and high temperature. However, the use of room-temperature brittleness, low immunity at high temperatures, and excessive cramping limits the use. Recently, several efforts have been devoted to improving the room-temperature characterization of intermediate systems (Cook, O'keefe, Smith, & Stobbs, 1983; Lin, Wu, Chou, & Kao, 1991; Piao, Miyazaki, & Otsuka, 1992; Piao, Otsuka, Miyazaki, & Horikawa, 1993; Scarsbrook, Cook, & Stobbs, 1984). Two important roads have been used to improve the room-temperature dilution: mixing with transition metals and modification of the stoichiometric relationship (Liancheng & Wei, 1997). A partial disorder of the naturally derived super-lattice lies in the transformation of the stoichiometric relation. It has also shown the addition of a third element to reduce the strong atomic order of intermediate alloys. We have carried some investigations out with conventional alloy casting methods with unmodified metal additives. Some researchers investigated that (Zhang, Zhao, Duerig, & Wayman, 1990) that Li AlCuFe and Mn_2FeAl improve environmental weakness in intermetallic alloys. Li additives enhance the compressibility of Mn_2FeAl and $NiFeAl_3$ alloys in intermetallic systems (Liu & Galvin, 1997). In the $Mn_{1.5}Fe_{1.5}Al$ alloy microstructure, many Mn_2FeAl matrix alloys and β-Fe occur. They have documented that Mn1.5Fe1.5Al can perform extensive hysteresis, for example, at 150oC, on decomposition in a manganese-rich Heusler alloy state or stress-induced manganese-rich Heusler alloy state (Zhang et al., 1990).

In this work, we investigated the effect of tensile water by adding different concentrations of Fe in Mn_2FeAl intermetallic matrices by phase transformation kinetics, DDT and CCD graphs. We observed that temperature-dependent thermodynamic properties could be formed in these materials such as density, Young's modulus, thermal expansion coefficient and thermal conductivity. Temperature with mechanical properties replaced the Heusler Alloys phase. In the present paper, it reports the new development computer programming simulation which can calculate the above material properties for general Heusler alloy. I base the result of the program on stress and a strain of Mn2FeAl major phase transformations taking place during heat treatment, along with the calculation of the properties of different phases designed in this material. This paper focused on preparing Al-based Heusler alloy through JMatPro software package, which is generating material data that has to be matched with experimental results, such as dilatometry, Jominy hardenability tests, and reliable testing. To make JMatPro's material data is easier to analyses with the developed model by arranged systematically in the format of FE simulation packages. We have successfully developed Heusler alloy with different Fe composition with formation/deformation simulation.

SIMULATION PROCEDURE

The software tool JMatPro, which is an acronym for Java-based Materials Properties software, is a suite of computer programs that have been developed by Thermotech Ltd. and Sente Software Ltd. for the prediction of a broad range of materials properties. JMatPro started in 1998 as a project funded by a consortium of companies and institutions interested in expanding their materials properties modelling capacity. The CALPHAD-book (Otsuka & Ren, 2005) can be seen as a starting point and scientific foundation of the system. The initial commercial release of JMatPro was available 2002, current release of JMatPro is V.6.2. Although engineers have the common and easy access to geometrical data for CAE and FEM software, there is a lack of accurate materials data for use with such software. However, these materials data is an important basis for the optimisation of products and processes. Because of the great effort involved in traditional testing, data is often scarce or inconsistent, especially for new products or multi-components alloys. JMatPro (Java-based Material Properties) closes these gaps in data. It combines solid and user-friendly thermodynamics (CalPhad) with the calculation of material properties based on physical models.

The material simulation software is available for many alloy systems, e.g. steels, nickel, aluminum, and titanium alloys. It calculates phase equilibria and Thermo-physical data, such as heat capacity and thermal expansion. Also, the software determines mechanical properties like strength and hardness and phase transformations like time-temperature transformation (TTT) diagrams. Version 6.0 of JMatPro is available since November 2010. The new version includes many enhancements, such as the calculation of flow curves as a function of temperature and strain rate.

The data export capabilities to CAE tools have been further expanded to Deform, Simufact, Forge and similar tools. For steels, re-austenitization calculations and time-temperature-austenitization (TTA) has also added diagrams in the new version. I have used successfully the software for many years in the areas of materials development, for steel-related industries, as well as in the areas of process modelling and product development like in the automotive industry. JMatPro was designed so that any engineer or scientist that requires materials can use it.

- Properties as part of their everyday work.
- ease of use because of an intuitive user interface
- fast and robust calculations, sound models for predictions of materials properties
- data management interface to browse through calculated properties.

The properties calculated by JMatPro were:

- Stable and metastable phase equilibria
- Solidification behaviour and properties, e.h. Scheil calculations
- Mechanical properties, e.g. tensile strength, hardness, flow curves
- Thermo-physical and physical properties, e.g., Young's (E), Shear (G) and Bulk (K) moduli, Poisson's ratio, Thermal conductivity, Thermal expansion coefficient, Density, Specific heat....

Phase transformations, e.g., TTT/CCT/TTA diagrams, Jominy diagrams JMatPro has different modules for each material group. Each module is pre-configured with certain preference settings so that the end-user can select only the relevant options for calculation of data. Thermodynamic calculations are a basis for the calculation of materials properties. JMatPro uses core minimization routines developed for the PMLFKT software programme by Lukas and extended by Kattner. Comprehensive new subroutines which provide a variety of benefits extended this basis, and it is including setting automatic start points, checking for miscibility gaps, finding phase boundaries.

Mn_2FeAl System

Figure 1 shows the Mn_2FeAl alloy with 5% of Fe. Here, there is a 4.6% of elongation and 1200 MPa of the ultimate tensile strength (UTS). Figure 3(b) shows that the heat treatment improves the mechanical properties of Mn_2FeAl. These specimens show 25% of elongation and 480 MPa of UTS. However, in relation with the yield strength, the best results were obtained with the lower element additions to the

Mn_2FeAl alloy. Thus, for example, the specimen with 3% of Fe additions reaches the best value of yield strength. We got similar results with the Fe additions. Table 1 shows the Rockwell C hardness measurements from the got alloys before and after the heat treatments. The general trend for the specimens without heat treatment is a decrement in the hardness values with the element additions to the Mn_2FeAl alloy. Increments in the Li additions give rise to the more pronounced decrement in the hardness values, getting the lowest value for the 5% addition of Fe. However, increments in the Fe content produce also increments in the hardness values. Heat treatments decrease the hardness values in comparison with those got without heat treatment. These hardness decrement values could be related to the reduction of the number of acting vacancies in the heat-treated alloys. We got similar experimental results for similar specimens under compression tests (Liu & Galvin, 1997).

Phase Transformation Involved Mechanical Behavior of Mn2feal Shape Memory Alloys

SMAs can hold different states at a temperature in MS and AS temperatures because of the transformation of the mechanical behavior of Heusler and semi-metallic alloys at the same temperature. The deformation behaviors of Heusler alloys in the state of Heusler and semi-metallic alloys can be investigated at the same temperature. Figure. 2 shows the stress-strain curves of Mn_2FeAl, $Mn_{1.5}Fe_{1.5}Al$ and Fe_2MnAl Heusler alloys in manganese-rich Heusler alloy and half-metallic Heusler alloy states, at the specific temperature for each alloy.

Take, for example, the Mn_2FeAl alloy whose MS and AS temperatures are 70°C and 75°C, respectively. For a comparative study of the mechanical behavior of a manganese-filled Heusler alloy or a semi-metal Heusler alloy at the same temperature, it cooled another sample to the same temperature at 56°C and another half-metal Heusler alloy. Subsequently, the stress-strain curves of Mn2FeAl and semi-metal Heusler alloy states were recorded at 56°C. Subsequent similar procedures are followed the decay behavior of the $Mn_{1.5}$ $Fe_{1.5}Al$ and Fe_2MnAl alloy crystal structure was characterized as 32 °C and 27 °C, respectively. The corresponding stress-strain curves are shown in Figure 2.

The stress-strain curves shown in Figure.2 show that the yield stress for the reconstruction of manganese-rich Heusler alloy types is 100 MPa higher than that of the semi-metal Heusler alloy decay, even if both samples are simultaneously decomposed. In addition, it can see that the manganese-rich Heusler alloy pressure results different from the semi-metal Heusler alloy. It followed on a previous yielding plateau and then shows the slope of the rising yield. At the initial stage of the decomposition process of the half-metal Heusler alloy yield decreases until it reaches 10%. These Mn_2FeAl SMAs have the same strength regardless of their initial state, and the different yield behaviors of manganese-rich Heusler alloy and semi-metal Heusler alloy at the same temperature.

As shown above, the reorganization of manganese-rich Heusler alloy variants even at the same temperature requires more stress than induction of C1b deformation in the semi-metal Heusler alloy structure. To further investigate the yield temperature of Mn_2FeAl alloys in manganese-rich Heusler alloy L21 and C1b half metal Heusler alloy, the Fe_2MnAl alloy was used to characterize the yield behaviour between manganese-rich Heusler alloy and semi-metal. Heusler Alloy. Considering MS (70 °C) and AS (93 °C) of the $MnFe_2Al$ alloy, the mechanical behavior of the alloy under 43°C and 150 °C under manganese-rich Heusler alloy and semi-metal Heusler alloy states with their corresponding results are shown in Figure 3.

Figure 1. General physical properties of Mn_2FeAl (a) Latent heat (b) Stress vs initial cooling and (c) Fe_2MnAl, at specific temperature for each alloy

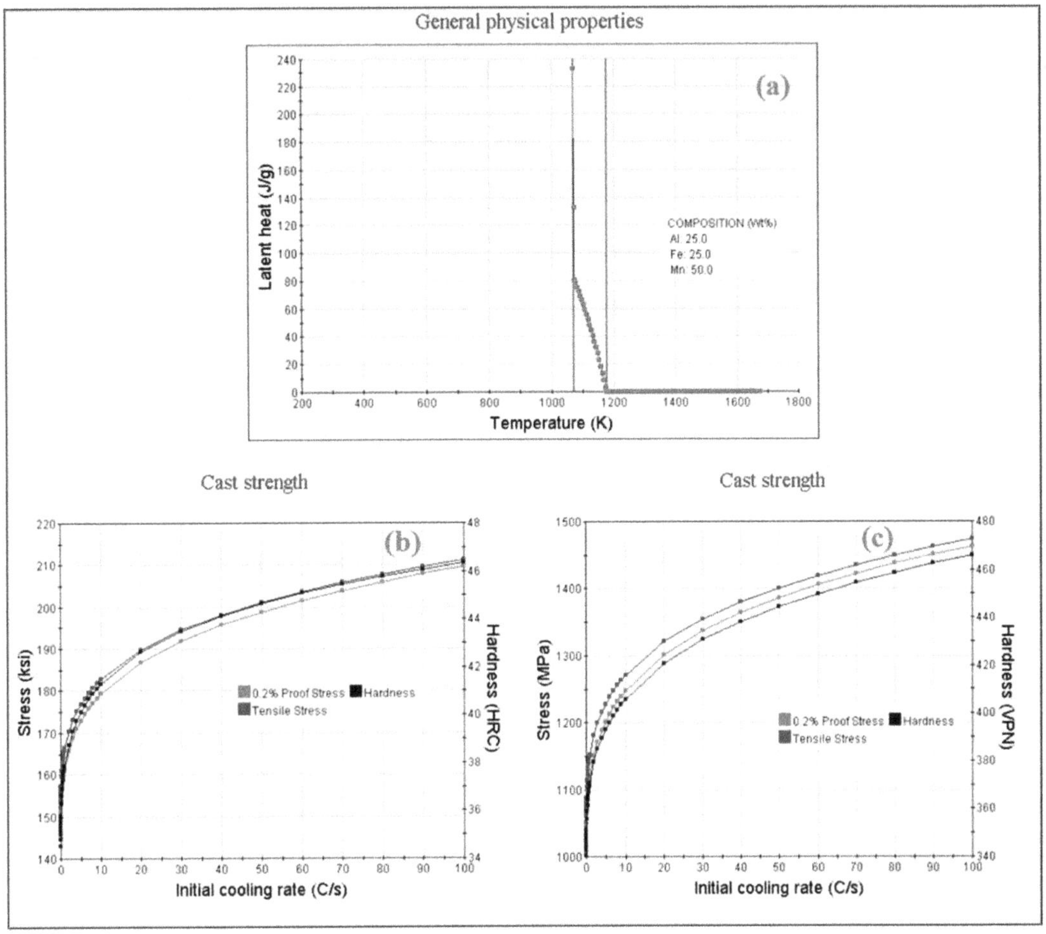

When the manganese-rich Heusler alloy and the semi-metal Heusler alloy of Fe_2MnAl were decomposed at low temperature, for example, we observed a significant difference in yield stress at 23 °C. As the temperature increases, the yield stresses of both manganese-rich Heusler alloy and semi-metal Heusler alloy will increase. The yield of a semi-metal Heusler alloy explodes with temperature compared to manganese-rich Heusler alloy. As a result, as the test temperature is high, the yield stress difference between the manganese-rich Heusler alloy and the semi-metal Heusler alloy weakens. In particular,

Table 1. Extreme intrinsic transformation hysteresis of alloys

Alloys	A_s °C	A_f °C	Elongation %	M_s °C	A_{sd} °C	M_f °C
Mn_2FeAl	93	45	40	70	180	30
$Mn_{1.5}Fe_{1.5}Al$	84	80	12	43	150	300
Fe_2MnAl	60	10	20	150	80	60

Figure 2. Stress–stain curves of Mn_2FeAl (a) Engineering Stress vs True strain (b) true Stress vs True strain and (c)Gibbs energy at specific temperature

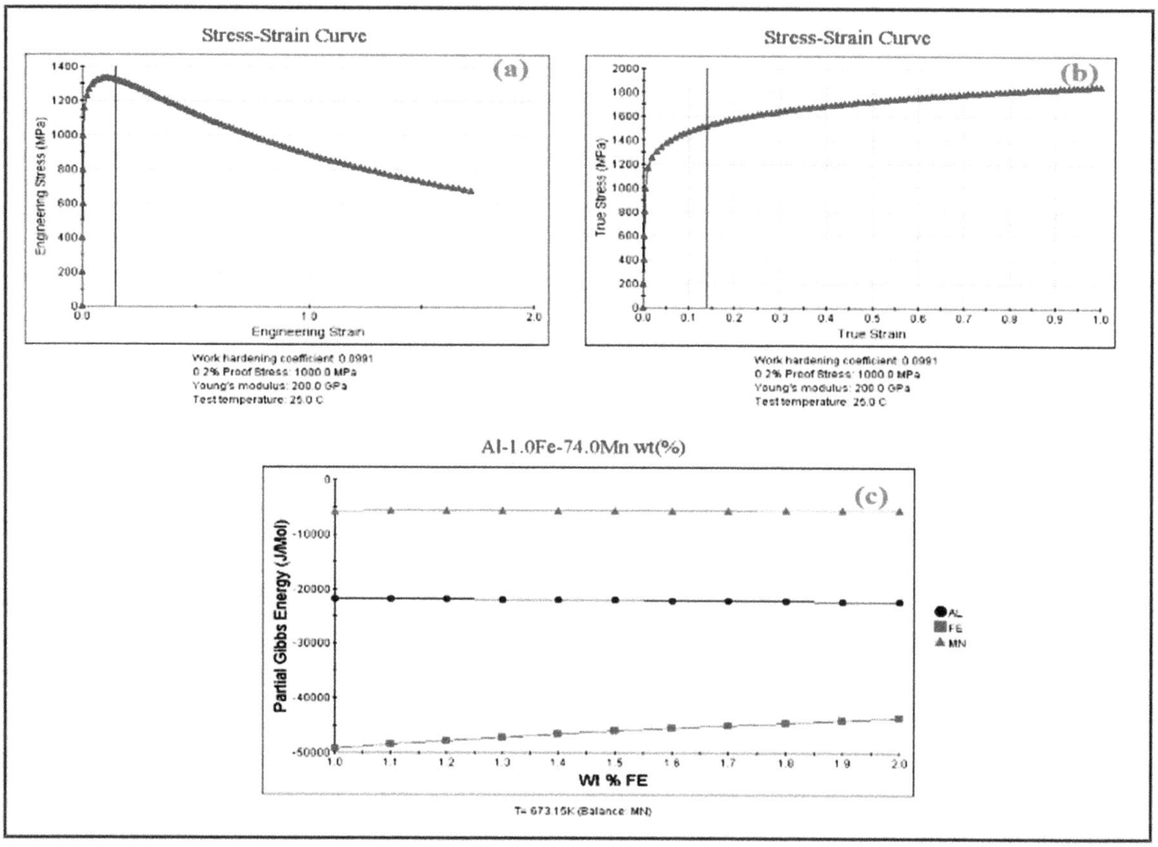

when the test temperature rises to 38°C, the difference between yield stresses for manganese-rich Heusler alloy and semi-metal Heusler alloy is negligible. This means that the stress for the reorganization of manganese-rich Heusler alloy types is almost identical to the stress induced by the conversion of half-metal Heusler alloy at 38°C.

Explanation of the Yield Stress Difference Between Manganese-Rich Heusler Alloy and Half Metallic Heusler Alloy

Because of thermoelastic deformation, Heusler alloys may have different phase conditions, such as a mixture of manganese-rich Heusler alloy and semi-metallic Heusler alloy. However, upon heating or cooling, Heusler alloys can keep their original position at a critical temperature between the MS and the AS, because of their amorphous hysteresis. This means that at the same temperature between the MS and the AS, the Heusler alloys can hold the manganese-rich Heusler alloy state or the semi-metal Heusler alloy state, by heating the sample in the whole manganese-filled Heusler alloy state by cooling. To illustrate the yield behavior of manganese-rich Heusler alloy and semi-metal Heusler alloy at different temperatures, Figure 4 represents a schematic of the Gibbs free energy of manganese-rich Heusler alloy and semi-metal Heusler alloy. From an energetic point of view, at a critical temperature between MS and

Figure 3. General physical properties of Heusler alloy is a) Young's Modulus b) Bulk modulus c) shear modulus d) poison's ratio vs Temperature

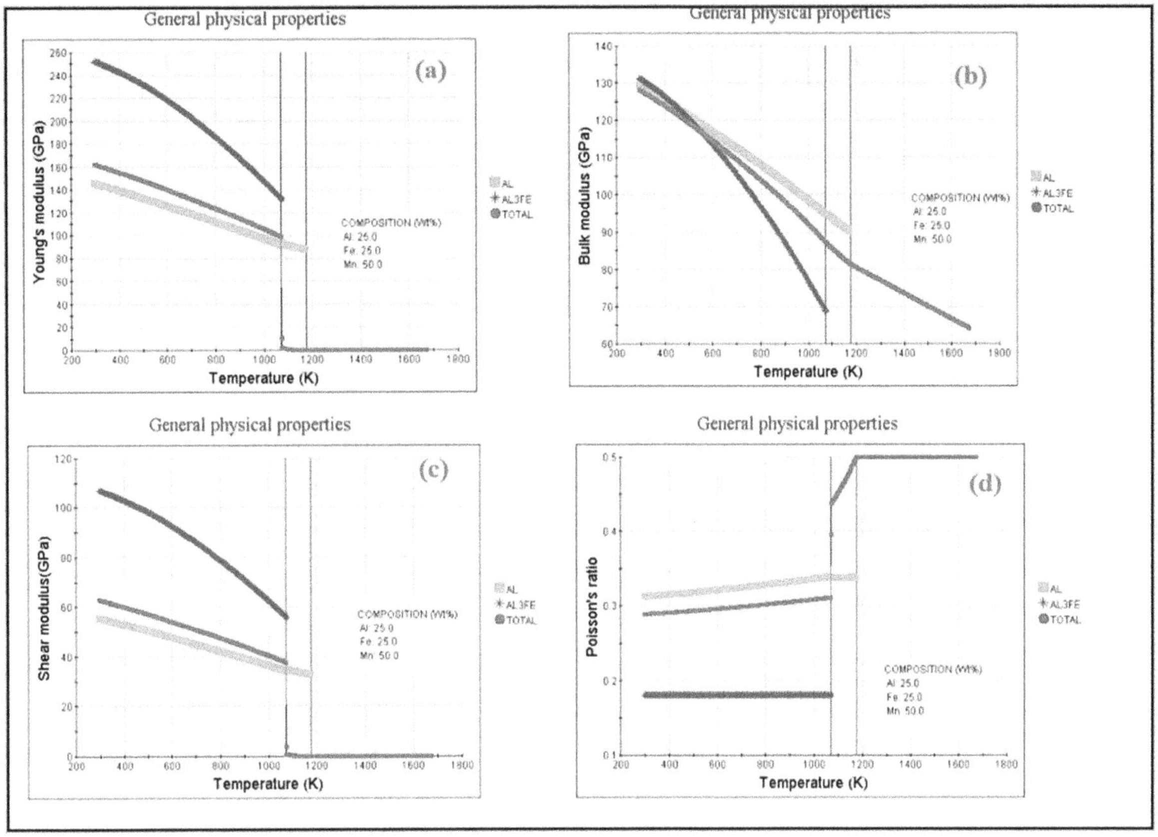

AS, the manganese Heusler alloy is in a super-heated state when the Heusler alloy is heated from the temperature at which the whole manganese is in a Heusler alloy state. Based on the same principle, the semi-metal Heusler alloy is in a super-cooled state while cools the Heusler alloy from the temperature of the full half metal Heusler alloy. To explain the temperature dependence of the yield stress difference between manganese-rich Heusler alloy and semi-metal Heusler alloy, two characteristic temperatures T1 and T2. According to Figure 4, the free energy conversion of the semi-metal Heusler alloy kips for the manganese-rich Heusler alloy conversion decreases with increasing temperature.

Under external stress the transformation from Half metallic Heusler alloy to manganese-rich Heusler alloy is more difficult to trigger at t2 than at t1. This might be the reason the yield stress for inducing martensitic transformation increases rapidly when the temperature increases between MS and AS. Contrary to usual belief, the experimental results of the present study showed that under external stress the Half metallic Heusler alloy is easier to yield than the manganese-rich Heusler alloy at the same temperature. About the lower stress for inducing Half metallic Heusler alloy to manganese-rich Heusler alloy transformation than for reorientation of manganese-rich Heusler alloy at the same specific temperature between MS and AS, the corresponding mechanism is not clear at present.

Figure 4. a) specific heat capacity b) Total diffusivity c) surface tension of manganese-rich d) Heusler alloy and Half metallic Heusler alloy as a function of Temperature.

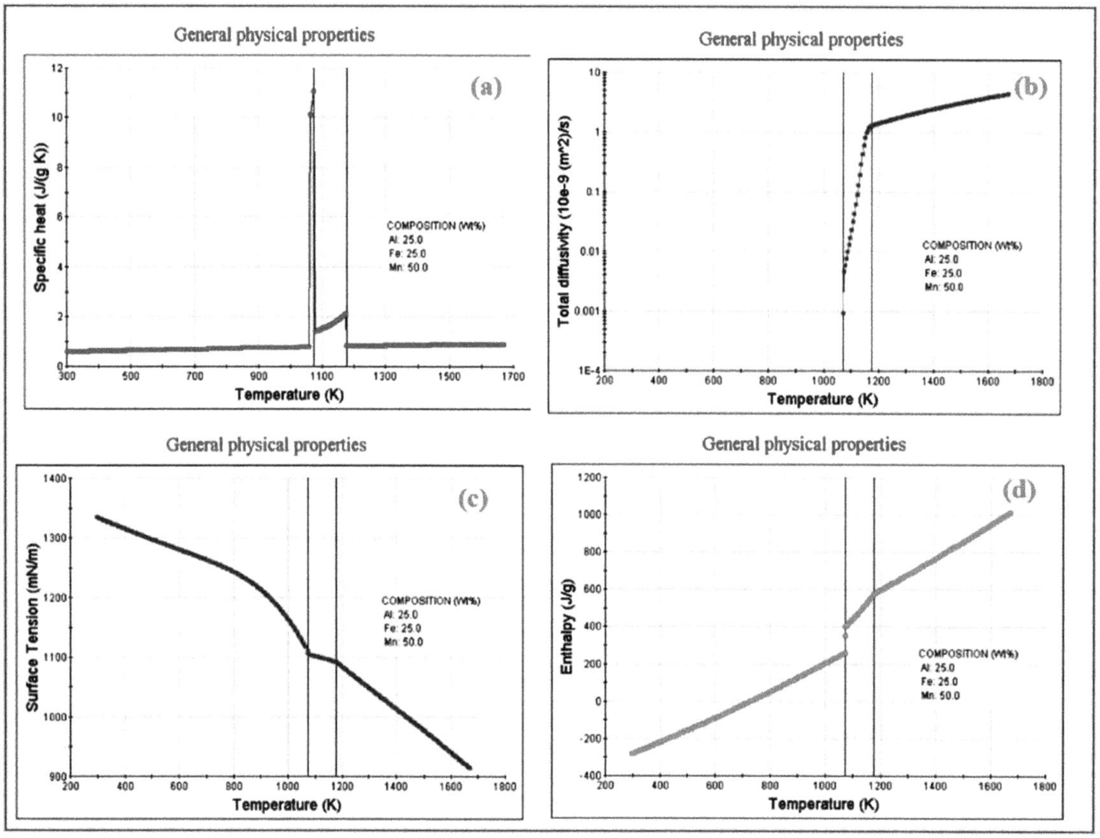

M_S Temperature Dependence of the Stress for a Reorientation of Manganese-Rich Heusler Alloy

Because of the thermoelastic Heusler alloy deformation, a specific always performs very low pressure to induce manganese-rich Heusler alloy deformation at MS temperature (Otsuka & Ren, 2005). However, for different rates of Heusler alloys, there is no investigation data on the fit between the MS temperature and the stress of the Heusler deformation and the reorganization of the manganese-rich Heusler alloy. In this study, we investigated the mechanical behavior of Mn_2FeAl alloys with different MS temperatures, showing that the yield stress for the reconstitution of manganese-rich Heusler alloy variants depends on the MS temperature of the Heusler alloys. Figure.5 illustrates spacing vs initial cooling rate for solidification of the samples at constant temperature. Temperature vs time graph can be explained it to stabilizing the micro crystal alloys. Fraction of except Mn weight reduced because of comparatively Fe and Al elements are increasing with respect to the fraction of solidification at constant temperature. Density of Aluminum elements were increasing in manganese-rich Heusler alloy temperature at 150°C below their corresponding MS temperatures. AS temperatures decrease, yield stresses decrease to restructure the manganese-rich Heusler alloy. This alloy has low MS temperature, and it shows low yield pressure to rearrange the manganese-rich Heusler alloy.

Figure 5. a) spacing vs initial cooling rate b) Temperature vs time for stabilizing the micro crystal alloys c) Wt% Vs fraction of all the elemental weight d) density of manganese-rich Heusler alloy and Half metallic Heusler alloy respect to temperature.

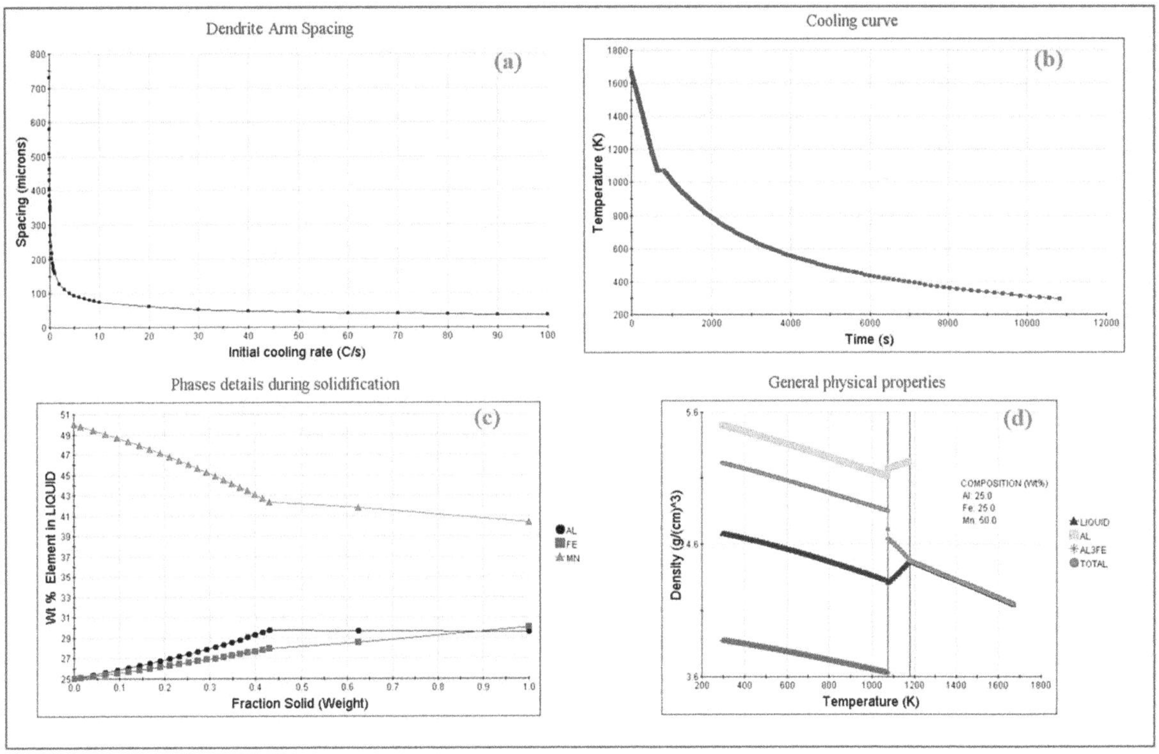

We achieved similar results in binary Mn-Fe alloys. In Figure 3. Shows the method of stress-strain curves of $Mn_{1.5}Fe_{1.5}Al$ and Fe_2MnAl alloys at their MS temperatures 28°C and 18°C. We find that the stress induced by martensitic deformation in $Mn_{1.5}Fe_{1.5}Al$ alloy is much lower than that of Fe_2MnAl alloy. The mechanism for this phenomenon is now under investigation, and we have provided such detailed explanations in this report.

Wide Hysteresis of Mn_2FeAl SMAs with Low Al Content

As mentioned above, the microstructure of Mn_2FeAl-shaped memory alloys are related to their Fe to Mn ratio and the content of Al. In the $Mn_{1.5}Fe_{1.5}Al$ alloy microstructure, many Mn_2FeAl matrix alloys and β-Fe occur. We have documented that Mn1.5Fe1.5Al can perform extensive hysteresis, for example, at 150°C, on decomposition in a manganese-rich Heusler alloy state or stress-induced manganese-rich Heusler alloy state (Zhang et al., 1990). Early researchers considered that the deformed Al particles in $Mn_{1.5}Fe_{1.5}Al$ inhibit reverse transformation, playing a key role in the achievement of wide hysteresis. Recent word by Liu et al. showed that low Al containing Mn_2FeAl alloys without the large amount of b-Fe phase in microstructure still show wide hysteresis upon a proper deformation (Wei & Xinqing, 2009). Figure.5 illustrates the dependence of the transformation hysteresis and recovery ratio of $MnFe_2Al$ and $Mn_{1.5}Fe_{1.5}Al$ Heusler alloys on the pre-deformation strain. We show it that both $Mn_{1.5}Fe_{1.5}Al$ and $MnFe_2Al$

Figure 6. shows the b) volume change vs Temperature and d) thermal conductivity of $Mn_{1.5}Fe_{1.5}Al$ alloys from the 300 to 1700°C.

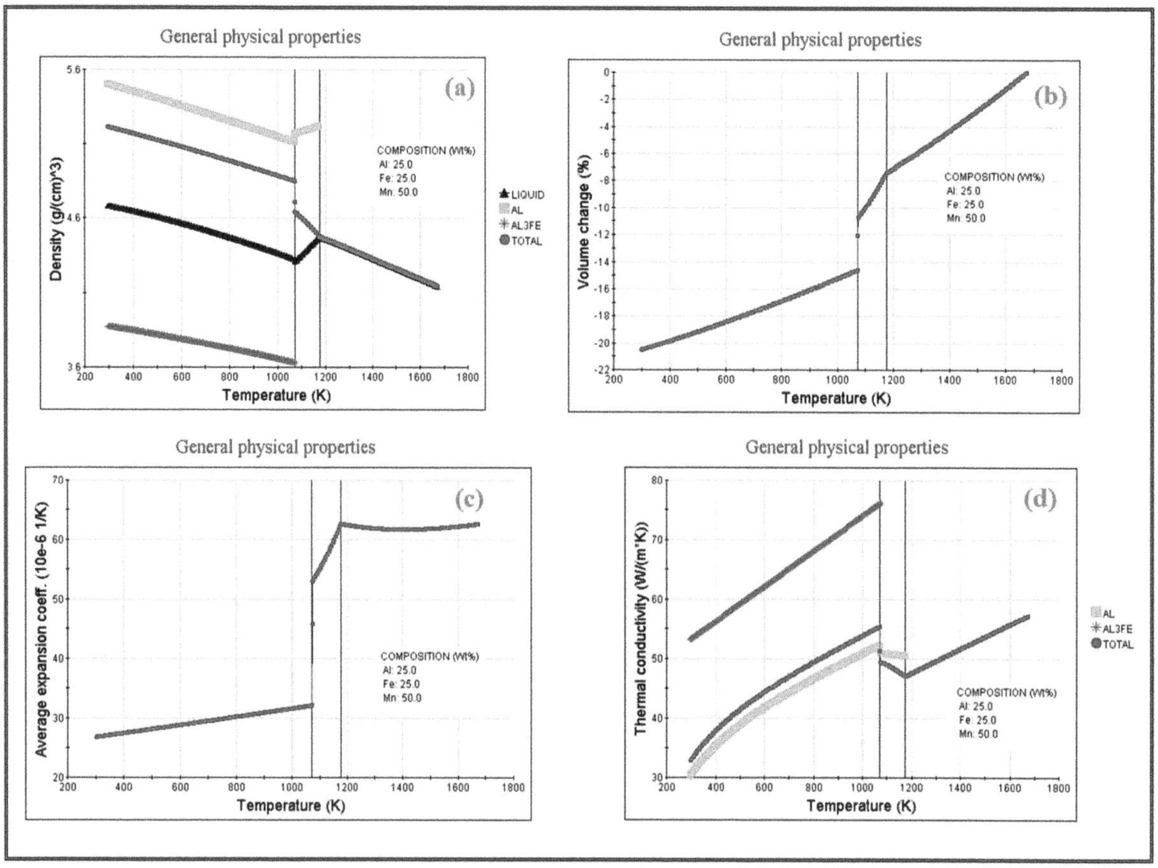

Heusler alloys perform wide hysteresis and similar recovery ratio upon a pre-deformation. $MnFe_2Al$ alloy possesses a wide hysteresis of 108°C by a pre-deformation of 16%, although a little smaller than that of $Mn_{1.5}Fe_{1.5}Al$ upon the same pre-deformation (136°C). Considering the difference in initial hysteresis between Mn_2FeAl (35⁰C) and $Mn_{1.5}Fe_{1.5}Al$ (48 K), $MnFe_2Al$ achieved similar net increase in hysteresis compared to $Mn_{2\text{-}0.5}Fe_{0.5}Al$, especially for small pre-deformation. Note that there are few isolated b-Fe phases in $Mn_{2\text{-}0.7}Fe_{0.7}Al$. Thus, one can conclude that large number of b-Al particles have the minor contribution to wide hysteresis, and the Al resolved in the FeAl matrix might be a predominant aspect. It is reasonable to speculate that the additive Al resolved in the matrix is considered changing the kinetics of thermoelastic martensitic transformation, resulting in the stabilization of manganese-rich Heusler alloy (Zhao, Yan, Yang, & Xu, 2006).

Figure. 6 shows the total volume change vs Temperature of Mn_2FeAl and $Mn_{1.5}Fe_{1.5}Al$ alloys from the 300 to 1700°C. It is found that when deformed at this temperature, Mn_2FeAl and $Mn_{1.5}Fe_{1.5}Al$ alloys perform similar deformation behavior, particularly almost the same yielding and reorientation behavior. As shown above, Mn_2FeAl contains no of b-Fe phase in microstructure. It suggests that the presence of b-Fe phase do not play a significant role in the deformation behavior.

Figure 7. a) Activity with Fe wt. % b) Heat capacity c) Schematic of the chemical (Gibbs) free energy of manganese-rich Heusler alloy and Half metallic Heusler alloy as a function of Fe wt %.

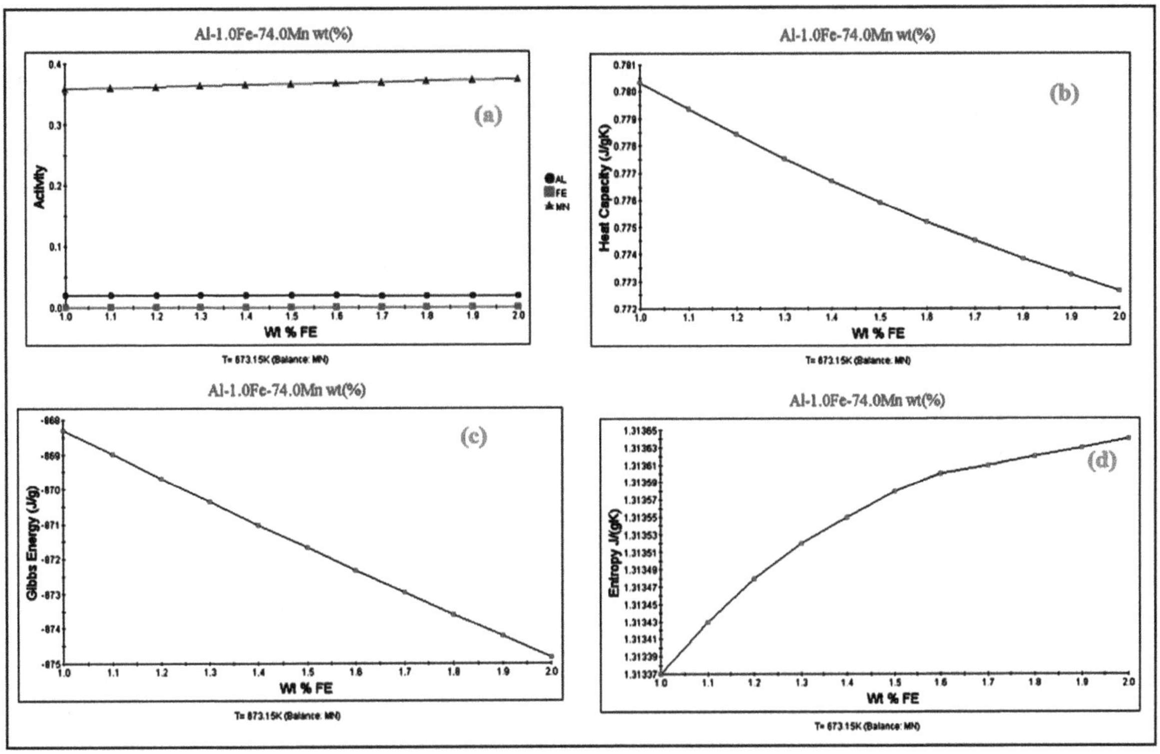

The Correlation of Elastic Energy Relaxation and Hysteresis Expansion

In thermoelastic martensitic transformations, the forward and reverse transformation temperatures start and finish in narrow temperature ranges due to their small chemical driving force. In the narrow temperature range, if the difference of specific heat between manganese-rich Heusler alloy and Half metallic Heusler alloy phases is not considered, the relationship between chemical free energy and temperature should be linear. Figure 7 schematically illustrates the chemical-free energy of manganese-rich Heusler alloy and Half metallic Heusler alloy phases as a function of temperature, in which T_0 denotes the temperature at which the Gibbs free energies of manganese-rich Heusler alloy and Half metallic Heusler alloy phases are equal.

$$\Delta G_{Chem}^{M\rightarrow P}(A_{sd}) = \Delta G_{Chem}^{M\rightarrow P}(A_{s}) + \Delta G_{el}^{P\rightarrow M}(M_{f}) \quad (1)$$

According to Tong and Wayman (Tong & Wayman, 1974), the chemical-free energy change at A_{Sd} and it can be expressed as the sum of the chemical-free energy change and the elastic energy change at temp (As) from the Figure 7, the relationships between the different temperature with corresponding chemical free energy changes can be estimated from the equation given below,

$$\frac{T_0 - M_f}{\Delta g_{M_f}} = \frac{A_s - T_0}{\Delta g_{As}} = \frac{A_{Sd} - T_0}{\Delta g_{A_{sd}}} \tag{2}$$

For thermos elastic martensitic transformations, the T_0 temperature could be represented by $T_0=(M_s+A_f)/2$. The increase in reverse deformation temperature was mathematically modeled as a function of T_0 and M_f, with data reported in some literature related to the hysteresis enhancement induced by mutations of MnFe and Mn_2FeAl alloys to verify the compatibility of A_{Sd} with the experimental results. Its values are tabulated in Table 1, where AS represents the experimentally measured reverse deformation initial temperature. In the case of the MnFe and FeAl alloy models, the hysteresis was defined as achieving its maximum. As Table 1 shows, the temperature of the A_{SD} after severe decomposition is always several degrees lower for MnFe than for A_S and about ten degrees lower for Mn_2FeAl alloys, although the A_{SD} temperature is close to that of the AS. In addition to relaxing elastic energy, there may be other factors involved in the hysteresis expansion, particularly Mn_2FeAl alloys, although these factors are not dominant.

CONCLUSION

A mathematical model of the solidification process on twin-roll continuous casting thin strip was developed based on the cellular automata (CA) theory. JMatPro numerically simulated the continuous solidification process in front of the nip point. The solidification microstructure, the grains size, the dendritic grain radius, and the secondary dendrite arm spacing were obtained to predict the mechanical properties of the wedge strip. The simulation results were verified by using the metallographic and tensile tests. The simulation results of the microstructure on the wedge strip rolled plain and cross-section, which was 1 mm away from the roll, were consistent with the experimental results. The simulation results of dendritic grain radius and the secondary dendrite arm spacing were, respectively, 1133 µm and 16 µm, which were consistent with the experimental results. The predicted yield strength of the aluminium alloy strip was 158 MPa, which was consistent with the experimental results (142.9 MPa).

- Manganese rich Heusler and Half metallic Heusler shape-memory alloys perform different mechanical behaviours even at the same temperature. The Half metallic Heusler alloy is easier to yield than the manganese-rich Heusler alloy under external stress.
- The yield stress for the reorientation of manganese-rich Heusler alloy variants depends on the MS temperature of these alloys. An alloy with lower MS temperature exhibits lower yield stress for reorientation of manganese-rich Heusler alloy.
- The presence of b-Fe phase does not have a considerable contribution to wide hysteresis and deformation behaviour of low Fe content could achieve wide transformation hysteresis upon pre-deformation.
- We established the relationship between the relaxation of elastic energy and the hysteresis expansion. The relaxation of elastic energy by plastic deformation plays a dominant role in the hysteresis expansion of Mn_2FeAl and Fe_2MnAl alloys.

The solidification process of twin-roll cast alloy was described accurately with the established mathematical model. The model can predict the solidification structure and mechanical properties of the strip under different process parameters and solidification parameters.

REFERENCES

Cook, J., O'keefe, M., Smith, D. J., & Stobbs, W. (1983). The high resolution electron microscopy of stacking defects in Cu–Zn–Al shape memory alloy. *Journal of Microscopy*, *129*(3), 295–306. doi:10.1111/j.1365-2818.1983.tb04186.x

Hiroki, C., Yamamoto, T., Takeda, Y., Suzuki, A., & Sakuma, T. (2010). Exploitation of shape memory alloy actuator using resistance feedback control and its development. *Progress in Natural Science: Materials International*, *20*, 97–103. doi:10.1016/S1002-0071(12)60013-6

Liancheng, Z., & Wei, C. (1997). Deformation-induced martensitic transformation and its reversibility in ni-ti-nb wide hysteresis shape memory alloys. *Chin Shu Hsueh Pao*, *33*(1), 90–98.

Lin, H., Wu, S.-K., Chou, T., & Kao, H. (1991). The effects of cold rolling on the martensitic transformation of an equiatomic TiNi alloy. *Acta Metallurgica et Materialia*, *39*(9), 2069–2080. doi:10.1016/0956-7151(91)90177-3

Liu, Y., & Galvin, S. (1997). Criteria for pseudoelasticity in near-equiatomic NiTi shape memory alloys. *Acta Materialia*, *45*(11), 4431–4439. doi:10.1016/S1359-6454(97)00144-4

Otsuka, K., & Ayman, C. M. W. (1998). *Shape Memory Materials*. Cambridge University Press.

Otsuka, K., & Ren, X. (2005). Physical metallurgy of Ti–Ni-based shape memory alloys. *Progress in Materials Science*, *50*(5), 511–678. doi:10.1016/j.pmatsci.2004.10.001

Piao, M., Miyazaki, S., & Otsuka, K. (1992). Characteristics of deformation and transformation in Ti-44Ni47Nb9 shape memory alloy. *Materials Transactions, JIM*, *33*(4), 346-353.

Piao, M., Otsuka, K., Miyazaki, S., & Horikawa, H. (1993). Mechanism of the As temperature increase by pre-deformation in thermoelastic alloys. *Materials Transactions, JIM*, *34*(10), 919-929.

Scarsbrook, G., Cook, J., & Stobbs, W. (1984). The stabilization of martensite in Cu-Zn-AI shape memory alloys. *Metallurgical Transactions. A, Physical Metallurgy and Materials Science*, *15*(11), 1977–1986. doi:10.1007/BF02646832

Tong, H., & Wayman, C. (1974). Characteristic temperatures and other properties of thermoelastic martensites. *Acta Metallurgica*, *22*(7), 887–896. doi:10.1016/0001-6160(74)90055-8

Wei, L., & Xinqing, Z. (2009). Mechanical properties and transformation behavior of NiTiNb shape memory alloys. *Chinese Journal of Aeronautics*, *22*(5), 540–543. doi:10.1016/S1000-9361(08)60138-7

Zhang, C., Zhao, L., Duerig, T., & Wayman, C. (1990). Effects of deformation on the transformation hysteresis and shape memory effect in a Ni47Ti44Nb9 alloy. *Scripta Metallurgica et Materialia*, *24*(9), 1807–1812. doi:10.1016/0956-716X(90)90550-Z

Zhao, X., Yan, X., Yang, Y., & Xu, H. (2006). Wide hysteresis NiTi (Nb) shape memory alloys with low Nb content (4.5 at.%). *Materials Science and Engineering*, *438*, 575–578. doi:10.1016/j.msea.2006.03.110

Compilation of References

Acharya, S., Moitra, A., Bysakh, S., Nanibabu, M., Krishanan, S. A., Mukhopadhyay, C. K., & Ghosh, K. S. (2019). Effect of high strain rate deformation on the properties of SS304L and SS316LN alloys. *Mechanics of Materials*, *136*, 103073. doi:10.1016/j.mechmat.2019.103073

Alberdi, A., Artaza, T., Suárez, A., Rivero, A., & Girot, F. (2016). An experimental study on abrasive waterjet cutting of CFRP/Ti6Al4V stacks for drilling operations. *International Journal of Advanced Manufacturing Technology*, *86*(1–4), 691–704. doi:10.100700170-015-8192-x

Aldrich, D. R., Ayranci, C., & Nobes, D. S. (2017). OSM-Classic: An optical imaging technique for accurately determining strain. *SoftwareX*, *6*, 225–230. doi:10.1016/j.softx.2017.08.007

Altpeter, I., Dobmann, G., Kröning, M., Rabung, M., & Szielasko, S. (2009). Micro-magnetic evaluation of micro residual stresses of the IInd and IIIrd order. *NDT & E International*, *42*(4), 283–290. doi:10.1016/j.ndteint.2008.11.007

Antonov, A. A. (1983). Development of the method and equipment for holographic inspection of residual-stresses in welded structures. *Welding Production*, *30*(12), 41–43.

Ao, J., Li, T., Liu, Y., Ding, Y., Wu, G., Hu, K., & Kong, L. (2010). Optimal design of thread height and width on an immediately loaded cylinder implant: A finite element analysis. *Computers in Biology and Medicine*, *40*(8), 681–686. doi:10.1016/j.compbiomed.2009.10.007 PMID:20599193

Apiwattanalunggarn, P., Shaw, S. W., Pierre, C., & Jiang, D. (2003). Finite-element-based nonlinear modal reduction of a rotating beam with large-amplitude motion. *Modal Analysis*, *9*(3-4), 235–263.

Arias, A., Rodríguez-Martínez, J. A., & Rusinek, A. (2008). Numerical simulations of impact behavior of thin steel plates subjected to cylindrical, conical and hemispherical non-deformable projectiles. *Engineering Fracture Mechanics*, *75*(6), 1635–1656. doi:10.1016/j.engfracmech.2007.06.005

Arrazola, P. J., & Ozel, T. (2010). Investigations on the effects of friction modeling in finite element simulation of machining. *International Journal of Mechanical Sciences*, *52*(1), 31–42. doi:10.1016/j.ijmecsci.2009.10.001

Arrazola, P. J., Ozel, T., Umbrello, D., Davies, M., & Jawahir, I. S. (2013). Recent advances in modeling of metal machining processes. *Journal of Manufacturing Technology*, *62*, 695–718.

Arshak, K. I., Ansari, F., Collins, D., & Perrem, R. (1994). Characterisation of a thin-film/thick-film strain gauge sensor on stainless steel. *Materials Science and Engineering B*, *26*(1), 13–17. doi:10.1016/0921-5107(94)90180-5

Arul, S., Vijayaraghavan, L., & Malhotra, S. K. (2007). Online monitoring of acoustic emission for quality control in drilling of polymeric composites. *Journal of Materials Processing Technology*, *185*(1–3), 184–190. doi:10.1016/j.jmatprotec.2006.03.114

Ashby, M., Shercliff, H., & Cebon, D. (2007). *Materials: Engineering, science, processing and design*. Amsterdam: Elsevier.

Ashtaputre, S. S., Deshpande, A., Marathe, S., Wankhede, M., Chimanpure, J., Pasricha, R., & Kulkarni, S. (2005). Synthesis and analysis of ZnO and CdSe nanoparticles. *Pramana*, *65*(4), 615–620. doi:10.1007/BF03010449

Badulescu, C., Grédiac, M., & Mathias, J. D. (2009). Investigation of the grid method for accurate in-plane strain measurement. *Measurement Science and Technology*, *20*(9), 1-17.

Badulescu, C., Grédiac, M., Mathias, J. D., & Roux, D. (2009). A procedure for accurate one-dimensional strain measurement using the grid method. *Experimental Mechanics*, *49*(6), 841–854. doi:10.100711340-008-9203-8

Bakshi, U. A., & Bakshi, A. V. (2008). *Electrical measurements and measuring instruments*. Pune: Technical Publications.

Balaji, P. S., Leblouba, M., Rahman, M. E., & Ho, L. H. (2016). Static lateral stiffness of wire rope isolators. *Mechanics Based Design of Structures and Machines*, *44*(4), 462–475. doi:10.1080/15397734.2015.1116996

Balaji, P. S., Moussa, L., Rahman, M. E., & Ho, L. H. (2016). An analytical study on the static vertical stiffness of wire rope isolators. *Journal of Mechanical Science and Technology*, *30*(1), 287–295. doi:10.100712206-015-1232-5

Balaji, P. S., Moussa, L., Rahman, M. E., & Vuia, L. T. (2015). Experimental investigation on the hysteresis behavior of the wire rope isolators. *Journal of Mechanical Science and Technology*, *29*(4), 1527–1536. doi:10.100712206-015-0325-5

Balaji, P. S., & Yadava, V. (2013). Three dimensional thermal finite element simulation of electro-discharge diamond surface grinding. *Simulation Modelling Practice and Theory*, *35*, 97–117. doi:10.1016/j.simpat.2013.03.007

Balanethiram, V. S., Hu, X., Altynova, M., & Daehn, G. S. (1994). Hyperplasticity: Enhanced formability at high rates. *Journal of Materials Processing Technology*, *45*(1-4), 595–600. doi:10.1016/0924-0136(94)90404-9

Baldi, A. (2014). Residual stress measurement using hole drilling and integrated digital image correlation techniques. *Experimental Mechanics*, *54*(3), 379–391. doi:10.100711340-013-9814-6

Balzar, D., & Ledbetter, H. (1993). Voigt-function modeling in Fourier analysis of size-and strain-broadened X-ray diffraction peaks. *Journal of Applied Crystallography*, *26*(1), 97–103. doi:10.1107/S0021889892008987

Bansal, R. K. (2016). *A textbook of strength of materials - Mechanics of solids*. Laxmi Publications.

Bass, J. D., Schmitt, D., & Ahrens, T. J. (1986). Holographic in situ stress measurements. *Geophysical Journal International*, *85*(1), 13–41. doi:10.1111/j.1365-246X.1986.tb05170.x

Bersuker, I. B. (1997). Limitations of density functional theory in application to degenerate states. *Journal of Computational Chemistry*, *18*(2), 260–267. doi:10.1002/(SICI)1096-987X(19970130)18:2<260::AID-JCC10>3.0.CO;2-M

Bertodo, R. (1959). Development of high-temperature strain gauges. *Proceedings - Institution of Mechanical Engineers*, *173*(1), 605–622. doi:10.1243/PIME_PROC_1959_173_052_02

Bhattacharya, S. K. (2011). *Basic electrical and electronics engineering*. New Delhi: Pearson Education India.

Bhattacharya, S. K. (2011). *Basic Electrical and Electronics Engineering*. Pearson Education India.

Bhavikatti, S. S. (2005). *Finite element analysis*. New Delhi: New Age International.

Bisht, A., Kumar, L., Subburaj, J., Jagadeesh, G., & Suwas, S. (2019). Effect of stacking fault energy on the evolution of microstructure and texture during blast assisted deformation of FCC materials. *Journal of Materials Processing Technology*, *271*, 568–583. doi:10.1016/j.jmatprotec.2019.04.029

Bjøntegaard, Ø., & Sellevold, E. J. (2004). The temperature-stress testing machine (TSTM): Capabilities and limitations. In *First International Rilem Symposium on Advances in Concrete through Science and Engineering. E&FN Spon.*

Boresi, A. P., & Chong, K. P. (2000). *Elasticity in engineering mechanics* (2nd ed.). Hoboken, NJ: Wiley.

Boyer, H. F. (1987). *Atlas of stress-strain curves.* Metals Park, OH: ASM International.

Branemark, P.-I. (1977). Osseointegrated implants in the treatment of the edentulous jaw. Experience from a 10-year period. *Scandinavian Journal of Plastic and Reconstructive Surgery. Supplementum*, 16. PMID:356184

Brinksmeier, E., & Janssen, R. (2002). Drilling of multi-layer composite materials consisting of carbon fiber reinforced plastics (CFRP), titanium and aluminum alloys. *CIRP Annals-Manufacturing Technology*, *51*(1), 87–90. doi:10.1016/S0007-8506(07)61472-3

Bronkhorst, C. A., Cerreta, E. K., Xue, Q., Maudlin, P. J., Mason, T. A., & Gray Iii, G. T. (2006). An experimental and numerical study of the localization behavior of tantalum and stainless steel. *International Journal of Plasticity*, *22*(7), 1304–1335. doi:10.1016/j.ijplas.2005.10.002

Broomhead, P., & Grieve, R. J. (1982). The effect of strain rate on the strain to fracture of a sheet steel under biaxial tensile stress conditions. *Journal of Engineering Materials and Technology*, *104*(2), 102–106. doi:10.1115/1.3225043

Byrne, G., Dornfeld, D., Inasaki, I., Ketteler, G., König, W., & Teti, R. (1995). Tool condition monitoring (TCM)— The status of research and industrial application. *CIRP Annals-Manufacturing Technology*, *44*(2), 541–567. doi:10.1016/S0007-8506(07)60503-4

Calamaz, M., Coupard, D., & Girot, F. (2008). A new material model for 2D numerical simulation of serrated chip formation when machining titanium alloy Ti–6Al–4V. *International Journal of Machine Tools & Manufacture*, *48*(3-4), 275–288. doi:10.1016/j.ijmachtools.2007.10.014

Campos, H. B., Butuc, M. C., Gracio, J. J., Rocha, J. E., & Duarte, J. M. F. (2006). Theoretical and experimental determination of the forming limit diagram for the AISI 304 stainless steel. *Journal of Materials Processing Technology*, *179*(1-3), 56–60. doi:10.1016/j.jmatprotec.2006.03.065

Cha, S. L., & Kim, J. K. (2014). Application of Thermal Stress Device for Measuring Thermal Stresses. *The 2014 World Congress on Advances in Civil, Environmental, and Materials Research (ACEM'14)*, 268-275.

Chen, Y. H., Chen, X., Xu, N., & Yang, L. (2014). *The digital image correlation technique applied to hole drilling residual stress measurement* (No. 2014-01-0825). SAE Technical Paper. doi:10.4271/2014-01-0825

Chen, A. Y., Ruan, H. H., Wang, J., Chan, H. L., Wang, Q., Li, Q., & Lu, J. (2011). The influence of strain rate on the microstructure transition of 304 stainless steel. *Acta Materialia*, *59*(9), 3697–3709. doi:10.1016/j.actamat.2011.03.005

Chen, C. H., Gau, J. T., & Lee, R. S. (2009). An experimental and analytical study on the limit drawing ratio of stainless steel 304 foils for microsheet forming. *Materials and Manufacturing Processes*, *24*(12), 1256–1265. doi:10.1080/10426910903129786

Cheng, Y.-C., Lin, D.-H., & Jiang, C.-P. (2016). Micromotion Improvement and Applications for Abutment-Implant System by Uniform Design and Kriging Interpolation. *International Journal of Bioscience, Biochemistry, Bioinformatics*, *6*(2), 41–49. doi:10.17706/ijbbb.2016.6.2.41-49

Cheng, Y.-C., Lin, D.-H., Jiang, C.-P., & Lee, S.-Y. (2015). Design improvement and dynamic finite element analysis of novel ITI dental implant under dynamic chewing loads. *Bio-Medical Materials and Engineering*, *26*(s1), S555–S561. doi:10.3233/BME-151346 PMID:26406049

Chen, J., Yu, X., Miller, R. G., & Feng, Z. (2015). In situ strain and temperature measurement and modelling during arc welding. *Science and Technology of Welding and Joining*, *20*(3), 181–188. doi:10.1179/1362171814Y.0000000270

Choi, J., & Bryant, M. (2002). Combining lumped parameter bond graphs with finite element shafts in a gearbox model. *Computer Modeling in Engineering & Sciences*, *3*(4), 431–446.

Chong, L., Khocht, A., Suzuki, J. B., & Gaughan, J. (2009). Effect of implant design on initial stability of tapered implants. *The Journal of Oral Implantology*, *35*(3), 130–135. doi:10.1563/1548-1336-35.3.130 PMID:19579524

Chopra, I. (1996). Review of current status of smart structures and integrated systems. *SPIE Smart Structures and Integrated Systems*, *17*, 20–62.

Cohodar, M. (2005). *Modelling and simulation of robot system with flexible links using bond graphs* (PhD thesis). University of Sarajevo.

Cohodar, M., Borutzky, W., & Damic, V. (2009). Comparison of different formulations of 2D beam elements based on bond graph technique. *Simulation Modelling Practice and Theory*, *17*(1), 107–124. doi:10.1016/j.simpat.2008.02.014

Cook, J., O'keefe, M., Smith, D. J., & Stobbs, W. (1983). The high resolution electron microscopy of stacking defects in Cu–Zn–Al shape memory alloy. *Journal of Microscopy*, *129*(3), 295–306. doi:10.1111/j.1365-2818.1983.tb04186.x

Courtney, T. H. (1990). *Mechanical behavior of materials*. New York, NY: McGraw-Hill.

Craig, R. G., & Peyton, F. A. (1965). Measurement of stresses in fixed-bridge restorations using a brittle coating technique. *Journal of Dental Research*, *44*(4), 756–762. doi:10.1177/00220345650440042201 PMID:14321410

Cullity, B. D., & Stock, S. R. (2001). *Elements of X-ray Diffraction* (Vol. 3). Prentice Hall.

Curie, J., & Curie, P. (1880). Development by pressure of polar electricity in hemihedral crystals with inclined faces. *Bull. soc. min. de France, 3*, 90.

Dally, J. W., & Rilley, W. F. Mc Connell and Kenneth G. (1993). Instrumentation for Engineering measurements (2nd ed.). John Wiley &Son.

Damić, V. (2006). Modelling flexible body systems: A bond graph component model approach. *Mathematical and Computer Modelling of Dynamical Systems*, *12*(2-3), 175–187. doi:10.1080/13873950500068757

Damic, V., & Cohodar, M. (2005). A bond graph approach to modelling of spatial flexible multibody system based on co-rotational formulation. *Proceedings of the 2005 International Conference on Bond Graph Modelling and Simulation.*

David, V. H. (2004). *Fundamentals of finite element analysis*. NewYork: McGraw-Hill.

Deheeger, A., Badulescu, C., Mathias, J. D., & Grédiac, M. (2009). Experimental study of thermal stresses in a bonded joint. *Journal of Physics: Conference Series*, *181*(1), 012041. doi:10.1088/1742-6596/181/1/012041

Depuydt, D., Hendrickx, K., Biesmans, W., Ivens, J., & Van Vuure, A. W. (2017). Digital image correlation as a strain measurement technique for fibre tensile tests. *Composites. Part A, Applied Science and Manufacturing*, *99*, 76–83. doi:10.1016/j.compositesa.2017.03.035

Derkaoui, A., Bideaux, E., & Scavarda, S. (2005). Finite element local structure assembly and shape functions influence on bond graph modelling. *International Conference on Bond Graph Modelling and Simulation (ICBGM'05).*

DeWald, A. T., & Hill, M. R. (2003). Improved data reduction for the deep-hole method of residual stress measurement. *Journal of Strain Analysis for Engineering Design*, *38*(1), 65–77. doi:10.1243/030932403762671908

Dhatrak, P., Shirsat, U., Sumanth, S., & Deshmukh, V. (2018). Finite element analysis and experimental investigations on stress distribution of dental implants around implant-bone interface. *Materials Today: Proceedings*, *5*(2), 5641–5648.

Dorf, R. C. (2003). *CRC Handbook of Engineering Tables*. CRC Press. doi:10.1201/9780203009222

Dour, G., Medjedoub, F., Leroux, S., Diaconu, G., & Rezai-Aria, F. (2004). Normalized thermal stresses analysis to design a thermal fatigue experiment. *Journal of Thermal Stresses*, *28*(1), 1–16. doi:10.1080/014957390523651

Doyle, J. F., Phillips, J. W., & Post, D. (1989). *Manual on Experimental Stress Analysis*. Society for Experimental Mechanics.

Duryea, G. R., & Martin, J. F. (1968). An improved piezoelectric balance for aerodynamic force. *IEEE Transactions on Aerospace and Electronic Systems*, *3*(3), 351–359. doi:10.1109/TAES.1968.5408988

Escamilla, I., Zapata, O., Gonzalez, B., Gamez, N., & Guerrero, M. (2010). 3D finite element simulation of the milling process of a Ti-6AL-4V alloy. *Simulia Customer Conference*.

Furgiuele, F. M., Pagnotta, L., & Poggialini, A. (1991). Measuring residual stresses by hole-drilling and coherent optics techniques: A numerical calibration. *Journal of Engineering Materials and Technology*, *113*(1), 41–50. doi:10.1115/1.2903381

Fu, Y., Ge, E., Su, H., Xu, J., & Li, R. (2015). Cold expansion technology of connection holes in aircraft structures: A review and prospect. *Chinese Journal of Aeronautics*, *28*(4), 961–973. doi:10.1016/j.cja.2015.05.006

Füzessy, Z. (1977). Methods of holographic interferometry for industrial measurements. *Periodica Polytechnica Mechanical Engineering*, *21*(3-4), 257–263.

Gao, C., & Zhang, L. (2013). Effect of cutting conditions on the serrated chip formation in high-speed cutting. *Machining Science and Technology: An International Journal*, *17*(1), 26–40. doi:10.1080/10910344.2012.747887

Gautschi, G. (2002). Piezoelectric sensors. In *Piezoelectric Sensorics* (pp. 73–91). Berlin: Springer. doi:10.1007/978-3-662-04732-3_5

Gerdeen, J. C., & Daudi, A. (1983). Formability Limits of Relatively Thick Sheet Metal as Determined by the Fukui Conical Cup Test. *Journal of Engineering for Industry*, *105*(4), 276–281. doi:10.1115/1.3185899

Gilat, A., Kuokkala, V. T., Seidt, J. D., & Smith, J. L. (2017). Full-field measurement of strain and temperature in quasi-static and dynamic tensile tests on stainless steel 316L. *Procedia Engineering*, *207*, 1994–1999. doi:10.1016/j.proeng.2017.10.1057

Goldrein, H. T., Palmer, S. J. P., & Huntley, J. M. (1995). Automated fine grid technique for measurement of large-strain deformation maps. *Optics and Lasers in Engineering*, *23*(5), 305–318. doi:10.1016/0143-8166(95)00036-N

Granda, J., Nguyen, L., & Hundal, S. (2011). Modeling the Completed Space Station a Three Dimensional Rigid-Flexible Dynamic Model to Predict Modes of Vibration and Stress Analysis. In Infotech@ Aerospace 2011 (p. 1535). doi:10.2514/6.2011-1535

Grediac, M., Sur, F., & Blaysat, B. (2016). The Grid Method for In-plane Displacement and Strain Measurement: A Review and Analysis. *Strain*, *52*(3), 205–243. doi:10.1111tr.12182

Grolleau, V., Gary, G., & Mohr, D. (2008). Biaxial testing of sheet materials at high strain rates using viscoelastic bars. *Experimental Mechanics*, *48*(3), 293–303. doi:10.100711340-007-9073-5

Gul, I., Maqsood, A., Naeem, M., & Ashiq, M. N. (2010). Optical, magnetic and electrical investigation of cobalt ferrite nanoparticles synthesized by co-precipitation route. *Journal of Alloys and Compounds*, *507*(1), 201–206. doi:10.1016/j.jallcom.2010.07.155

Gupta, N. K., Iqbal, M. A., & Sekhon, G. S. (2006). Experimental and numerical studies on the behavior of thin aluminum plates subjected to impact by blunt-and hemispherical-nosed projectiles. *International Journal of Impact Engineering*, *32*(12), 1921–1944. doi:10.1016/j.ijimpeng.2005.06.007

Gupta, T. K. (1990). Application of zinc oxide varistors. *Journal of the American Ceramic Society*, *73*(7), 1817–1840. doi:10.1111/j.1151-2916.1990.tb05232.x

Haines, D. J., & Wright, G. P. (1969). An experimental method of determining thermal strains. *Experimental Mechanics*, *9*(7), 327–331. doi:10.1007/BF02325139

Hannah, R. L., & Reed, S. E. (1992). *Strain Gage users Manual*. Society for Experimental Mechanics.

Hannah, R. L., & Reed, S. E. (Eds.). (1992). *Strain gauge users' handbook*. Springer Science & Business Media.

Harbour, J. R., & Hair, M. L. (1979). Radical intermediates in the photosynthetic generation of hydrogen peroxide with aqueous zinc oxide dispersions. *Journal of Physical Chemistry*, *83*(6), 652–656. doi:10.1021/j100469a003

Hearn, E. J. (1971). *Brittle Lacquers for Strain Measurement*. Watford, UK: Merrow Publishing Co.

Hedayati, A., Najafizadeh, A., Kermanpur, A., & Forouzan, F. (2010). The effect of cold rolling regime on microstructure and mechanical properties of AISI 304L stainless steel. *Journal of Materials Processing Technology*, *210*(8), 1017–1022. doi:10.1016/j.jmatprotec.2010.02.010

Hensley, S., Christensen, M., Small, S., Archer, D., Lakes, E., & Rogge, R. (2017). Digital image correlation techniques for strain measurement in a variety of biomechanical test models. *Acta of Bioengineering and Biomechanics*, *19*(3). PMID:29205227

Hertl, M., Fayolle, R., Weidmann, D., & Lecomte, J. C. (2006). Thermal Stress Failures: A New Experimental Approach For Prediction and Prevention. THERMINIC 2006, 169-174.

Hetnarski, R. B., Eslami, M. R., & Gladwell, G. M. L. (2009). *Thermal stresses: Advanced theory and applications* (Vol. 41). New York, NY: Springer.

Higson, G. (1964). Recent advances in strain gauges. *Journal of Scientific Instruments*, *41*(7), 405–414. doi:10.1088/0950-7671/41/7/301

Hiroki, C., Yamamoto, T., Takeda, Y., Suzuki, A., & Sakuma, T. (2010). Exploitation of shape memory alloy actuator using resistance feedback control and its development. *Progress in Natural Science: Materials International*, *20*, 97–103. doi:10.1016/S1002-0071(12)60013-6

Holister G. S. (1967). *Experimental Stress Analysis* (C.V.P) 63rd yr. 2, 33-41,

Holman, J. P. (1966). *Experimental methods for engineers*. McGraw Hill.

Hosford, W. F., & Caddell, R. M. (2011). *Metal forming: mechanics and metallurgy*. Cambridge University Press. doi:10.1017/CBO9780511976940

Hotchandani, S., & Kamat, P. V. (1992). Photoelectrochemistry of semiconductor ZnO particulate films. *Journal of the Electrochemical Society*, *139*(6), 1630–1634. doi:10.1149/1.2069468

Huang, G. L., Matlock, D. K., & Krauss, G. (1989). Martensite formation, strain rate sensitivity, and deformation behavior of type 304 stainless steel sheet. *Metallurgical Transactions. A, Physical Metallurgy and Materials Science*, *20*(7), 1239–1246. doi:10.1007/BF02647406

Huang, X., Liu, Z., & Xie, H. (2013). Recent progress in residual stress measurement techniques. *Guti Lixue Xuebao*, *26*(6), 570–583.

Hung, Y. Y. (1982). Shearography: A new optical method for strain measurement and nondestructive testing. *Optical Engineering (Redondo Beach, Calif.)*, *21*(3), 391–395. doi:10.1117/12.7972920

Hwang, Y. L. (2005). A new approach for dynamic analysis of flexible manipulator systems. *International Journal of Non-linear Mechanics*, *40*(6), 925–938. doi:10.1016/j.ijnonlinmec.2004.12.001

Inman, D. J., Dosch, J. J., & Garcia, E. (1992). A self-sensing piezoelectric actuator for collocated control. *Journal of Intelligent Material Systems and Structures*, *3*(1), 166–185. doi:10.1177/1045389X9200300109

Irfan, H., Racik, K. M., & Anand, S. (2018). Microstructural evaluation of CoAl2O4 nanoparticles by Williamson–Hall and size–strain plot methods. *Journal of Asian Ceramic Societies*, *6*(1), 54–62. doi:10.1080/21870764.2018.1439606

Iwata, K., & Moriwaki, T. (1978). Cutting state identification and in-process tool wear sensing by acoustic emission analysis. *Bulletin of the Japan Society of Precision Engineering*, *12*(4), 213–214.

Jacquot, P. (2008). Speckle interferometry: A review of the principal methods in use for experimental mechanics applications. *Strain*, *44*(1), 57–69. doi:10.1111/j.1475-1305.2008.00372.x

Jamal, F., Ahmadi, F., Khanzadeh, M., & Malekzadeh, S. (2019). Application of image processing in optical method, moire deflectometry for investigating the optical properties of zinc oxide nanoparticle. *Optic and Laser Technology.* doi:10.13140/RG.2.2.23059.32804

Jenkins, C., & Khanna, S. (2005). *Mechanics of materials*. Amsterdam: Elsevier.

Jerrett, R. S. (1945). The acoustic strain gauge. *Journal of Scientific Instruments*, *22*(2), 29–34. doi:10.1088/0950-7671/22/2/303

Jiang, T., Ryu, S. K., Zhao, Q., Im, J., Huang, R., & Ho, P. S. (2013). Measurement and analysis of thermal stresses in 3D integrated structures containing through-silicon-vias. *Microelectronics and Reliability*, *53*(1), 53–62. doi:10.1016/j.microrel.2012.05.008

Jin, X., & Altintas, Y. (2012). Prediction of micro-milling forces with finite element method. *Journal of Materials Processing Technology*, *212*(3), 542–552. doi:10.1016/j.jmatprotec.2011.05.020

John, M. R. S., Shrivastava, K., Banerjee, N., Madhukar, P. D., & Vinayagam, B. K. (2013). Finite element method-based machining simulation for analyzing surface roughnessduring turning operation with HSS and carbide insert tool. *Arabian Journal for Science and Engineering*, *38*(6), 1615–1623. doi:10.100713369-013-0541-1

Johns, D. J. (2013). *Thermal stress analyses* (Vol. 2223). Elsevier.

Justusson, B., Pankow, M., Heinrich, C., Rudolph, M., & Waas, A. M. (2013). Use of a shock tube to determine the biaxial yield of an aluminum alloy under high rates. *International Journal of Impact Engineering*, *58*, 55–65. doi:10.1016/j.ijimpeng.2013.01.012

Karl, H. (2000). *An Introduction to Measurements using strain gauges*. Druckerei Drach Press.

Karl, M., Graef, F., & Winter, W. (2015). *Determination of Micromotion at the Implant Bone Interface–An: Vitro.* Academic Press.

Karthik Selva Kumar, K., & Kumaraswamidhas, L. A. (2018). Investigation on stability of an elastically mounted circular tube under cross flow in inline square arrangement. *Iranian Journal of Science and Technology, Transactions of Mechanical Engineering,* 1-13.

Karthik Selva Kumar, K., & Kumaraswamidhas, L. A. (2018). Investigation on stability of an elastically mounted circular tube under cross flow in inline square arrangement. *Iranian Journal of Science and Technology: Transactions of Mechanical Engineering,* 1-13.

Karthik Selva Kumar, K., & Kumaraswamidhas, L.A. (2018). Investigation on Stability of an Elastically Mounted Circular Tube under Cross Flow in Inline Square Arrangement. *Iranian Journal of Science and Technology, Transactions of Mechanical Engineering,* 1-13.

Karthik Selva Kumar, K., & Kumaraswamidhas, L. A. (2015). Experimental investigation on flow-induced vibration excitation in an elastically mounted circular cylinder in arrays. *Fluid Dynamics Research, 47*(1), 015508. doi:10.1088/0169-5983/47/1/015508

Karthik Selva Kumar, K., & Kumaraswamidhas, L. A. (2015). Experimental investigation on flow-induced vibration excitation in an elastically mounted square cylinder. *Journal of Vibroengineering, 17*(1), 468–477.

Karthik Selva Kumar, K., & Kumaraswamidhas, L. A. (2015). Experimental investigation on Flow-Induced Vibration Excitation in an Elastically Mounted Square Cylinder. *Journal of Vibroengineering, 17*(1), 468–477.

Karthik Selva Kumar, K., & Kumaraswamidhas, L. A. (2015b). Experimental investigation on flow-induced vibration excitation in an elastically mounted square cylinder. *Journal of Vibroengineering, 17*(1), 468–477.

Kasahara, N., Jinbo, M., & Hosogai, H. (2003). *Mitigation method of thermal transient stress by thermalhydraulic-structure total analysis.* Academic Press.

Kasani, R., Attili, B. K. R. S., Dommeti, V. K., Merdji, A., Biswas, J. K., & Roy, S. (2019). Stress distribution of overdenture using odd number implants–A Finite Element Study. *Journal of the Mechanical Behavior of Biomedical Materials, 98*, 369–382. doi:10.1016/j.jmbbm.2019.06.030 PMID:31326699

Kayabaşı, O., Yüzbasıoğlu, E., & Erzincanlı, F. (2006). Static, dynamic and fatigue behaviors of dental implant using finite element method. *Advances in Engineering Software, 37*(10), 649–658. doi:10.1016/j.advengsoft.2006.02.004

Kaya, S. (2016). Nonisothermal warm deep drawing of SS304: FE modeling and experiments using servo press. *International Journal of Advanced Manufacturing Technology, 83*(5-8), 1047–1056. doi:10.100700170-015-7620-2

Keeler, S. P. (1989). On the Origins of the FLD. *The Minerals, Metals & Materials Society, Forming Limit Diagrams: Concepts, Methods, and Applications*, 3-8.

Keil, S. (2017). *Technology and practical use of strain gauges: with particular consideration of stress analysis using strain gauges.* John Wiley & Sons. doi:10.1002/9783433606667

Kemper, B., & Von Bally, G. (2008). Digital holographic microscopy for live cell applications and technical inspection. *Applied Optics, 47*(4), A52–A61. doi:10.1364/AO.47.000A52 PMID:18239699

Khanna, S. K., He, C., & Agrawal, H. N. (2001). Residual stress measurement in spot welds and the effect of fatigue loading on redistribution of stresses using high sensitivity moiré interferometry. *Journal of Engineering Materials and Technology, 123*(1), 132–138. doi:10.1115/1.1286218

Kim, D., Beal, A., & Kwon, P. (2016). Effect of tool wear on hole quality in drilling of carbon fiber reinforced plastic–titanium alloy stacks using tungsten carbide and polycrystalline diamond tools. *Journal of Manufacturing Science and Engineering*, *138*(3), 031006. doi:10.1115/1.4031052

Kim, J. H. J., Jeon, S. E., & Kim, J. K. (2002). Development of new device for measuring thermal stresses. *Cement and Concrete Research*, *32*(10), 1645–1651. doi:10.1016/S0008-8846(02)00842-6

Kim, S. B., Huh, H., Bok, H. H., & Moon, M. B. (2011). Forming limit diagram of auto-body steel sheets for high-speed sheet metal forming. *Journal of Materials Processing Technology*, *211*(5), 851–862. doi:10.1016/j.jmatprotec.2010.01.006

Kim, S. H., Kim, J. B., & Lee, W. J. (2009). Numerical prediction and neutron diffraction measurement of the residual stresses for a modified 9Cr–1Mo steel weld. *Journal of Materials Processing Technology*, *209*(8), 3905–3913. doi:10.1016/j.jmatprotec.2008.09.012

Kistler Instrument Corporation. (2007). *Acceleration. Piezotron sensor. Acoustic emission sensor.* Accessed August 1, 2019, from http://www.intertechnology.com/Kistler/pdfs/Accelerometer_Model_8152B.pdf

Koç, M., Billur, E., & Cora, Ö. N. (2011). An experimental study on the comparative assessment of hydraulic bulge test analysis methods. *Materials & Design*, *32*(1), 272–281. doi:10.1016/j.matdes.2010.05.057

Kreis, T. (2006). *Handbook of holographic interferometry: Optical and digital methods*. John Wiley & Sons.

Krishnaraj, V., Prabukarthi, A., Santhosh, M., Senthilkumar, M., & Zitoune, R. (2012b, June). Optimization of machining parameters in CFRP/Ti stacks drilling. In *ASME 2012 International Manufacturing Science and Engineering Conference collocated with the 40th North American Manufacturing Research Conference and in participation with the International Conference on Tribology Materials and Processing* (pp. 269–275). American Society of Mechanical Engineers. 10.1115/MSEC2012-7216

Krishnaraj, V., Prabukarthi, A., Ramanathan, A., Elanghovan, N., Kumar, M. S., Zitoune, R., & Davim, J. P. (2012a). Optimization of machining parameters at high-speed drilling of carbon fiber reinforced plastic (CFRP) laminates. *Composites. Part B, Engineering*, *43*(4), 179–1799. doi:10.1016/j.compositesb.2012.01.007

Kug, W. K., Woo, Y. L., & Hyo-chol, S. (1999). A finite element analysis for the characteristics of temperature and stress in micro-machining considering the size effect. *International Journal of Machine Tools & Manufacture*, *39*(9), 1507–1524. doi:10.1016/S0890-6955(98)00071-6

Kumar, K. S., & Kumaraswamidhas, L. A. (2018). Investigation on Stability of an Elastically Mounted Circular Tube under Cross Flow in Inline Square Arrangement. *Iranian Journal of Science and Technology, Transactions of Mechanical Engineering,* 1-13.

Kumar, P., LeBlanc, J., & Shukla, A. (2011). Effect of curvature on shock loading response of aluminum panels. In Dynamic Behavior of Materials (vol. 1, pp. 369-374). doi:10.1007/978-1-4614-0216-9_51

Kumaraswamidhas, L. A., & Karthik Selva Kumar, K. (2016). Experimental investigation on stability of an elastically mounted circular tube under cross flow in normal triangular arrangement. *Journal of Vibroengineering*, *18*(3), 1824–1838. doi:10.21595/jve.2016.16708

Leblouba, M., Altoubat, S., Ekhlasur Rahman, M., & Palani Selvaraj, B. (2015). Elliptical leaf spring shock and vibration mounts with enhanced damping and energy dissipation capabilities using lead spring. *Shock and Vibration*, *12*. doi:10.1155/2015/482063

Lee, C-K., & O'Sullivan, T., & Chiang. (1991). Piezoelectric strain rate sensor and actuator designs for active vibration control. *Structures, Structural Dynamics, and Materials Conference, 132*, 1064.

Lee, R. S., & Hsu, Q. C. (1994). Image-processing system for circular-grid analysis in sheet-metal forming. *Experimental Mechanics, 34*(2), 108–115. doi:10.1007/BF02325707

Lee, W. S., & Lin, C. F. (2001). Effects of prestrainning on the impact response and twinning structure of 304L stainless steel. *Materials Transactions, 42*(10), 2080–2086. doi:10.2320/matertrans.42.2080

Leggatt, R. H., Smith, D. J., Smith, S. D., & Faure, F. (1996). Development and experimental validation of the deep hole method for residual stress measurement. *Journal of Strain Analysis for Engineering Design, 31*(3), 177–186. doi:10.1243/03093247V313177

Lei, J. F., & Will, H. A. (1998). Thin-film thermocouples and strain-gauge technologies for engine applications. *Sensors and Actuators. A, Physical, 65*(2-3), 187–193. doi:10.1016/S0924-4247(97)01683-X

Leo Kumar, S. P., Jerald, J., Kumanan, S., & Prabakaran, R. (2014). Review on current research aspects in tool based micromachining processes. *Materials and Manufacturing Processes, 2*(11-12), 1291–1337. doi:10.1080/10426914.2014.952037

Liancheng, Z., & Wei, C. (1997). Deformation-induced martensitic transformation and its reversibility in ni-ti-nb wide hysteresis shape memory alloys. *Chin Shu Hsueh Pao, 33*(1), 90–98.

Lichtenfeld, J. A., Van Tyne, C. J., & Mataya, M. C. (2006). Effect of strain rate on stress-strain behavior of alloy 309 and 304L austenitic stainless steel. *Metallurgical and Materials Transactions. A, Physical Metallurgy and Materials Science, 37*(1), 147–161. doi:10.100711661-006-0160-5

Limbert, G., van Lierde, C., Muraru, O. L., Walboomers, X. F., Frank, M., Hansson, S., ... Jaecques, S. (2010). Trabecular bone strains around a dental implant and associated micromotions—A micro-CT-based three-dimensional finite element study. *Journal of Biomechanics, 43*(7), 1251–1261. doi:10.1016/j.jbiomech.2010.01.003 PMID:20170921

Lin, H., Wu, S.-K., Chou, T., & Kao, H. (1991). The effects of cold rolling on the martensitic transformation of an equiatomic TiNi alloy. *Acta Metallurgica et Materialia, 39*(9), 2069–2080. doi:10.1016/0956-7151(91)90177-3

List, G., Sutter, G., & Bouthiche, A. (2012). Cutting temperature prediction in high speedmachining by numerical modeling of chip formation and its dependence with craterwear. *International Journal of Machine Tools & Manufacture, 54*, 1–9. doi:10.1016/j.ijmachtools.2011.11.009

Li, T., Hu, K., Cheng, L., Ding, Y., Ding, Y., Shao, J., & Kong, L. (2011). Optimum selection of the dental implant diameter and length in the posterior mandible with poor bone quality–A 3D finite element analysis. *Applied Mathematical Modelling, 35*(1), 446–456. doi:10.1016/j.apm.2010.07.008

Liu, D. H., Yu, H. P., & Li, C. F. (2012). Comparative study of the microstructure of 5052 aluminum alloy sheets under quasi-static and high-velocity tension. *Materials Science and Engineering A, 551*, 280–287. doi:10.1016/j.msea.2012.05.018

Liu, Y., & Galvin, S. (1997). Criteria for pseudoelasticity in near-equiatomic NiTi shape memory alloys. *Acta Materialia, 45*(11), 4431–4439. doi:10.1016/S1359-6454(97)00144-4

Li, X., & Dong, F. (2017). Three-dimensional finite element stress analysis of uneven-threaded ti dental implant. *International Journal of Clinical and Experimental Medicine, 10*(1), 307–315.

Logesh, K., Raja, V. B., & Velu, R. (2015). Experimental investigation for characterization of formability of epoxy based fiber metal laminates using Erichsen cupping test method. *Indian Journal of Science and Technology*, *8*(33), 1–7. doi:10.17485/ijst/2015/v8i33/72244

Lord, J. D., Penn, D., & Whitehead, P. (2008). The application of digital image correlation for measuring residual stress by incremental hole drilling. In Applied Mechanics and Materials (Vol. 13, pp. 65-73). Trans Tech Publications. doi:10.4028/www.scientific.net/AMM.13-14.65

Macedo, J., Pereira, J., Faria, J., Pereira, C., Alves, J., Henriques, B., ... López-López, J. (2017). Finite element analysis of stress extent at peri-implant bone surrounding external hexagon or Morse taper implants. *Journal of the Mechanical Behavior of Biomedical Materials*, *71*, 441–447. doi:10.1016/j.jmbbm.2017.03.011 PMID:28499606

MacLauchlan, D. T., & Alers, G. A. (1987). Measurement of thermal stress in railroad rails using ultrasonic SH waves. In *Review of Progress in Quantitative Nondestructive Evaluation*. Boston, MA: Springer. doi:10.1007/978-1-4613-1893-4_175

Makino, A., Nelson, D. V., Fuchs, E. A., & Williams, D. R. (1996). Determination of biaxial residual stresses by a holographic-hole drilling technique. *Journal of Engineering Materials and Technology*, *118*(4), 583–588. doi:10.1115/1.2805960

Makkouk, R., Bourgeois, N., Serri, J., Bolle, B., Martiny, M., Teaca, M., & Ferron, G. (2008). Experimental and theoretical analysis of the limits to ductility of type 304 stainless steel sheet. *European Journal of Mechanics. A, Solids*, *27*(2), 181–194. doi:10.1016/j.euromechsol.2007.04.004

Martínez-García, V., Wenzelburger, M., Killinger, A., Pedrini, G., Gadow, R., & Osten, W. (2014). Residual Stress Measurement with Laser-Optical and Mechanical Methods. *Advanced Materials Research*, 996.

Maurizio, C., El Habra, N., Rossetto, G., Merlini, M., Cattaruzza, E., Pandolfo, L., & Casarin, M. (2010). XAS and GIXRD study of Co sites in CoAl2O4 layers grown by MOCVD. *Chemistry of Materials*, *22*(5), 1933–1942. doi:10.1021/cm9018106

McDonach, A., McKelvie, J., MacKenzie, P., & Walker, C. A. (1983). Improved moiré interferometry and applications in fracture mechanics, residual stress. and damaged composites. *Experimental Techniques*, *7*(6), 20–24. doi:10.1111/j.1747-1567.1983.tb01766.x

McGee, T. D. (1988). *Principles and methods of temperature measurement*. John Wiley & Sons.

Merzouki, R., Samantaray, A. K., Pathak, P. M., & Bouamama, B. O. (2012). *Intelligent mechatronic systems: Modelling, control and diagnosis*. Springer Science & Business Media.

Meuleman, D. J., Siles, J. L., & Zoldak, J. J. (1985). *The limiting dome height test for assessing the formability of sheet steel* (No. 850005). SAE Technical Paper.

Mimani, T. (2001). Instant synthesis of nanoscale spinel aluminates. *Journal of Alloys and Compounds*, *315*(1-2), 123–128. doi:10.1016/S0925-8388(00)01262-7

Min, F., Yang, Y., Qiu, H., Zhong, S., & Pi, X. (2018). Hypersonic aerodynamics measurement with fiber-optic balance based on MEMS Fabry-Perot strain sensor. *Fiber Optic Sensing and Optical Communication*, 10849.

Mitra, P., Chatterjee, A. P., & Maiti, H. S. (1998). ZnO thin film sensor. *Materials Letters*, *35*(1-2), 33–38. doi:10.1016/S0167-577X(97)00215-2

Molinari, A., Cheriguene, R., & Miguelez, H. (2011). Numerical and analytical modeling oforthogonal cutting: The link between local variables and global contact characteristics. *International Journal of Mechanical Sciences*, *53*(3), 183–206. doi:10.1016/j.ijmecsci.2010.12.007

Morimoto, Y., Seguchi, Y., & Inatani, T. (1988). Strain Analysis by Grid Method and its Automization Using Image Processing. *Nondestructive Testing Communications*, *4*(2-3), 74. doi:10.1080/02780898808962125

Mosavar, A., Ziaei, A., & Kadkhodaei, M. (2015). The effect of implant thread design on stress distribution in anisotropic bone with different osseointegration conditions: A finite element analysis. *International Journal of Oral & Maxillofacial Implants*, *30*(6), 1317–1326. doi:10.11607/jomi.4091 PMID:26478976

Mote, V., Purushotham, Y., & Dole, B. (2012). Williamson-Hall analysis in estimation of lattice strain in nanometer-sized ZnO particles. *Journal of Theoretical and Applied Physics*, *6*(1), 6. doi:10.1186/2251-7235-6-6

Motra, H. B., Hildebrand, J., & Dimmig-Osburg, A. (2014). Assessment of strain measurement techniques to characterise mechanical properties of structural steel. *Engineering Science and Technology, an International Journal, 17*(4), 260-269.

Moyer, R. G., & Gillespie, G. E. (1972). *Displacement measurement by holographic interferometry* (No. AECL--4129). Academic Press.

Mukherjee, A., & Karmakar, R. (2000). *Modelling and simulation of engineering systems through bondgraphs*. Alpha Science International Ltd.

Munoz-Sanchez, A., Canteli, J. A., Cantero,J. L., &Miguelez, M. H. (2011). Numericalanalysis of the tool wear effect in the machining induced residual stress. *Journal of Simulation Modeling Practice and Theory, 19*, 872–886.

Mynors, D. J., & Zhang, B. (2002). Applications and capabilities of explosive forming. *Journal of Materials Processing Technology*, *125*, 1–25. doi:10.1016/S0924-0136(02)00413-2

Nanda, S. R., Agarwal, S., Kulkarni, V., & Sahoo, N. (2017). Shock Tube as an Impulsive Application Device. *International Journal of Aerospace Engineering*.

Nanda, S. R., Kulkarni, V., & Sahoo, N. (2017). Apt strain measurement technique for impulsive loading applications. *Measurement Science & Technology*, *28*(3). doi:10.1088/1361-6501/aa577a

Nasr, M. N. A., & Elbestawi, M. A. (2008). A modified time-efficient FE approach for predicting machining induced residual stress. *Journal of Finite Element in Analysis and Design*, *44*(4), 149–161. doi:10.1016/j.finel.2007.11.005

Nelson, D. V., Makino, A., & Schmidt, T. (2006). Residual stress determination using hole drilling and 3D image correlation. *Experimental Mechanics*, *46*(1), 31–38. doi:10.100711340-006-5859-0

Nelson, D. V., & McCrickerd, J. T. (1986). Residual-stress determination through combined use of holographic interferometry and blind-hole drilling. *Experimental Mechanics*, *26*(4), 371–378. doi:10.1007/BF02320153

Neubert, H. K. P. (2000). *Strain Guages: Kinds and Uses*. Macmillian.

Neugebauer, R., Ben-Hanan, U., Ihlenfeldt, S., Wabner, M., & Stoll, A. (2012). Acoustic emission as a tool for identifying drill position in fiber-reinforced plastic and aluminum stacks. *International Journal of Machine Tools & Manufacture*, *57*, 20–26. doi:10.1016/j.ijmachtools.2012.01.013

Nicoletto, G. (1988). Theoretical fringe analysis for a coherent optics method of residual stress measurement. *Journal of Strain Analysis for Engineering Design*, *23*(4), 169–178. doi:10.1243/03093247V234169

Noda, N. (2018). *Thermal stresses*. Routledge. doi:10.1201/9780203735831

O'Sullivan, D., Sennerby, L., & Meredith, N. (2004). Influence of implant taper on the primary and secondary stability of osseointegrated titanium implants. *Clinical Oral Implants Research*, *15*(4), 474–480. doi:10.1111/j.1600-0501.2004.01041.x PMID:15248883

Okochi, M., Hashimoto, S., & Matsui, S. (1925). Tokyo. High-Speed Internal Combustion Engine and Piezoelectric Pressure Indicator. *Inst. Phys. Chem. Research*, *4*, 85–97.

Oliveira, D. A., Worswick, M. J., Finn, M., & Newman, D. (2005). Electromagnetic forming of aluminum alloy sheet: Free-form and cavity fill experiments and model. *Journal of Materials Processing Technology*, *170*(1-2), 350–362. doi:10.1016/j.jmatprotec.2005.04.118

Omega. (2018). *Practical Strain Gage Measurements*. Retrieved May 24, 2019, from https://www.omega.co.uk/techref/pdf/StrainGage_Measurement.pdf

Otsuka, K., & Ayman, C. M. W. (1998). *Shape Memory Materials*. Cambridge University Press.

Otsuka, K., & Ren, X. (2005). Physical metallurgy of Ti–Ni-based shape memory alloys. *Progress in Materials Science*, *50*(5), 511–678. doi:10.1016/j.pmatsci.2004.10.001

Ozturk, F., Dilmec, M., Turkoz, M., Ece, R. E., & Halkaci, H. S. (2009, June). Grid marking and measurement methods for sheet metal formability. *5th International Conference and Exhibition on Design and Production of MACHINES and DIES/MOLDS*, 41-49.

Paynter, H. (1961). Analysis and design of engineering systems. *Class Notes for MIT Course*, *2*, 751.

Pedrini, G., Pfister, B., & Tiziani, H. (1993). Double pulse-electronic speckle interferometry. *Journal of Modern Optics*, *40*(1), 89–96. doi:10.1080/09500349314550111

Piao, M., Miyazaki, S., & Otsuka, K. (1992). Characteristics of deformation and transformation in Ti44Ni47Nb9 shape memory alloy. *Materials Transactions, JIM, 33*(4), 346-353.

Piao, M., Otsuka, K., Miyazaki, S., & Horikawa, H. (1993). Mechanism of the As temperature increase by pre-deformation in thermoelastic alloys. *Materials Transactions, JIM, 34*(10), 919-929.

Polyaev, V. M., Genbach, A. N., & Genbach, A. A. (1992). An experimental study of thermal stress in porous materials by methods of holography and photoelasticity. *Experimental Thermal and Fluid Science*, *5*(6), 697–702. doi:10.1016/0894-1777(92)90113-J

Poutord, A., Rossi, F., Poulachon, G., M'Saoubi, R., & Abrivard, G. (2013). Local approach of wear in drilling Ti6Al4V/CFRP for stack modelling. *Procedia CIRP, 8*(1), 316–321.

Prabukarthi, A., Senthilkumar, M., & Krishnaraj, V. (2016). Study on drilling of CFRP/Ti6Al4V stack with modified twist drills using acoustic emission technique. *Steel and Composite Structures*, *21*(3), 573–588. doi:10.12989cs.2016.21.3.573

Prime, M., & Gonzales, A. (2000). The contour method: Simple 2-D mapping of residual stresses. *6th International Conference on Residual Stress Proceedings*, *1*, 617-624.

Qin, W., & Szpunar, J. (2005). Origin of lattice strain in nanocrystalline materials. *Philosophical Magazine Letters*, *85*(12), 649–656. doi:10.1080/09500830500474339

Ramakanth, K. (2007). *Basics of X-ray Diffraction and its Application.* IK.

Ramesh, K. (2018). *Experimental stress analysis*. Retrieved June 20, 2019, from https://nptel.ac.in/courses/112106068/

Ramesh, K. (2018). *Experimental Stress Analysis*. IIT Madras.

Ramezani, M., & Ripin, Z. M. (2010). Combined experimental and numerical analysis of bulge test at high strain rates using split Hopkinson pressure bar apparatus. *Journal of Materials Processing Technology*, *210*(8), 1061–1069. doi:10.1016/j.jmatprotec.2010.02.016

Ramulu, M., Branson, T., & Kim, D. (2001). A study on the drilling of composite and titanium stacks. *Composite Structures*, *54*(1), 67–77. doi:10.1016/S0263-8223(01)00071-X

Ravindra, H. V., Srinivasa, Y. G., & Krishnamurthy, R. (1997). Acoustic emission for tool condition monitoring in metal cutting. *Wear*, *212*(1), 78–84. doi:10.1016/S0043-1648(97)00137-3

Ravishankar, S. R., & Murthy, C. R. L. (2000). Application of acoustic emission in drilling of composite laminates. *NDT & E International*, *33*(6), 429–435. doi:10.1016/S0963-8695(00)00014-1

Raymond, C., &Granda, J. (2003). Using bond graphs for articulated, flexible multi-bodies, sensors. *Actuators, and Controllers with Application to the International Space Station. ICBG03.*

Ray, N., Jagadeesh, G., & Suwas, S. (2015). Response of shock wave deformation in AA5086 aluminum alloy. *Materials Science and Engineering A*, *622*, 219–227. doi:10.1016/j.msea.2014.10.010

Ribeiro, J., Monteiro, J., Lopes, H., & Vaz, M. (2011). Moire interferometry assessement of residual stress variation in depth on a shot peened surface. *Strain*, *47*, e542–e550. doi:10.1111/j.1475-1305.2009.00653.x

Robinson, M. J., Martinez Schramm, J., & Hannemann, K. (2011). Design and implementation of an internal stress wave force balance in a shock tunnel. *CEAS Space Journal*, *1*(1-4), 45–57. doi:10.100712567-010-0003-5

Rohatgi, A., Stephens, E. V., Davies, R. W., Smith, M. T., Soulami, A., & Ahzi, S. (2012). Electro-hydraulic forming of sheet metals: Free-forming vs. conical-die forming. *Journal of Materials Processing Technology*, *212*(5), 1070–1079. doi:10.1016/j.jmatprotec.2011.12.014

Roosbroeck, J. V., Vlekken, J., Voet, E., & Voet, M. (2009). A new methodology for fiber optic strain gauge measurements and its characterization. *Proceedings of the SENSOR+ TEST conference Opto,* 59-64.

Rosa, P. A. R., Martins, P. A. F., & Atkins, A. G. (2007). Revisiting the fundamentals of metalcutting by means of finite elements and ductile fracture mechanics. *International Journal of Machine Tools & Manufacture*, *47*(3–4), 607–617. doi:10.1016/j.ijmachtools.2006.05.003

Rossini, N. S., Dassisti, M., Benyounis, K. Y., & Olabi, A. G. (2012). Methods of measuring residual stresses in components. *Materials & Design*, *35*, 572–588. doi:10.1016/j.matdes.2011.08.022

Roy, S., Das, M., Chakraborty, P., Biswas, J. K., Chatterjee, S., & Khutia, N., & RoyChowdhury, A. (2017). Optimal selection of dental implant for different bone conditions based on the mechanical response. *Acta of Bioengineering and Biomechanics*, *19*(2). PMID:28869633

Roy, S., Dey, S., Khutia, N., Chowdhury, A. R., & Datta, S. (2018). Design of patient specific dental implant using FE analysis and computational intelligence techniques. *Applied Soft Computing*, *65*, 272–279. doi:10.1016/j.asoc.2018.01.025

Roy, S., Panda, D., Khutia, N., & Chowdhury, A. R. (2014). Pore geometry optimization of titanium (Ti6Al4V) alloy, for its application in the fabrication of customized hip implants. *International Journal of Biomaterials*. PMID:25400663

Rusinek, A., Rodríguez-Martínez, J. A., Arias, A., Klepaczko, J. R., & López-Puente, J. (2008). Influence of conical projectile diameter on perpendicular impact of thin steel plate. *Engineering Fracture Mechanics*, *75*(10), 2946–2967. doi:10.1016/j.engfracmech.2008.01.011

Sainthuile, T., Grondel, S., Delebarre, C., Godts, S., & Paget, C. (2012). Energy harvesting process modelling of an aeronautical structural health monitoring system using a bond-graph approach. *International Journal of Aerospace Sciences*, *1*(5), 2.

Salavati-Niasari, M., Davar, F., & Emadi, H. (2010). Hierarchical nanostructured nickel sulfide architectures through simple hydrothermal method in the presence of thioglycolic acid. *Chalcogenide Letters*, *7*(12), 647–655.

Saleem, Q., Wildman, R. D., Huntley, J. M., & Whitworth, M. B. (2003). A novel application of speckle interferometry for the measurement of strain distributions in semi-sweet biscuits. *Measurement Science & Technology*, *14*(12), 2027–2033. doi:10.1088/0957-0233/14/12/001

Scarsbrook, G., Cook, J., & Stobbs, W. (1984). The stabilization of martensite in Cu-Zn-AI shape memory alloys. *Metallurgical Transactions. A, Physical Metallurgy and Materials Science*, *15*(11), 1977–1986. doi:10.1007/BF02646832

Schelkens, P., Ebrahimi, T., Gilles, A., Gioia, P., Oh, K. J., Pereira, F., & Pinheiro, A. M. (2019). JPEG Pleno: Providing representation interoperability for holographic applications and devices. *ETRI Journal*, *41*(1), 93–108. doi:10.4218/etrij.2018-0509

Schiavo, F., Vigano, L., & Ferretti, G. (2006). Object-oriented modelling of flexible beams. *Multibody System Dynamics*, *15*(3), 263–286. doi:10.100711044-006-9012-8

Schulze, V., Autenrieth, H., Deuchert, M., & Weule, H. (2010). Investigation of surface near residual stress after microcutting by finite element simulation. *CIRP Annals–Manufacturing Technology, 59*, 117–120.

Sebastian, J. R., & Stubbs, D. A. (2001). Strain measurement using surface acoustic waves. AIP Conference Proceedings, 557(1), 1467-1471. doi:10.1063/1.1373926

Senthilkumar, M., Prabukarthi, A., & Krishnaraj, V. (2017). Multi-response optimisation in drilling of CFRP/Ti6Al4V stacks: A degree of similarity approach. *International Journal of Materials Engineering Innovation*, *8*(3–4), 222–249. doi:10.1504/IJMATEI.2017.090235

Senthilkumar, M., Prabukarthi, A., & Krishnaraj, V. (2018). Machining of CFRP/Ti6Al4V stacks under minimal quantity lubricating condition. *Journal of Mechanical Science and Technology*, *32*(8), 3787–3796. doi:10.100712206-018-0731-6

Seo, C. H., Shi, Y., Huang, S. W., Kim, K., & O'Donnell, M. (2011). Thermal strain imaging: A review. *Interface Focus*, *1*(4), 649–664. doi:10.1098/rsfs.2011.0010 PMID:22866235

Shadley, J. R., Rybicki, E. F., & Shealy, W. S. (1987). Application guidelines for the parting out step in a through thickness residual stress measurement procedure. *Strain*, *23*(4), 157–166. doi:10.1111/j.1475-1305.1987.tb00640.x

Shafi, A. A., Kadir, M. R. A., Sulaiman, E., Kasim, N. H. A., & Kassim, N. L. A. (2013). The effect of dental implant materials and thread profiles—A finite element and statistical study. *Journal of Medical Imaging and Health Informatics*, *3*(4), 509–513. doi:10.1166/jmihi.2013.1199

Shams, A., & Mashayekhi, M. (2012). Improvement of orthogonal cutting simulation with non-local damage model. *International Journal of Mechanical Sciences*, *61*(1), 88–96. doi:10.1016/j.ijmecsci.2012.05.008

Sharpe, J., & William, N. (Eds.). (2008). *Springer handbook of experimental solid mechanics.* Berlin: Springer. doi:10.1007/978-0-387-30877-7

Sharpe, W. N. Jr. (1968). The interferometric strain gauge. *Experimental Mechanics*, *8*(4), 164–170. doi:10.1007/BF02326343

Showa Measuring Instruments. (2018). *Strain Gauges*. Retrieved April 14, 2019, from http://www.showa-sokki.co.jp/English/index_e.html

Shrot, A., & Baker, M. (2012). Determination of Johnson-Cook parameters from machining simulations. *Journal of Computational Materials Science*, *52*(1), 298–304. doi:10.1016/j.commatsci.2011.07.035

Shyha, I., Soo, S. L., Aspinwall, D. K., Bradley, S., Dawson, S., & Pretorius, C. J. (2010). Drilling of titanium/CFRP/aluminum stacks. *Key Engineering Materials*, *447*, 624–633. doi:10.4028/www.scientific.net/KEM.447-448.624

Sima, M., & Ozel, T. (2010). Modified material constitutive models for serrated chipformation simulations and experimental validation in machining of titanium alloy Ti–6Al–4V. *International Journal of Machine Tools & Manufacture*, *50*(11), 943–960. doi:10.1016/j.ijmachtools.2010.08.004

Sirkis, J. S., & Lim, T. J. (1991). Displacement and strain measurement with automated grid methods. *Experimental Mechanics*, *31*(4), 382–388. doi:10.1007/BF02325997

Sirohi, J., & Chopra, I. (2000). Fundamental understanding of piezoelectric strain sensors. *Journal of Intelligent Material Systems and Structures*, *11*(4), 246–257. doi:10.1106/8BFB-GC8P-XQ47-YCQ0

Souza, N. E. A. (2008). *Computational methods for plasticity: theory and applications*. Willey Publications.

Spillman, W. B., Jr., & Weissman, E. M. (1998). *U.S. Patent No. 5,747,698*. Washington, DC: U.S. Patent and Trademark Office.

Stachowicz, F., Trzepieciński, T., & Pieja, T. (2010). Warm forming of stainless steel sheet. *Archives of Civil and Mechanical Engineering*, *10*(4), 85–94. doi:10.1016/S1644-9665(12)60034-X

Stern, F. B. (1979). Brittle coatings. *Experimental Mechanics*, *19*(6), 221–224. doi:10.1007/BF02324986

Stetson, K. A., & Brohinsky, W. R. (1985). Electrooptic holography and its application to hologram interferometry. *Applied Optics*, *24*(21), 3631–3637. doi:10.1364/AO.24.003631 PMID:18224099

Stoffel, M. (2007). Experimental validation of anisotropic ductile damage and failure of shock wave-loaded plates. *European Journal of Mechanics. A, Solids*, *26*(4), 592–610. doi:10.1016/j.euromechsol.2006.12.002

Stoffel, M., Schmidt, R., & Weichert, D. (2001). Shock wave-loaded plates. *International Journal of Solids and Structures*, *38*(42-43), 7659–7680. doi:10.1016/S0020-7683(01)00038-5

St-ograve, V. (1998). Force measurements in hypersonic impulse facilities. *AIAA Journal*, *36*(3), 342–348. doi:10.2514/2.402

Stout, M. B. (1960). *Basic electrical measurements*. Englewood Cliffs, NJ: Prentice-Hall.

Sugiura, T., Yamamoto, K., Horita, S., Murakami, K., & Kirita, T. (2018). Micromotion analysis of different implant configuration, bone density, and crestal cortical bone thickness in immediately loaded mandibular full-arch implant restorations: A nonlinear finite element study. *Clinical Implant Dentistry and Related Research*, *20*(1), 43–49. doi:10.1111/cid.12573 PMID:29214714

Suryanarayana, C. (2004). *Mechanical Alloying and Milling*. Marcel Dekker.

Suryanarayana, C., & Norton, M. G. (1998). *Practical aspects of X-ray diffraction X-Ray Diffraction*. Springer.

Sutton, M. A. (2008). Digital image correlation for shape and deformation measurements. In W. N. Sharpe (Ed.), *Springer handbook of experimental solid mechanics* (pp. 565–600). Berlin: Springer. doi:10.1007/978-0-387-30877-7_20

Takayama, K. (1983, October). Application of holographic interferometry to shock wave research. In *Industrial Applications of Laser Technology* (Vol. 398, pp. 174–181). International Society for Optics and Photonics. doi:10.1117/12.935372

Talyan, V., Wagoner, R. H., & Lee, J. K. (1998). Formability of stainless steel. *Metallurgical and Materials Transactions. A, Physical Metallurgy and Materials Science*, *29*(8), 2161–2172. doi:10.100711661-998-0041-1

Tebedge, N., Alpsten, G., & Tall, L. (1973). Residual-stress measurement by the sectioning method. *Experimental Mechanics*, *13*(2), 88–96. doi:10.1007/BF02322389

Tekalur, S. A., Bogdanovich, A. E., & Shukla, A. (2009). Shock loading response of sandwich panels with 3-D woven E-glass composite skins and stitched foam core. *Composites Science and Technology*, *69*(6), 736–753. doi:10.1016/j.compscitech.2008.03.017

Thalmann, R., & Dändliker, R. (1987). Strain measurement by heterodyne holographic interferometry. *Applied Optics*, *26*(10), 1964–1971. doi:10.1364/AO.26.001964 PMID:20454429

Thanongsak, T., & Ozel, T. (2013). Experimental and finite element simulation based investigation on micro-milling Ti-6Al-4V titanium alloy: Effects of cBN coating on tool wear. *Journal of Materials Processing Technology*, *213*, 535–542.

Toal, V. (2011). *Introduction to holography*. CRC Press. doi:10.1201/b11706

Tong, H., & Wayman, C. (1974). Characteristic temperatures and other properties of thermoelastic martensites. *Acta Metallurgica*, *22*(7), 887–896. doi:10.1016/0001-6160(74)90055-8

Ugural, A. C. (2010). *Stresses in beams, plates, and shells* (3rd ed.). Boca Raton, FL: CRC Press.

Umbrello, D. (2008). Finite element simulation of conventional and high speed machiningof Ti6Al4V alloy. *Journal of Materials Processing Technology*, *196*(1-3), 79–87. doi:10.1016/j.jmatprotec.2007.05.007

Ungár, T. (2007). Characterization of nanocrystalline materials by X-ray line profile analysis. *Journal of Materials Science*, *42*(5), 1584–1593. doi:10.100710853-006-0696-1

Vandenrijt, J. F., & Georges, M. P. (2010). Electronic speckle pattern interferometry and digital holographic interferometry with micro-bolometer arrays at 10.6 μm. *Applied Optics*, *49*(27), 5067–5075. doi:10.1364/AO.49.005067 PMID:20856279

Vavourakis, V., Loukidis, D., Charmips, D. C., & Panastasiou, P. (2013). Assessment of remeshing and remapping strategies for large deformation elastoplastic finite element analysis. *Journal of Computers and Structures*, *114-115*, 133–146. doi:10.1016/j.compstruc.2012.09.010

Vaziri, M. R., Salimi, M., & Mashayekhi, M. (2010). A new calibration method for ductile fracture models as chip separation criteria in machining. *Journal of Simulation Modeling Practice and Theory*, *18*(9), 1286–1296. doi:10.1016/j.simpat.2010.05.003

Vaziri, M. R., Salimi, M., & Mashayekhi, M. (2011). Evaluation of chip formation simulation models for material separation in the presence of damage models. *Journal of Simulation Modeling Practice and Theory*, *19*(2), 718–733. doi:10.1016/j.simpat.2010.09.006

Vaz, M. Jr, Owen, D. J. R., Kalhori, V., Lundblad, M., & Lindgren, L. E. (2007). Modeling and simulation of machining processes. *Archives of Computational Methods in Engineering*, *14*(2), 173–204. doi:10.100711831-007-9005-7

Velayudham, A., Krishnamurthy, R., & Soundarapandian, T. (2005). Acoustic emission based drill condition monitoring during drilling of glass/phenolic polymeric composite using wavelet packet transform. *Materials Science and Engineering A*, *412*(1–2), 141–145. doi:10.1016/j.msea.2005.08.036

Venkateshan, S. P. (2015). *Mechanical measurements*. John Wiley & Sons.

Verleysen, P., Peirs, J., Van Slycken, J., Faes, K., & Duchene, L. (2011). Effect of strain rate on the forming behaviour of sheet metals. *Journal of Materials Processing Technology*, *211*(8), 1457–1464. doi:10.1016/j.jmatprotec.2011.03.018

Vijayasankar, A., Garg, H., & Karar, V. (2016). A Novel Experimental Approach for Thermal Stress Measurement in Elastomerically Mounted Optics of Head-Up Display System. *Journal of Display Technology*, *12*(8), 859–868. doi:10.1109/JDT.2016.2540720

Walker, D. (2001). Residual stress measurement techniques. *Advanced Materials & Processes*, *159*(8), 30–33.

Wang, E., & Shukla, A. (2010). Analytical and experimental evaluation of energies during shock wave loading. *International Journal of Impact Engineering*, *37*(12), 1188–1196. doi:10.1016/j.ijimpeng.2010.07.003

Wang, Y., Liu, Y., & Jiang, Z. (2016). Design of a pulse-type strain gauge balance for a long-test-duration hypersonic shock tunnel. *Shock Waves*, *6*(6), 835–844. doi:10.100700193-015-0616-x

Warren, B., & Averbach, B. (1950). The effect of cold-work distortion on X-ray patterns. *Journal of Applied Physics*, *21*(6), 595–599. doi:10.1063/1.1699713

Warren, B., & Averbach, B. (1952). The separation of cold-work distortion and particle size broadening in X-ray patterns. *Journal of Applied Physics*, *23*(4), 497–497. doi:10.1063/1.1702234

Wasa, K., Kitabatake, M., & Adachi, H. (2004). *Thin film materials technology: sputtering of control compound materials*. Springer Science & Business Media.

Wei, L., & Xinqing, Z. (2009). Mechanical properties and transformation behavior of NiTiNb shape memory alloys. *Chinese Journal of Aeronautics*, *22*(5), 540–543. doi:10.1016/S1000-9361(08)60138-7

Wu, Z., Lu, J., & Han, B. (1998). Study of residual stress distribution by a combined method of Moire interferometry and incremental hole drilling, Part I: Theory. *Journal of Applied Mechanics*, *65*(4), 837–843. doi:10.1115/1.2791919

Wyant, J. C. (2002, July). White light interferometry (Vol. 4737). SPIE.

Xiaoli, L., & Zhejun, Y. (1998). Tool wear monitoring with wavelet packet transform—Fuzzy clustering method. *Wear*, *219*(2), 145–154. doi:10.1016/S0043-1648(98)00165-3

Xing, Y., Pedersen, E., & Moan, T. (2011). An inertia-capacitance beam substructure formulation based on the bond graph method with application to rotating beams. *Journal of Sound and Vibration*, *330*(21), 5114–5130. doi:10.1016/j.jsv.2011.05.025

Xu, L., Peng, X., Miao, J., & Asundi, A. K. (2001). Studies of digital microscopic holography with applications to microstructure testing. *Applied Optics*, *40*(28), 5046–5051. doi:10.1364/AO.40.005046 PMID:18364784

Yang, J., Jiang, L., & Chen, D. C. (2004). Dynamic modelling and control of a rotating Euler–Bernoulli beam. *Journal of Sound and Vibration*, *274*(3-5), 863–875. doi:10.1016/S0022-460X(03)00611-4

Yildiz, R. A., & Yilmaz, S. (2017). The verification of strains obtained by grid measurements using digital image processing for sheet metal formability. *Journal of Strain Analysis for Engineering Design*, *52*(8), 506–514. doi:10.1177/0309324717734669

Yoshida, S., Sasaki, T., Usui, M., Sakamoto, S., Gurney, D., & Park, I. K. (2016). Residual stress analysis based on acoustic and optical methods. *Materials (Basel)*, *9*(2), 112. doi:10.3390/ma9020112 PMID:28787912

Zhang, C., Zhao, L., Duerig, T., & Wayman, C. (1990). Effects of deformation on the transformation hysteresis and shape memory effect in a Ni47Ti44Nb9 alloy. *Scripta Metallurgica et Materialia*, *24*(9), 1807–1812. doi:10.1016/0956-716X(90)90550-Z

Zhang, J. (1998). Two-dimensional in-plane electronic speckle pattern interferometer and its application to residual stress determination. *Optical Engineering (Redondo Beach, Calif.)*, *37*(8), 2402–2410. doi:10.1117/1.602007

Zhang, J. M., Zhang, Y., Xu, K.-W., & Ji, V. (2006). General compliance transformation relation and applications for anisotropic hexagonal metals. *Solid State Communications*, *139*(3), 87–91. doi:10.1016/j.ssc.2006.05.026

Zhang, Y. C., Mabrouki, T., Nelias, D., & Gong, Y. D. (2011). Chip formation in orthogonal cutting considering interface limiting shear stress and damage evolution based on fracture energy approach. *Journal of Finite Elements in Analysis and Design*, *47*(7), 850–863. doi:10.1016/j.finel.2011.02.016

Zhao, B., Asundi, A. K., & Oh, K. E. (1999). Grid method for strain measurement in electronic packaging using optical, electronic, and atomic force microscope. Advanced Photonic Sensors and Applications, 3897, 260-271. doi:10.1117/12.369315

Zhao, X., Yan, X., Yang, Y., & Xu, H. (2006). Wide hysteresis NiTi (Nb) shape memory alloys with low Nb content (4.5 at.%). *Materials Science and Engineering*, *438*, 575–578. doi:10.1016/j.msea.2006.03.110

Zhu, X., & di Scalea, F. L. (2017). Thermal stress measurement in continuous welded rails using the hole-drilling method. *Experimental Mechanics*, *57*(1), 165–178. doi:10.100711340-016-0204-8

Zhu, X., Lanza di Scalea, F., & Fateh, M. (2015). On the study of the in-situ thermal stress measurement using a Hole-Drilling method. In *11th International Workshop on Advanced Materials and Smart Structures Technology*. University of Illinois Urbana-Champaign.

Zitoune, R., Krishnaraj, V., & Collombet, F. (2010). Study of drilling of composite material and aluminum stack. *Composite Structures*, *92*(5), 1246–1255. doi:10.1016/j.compstruct.2009.10.010

About the Contributors

Karthik Selva Kumar Karuppasmy, amid a healthy and intellectual upbringing, obtained his Bachelor Degree in Mechanical Engineering from Anna University. Subsequently, he obtained his Master Degree in Production Engineering from Thiagarajar College of Engineering/ Anna University. Besides having the strong technical expertise and analytical skills, he acquired his PhD degree from Indian Institute of Technology IIT, Dhanbad. He started his career to satisfy his passion for teaching and research by working at the National Institute of Technology (NIT)-Jamshedpur from 2016-2018. At present, he is working as a Post-Doctoral Research Fellow at IIT-Guwahati. In addition to teaching at engineering colleges, he has also been working as an administrator/coordinator for student associations and laboratories. As an active researcher, he is associated with many foreign Universities, along with that, he is a reviewer of 8 International Peer-Reviewed Journals. Consecutively he has contributed to the areas of Automobiles, Vibration, Engineering Design and Computational Fluid Dynamics with his 17 research publications in highly reputed International Journals and Conferences. He also published a book and a book chapter in the area of Fluid-Structure Interaction through international publishing houses.

Balaji P. S. received his PhD in vibration control from Curtin University, Australia and Master in Technology in Mechanical Engineering from National Institute of Technology Allahabad. He did his Bachelors of Engineering in Aeronautical Engineering from Anna University Chennai. His area of research is Vibration, Composite materials and Advanced Machining Process. He is actively involved in the consultancy projects on vibrations and stress analysis of mechanical components. He has also undertaken funded projects in the field of vibrations and composite materials. He is currently, associated with National Institute of Technology Rourkela in the Department of Mechanical Engineering. He has also authored patents on the vibration control devices. He also a reviewer for reputed International and National journals.

* * *

Avinash is a Research Scholar, PSG College of Technology, Coimbatore, Tamil Nadu, India.

Payal Banerjee is a Doctoral Research Scholar in the Department of Mechanical Engineering, National Institute of Technology Rourkela, Odisha.

Saibal Kanchan Barik is a Doctoral Research Scholar in the Department of Mechanical Engineering at IIT Guwahati.

K. V. J. Bhargav is pursuing Ph.D. from the National Institute of Technology, Rourkela. His research area is micromachining. He has pursued his B. Tech (ME) and M. Tech (Manufacturing) from Jawaharlal Nehru Technological University, Hyderabad. Even though his research area is machining, he is fond of design and its related areas.

Manoj Kumar Bhowmik is an Undergraduate Student in the Department of Civil Engineering at Tripura Institute of Technology. He is a dynamic candidate in the field of structural material and geothermal.

Mrinal Bhowmik received B.E degree in Mechanical Engineering from Tripura Institute of Technology, India. He obtained his M.Tech in Thermal Science and Engineering under Mechanical Engineering department at National Institute of Technology (NIT), Agartala. He worked as a faculty member in Mechanical Engineering department, National Institute of Technology Warangal. Presently, he is a doctoral research scholar in Centre for Energy at IIT Guwahati. His field of research is Non-conventional energy, Thermal science and IC Engine. He is a life member of Institution of Engineers (India) (IEI), Combustion Institute Indian Section (CI-IS) and Indian Society of Heating, Refrigerating and Air Conditioning Engineers (ISHRAE).

Srajan Dalela is pursuing Ph.D. from the National Institute of Technology, Rourkela. His research area is Metamaterials. He did his B. Tech (ME) from BIT Durg and M. Tech (Machine Design) from NIT Bhopal. He is actively involved in the design of metamaterials for the vibration isolation applications.

Vamsi Krishna Dommeti currently associated with SRM Institute of Science and Technology as Assistant professor in Department of Mechanical Engineering. The area of my research interest is dental biomechanics, dental materials.

Nuţu Emil is an Associate professor at the University Politehnica of Bucharest, Faculty of Industrial Engineering and Robotics, Strength of Materials Department. Between 2005 and 2013 he was an assistant professor at the same institution and lecturer between 2013 and 2019. He is experienced in numerical modeling of mechanical structures and in structural biomechanics. He develops models and algorithms to simulate the behavior of mechanical and biomechanical structures under different loading conditions and material constitutive laws. Within his doctoral thesis, entitled "Algorithms for studying bone remodeling with applicability to the design, mechanical analysis and optimization of prosthetic implants", he has developed algorithms that allow the computer implementation of most common adaptive bone remodeling theories using a general form and implemented them within a program that couples MATLAB and ANSYS simulation environments. He applied the program during a postdoctoral research project, entitled "Application of bone remodeling simulations to the optimization of dental implants". Also, several journal and conference publications emerged from simulations with his program. The scientific activity: coauthor to 2 national patents, 32 papers in scientific reviews or proceedings of different international and national scientific meeting volumes (congresses, conferences, symposiums); 14 research projects (as researcher); one third party research project as a project leader.

Abhishek Kamal is a Doctoral Research Scholar in the Department of Mechanical Engineering at IIT Guwahati.

Krishnaraj is currently working as an Associate Professor in the Department of Production Engineering at PSG College of Technology, India. He obtained his PhD in Machining of Composite Materials. He has authored about 80 journal papers, about 50 conference papers. He has successfully completed many sponsored research projects. His fields of interest include CNC machining and composite materials.

Vinayak Kulkarni is currently working as Professor in the department of Mechanical Engineering, IIT Guwahati. He received his Ph. D Degree from IISC Bangalore, India. He has expertise in the field of high speed aerodynamic for force and heat transfer measurement, force prediction algorithms and simulation of high-speed flows.

Senthil Kumar is currently working as an Associate Professor in the Department of Production Engineering at PSG College of Technology. He has authored more than 100 journal papers and about 50 conference papers. He has received an ISTE Award in 2006 and AICTE Career Award in 2009 and Outstanding Academician Award in 2011. He has completed many sponsored research projects and consultancy works. His fields of interest include vibration control, composites and smart structures.

Manikandan Padinjare Kunnath is a scholar from the Federation University Australia.

Malaidurai is a Doctoral Research Scholar in the Department of Applied Physics at Indian Institute of Technology (ISM) Dhanbad.

Manimaran is currently associated with SRM Institute of Science and Technology. He has completed his M-Tech from SRM Institute of Science and Technology. The area of his research interest is dental biomechanics.

Prabukarthi is currently working as an Assistant Professor in the Department of Mechanical Engineering at PSG College of Technology, India. He obtained his BE degree in Mechanical Engineering, ME in Manufacturing System and Management and obtained his PhD in the drilling of stacked materials. He has about nine years of teaching experience and two-year of industrial experience. He has authored 15 journal papers and about 25 national/ International conference papers. He has been actively involved in sponsored research projects and consultancy works. His field of interest includes supply chain management, lean manufacturing and machining of composite materials.

A. Karpaga Raj received his Bachelor Degree in Mechanical Engineering from the Institute of Road and Transport Technology (IRTT- Erode) under Anna University, in the year of 2010 and obtained his Master Degree in Manufacturing Engineering from Alagappa Chettiar College of Engineering and Technology (ACCET) under Anna University Chennai, in the year 2012. He joined as a research scholar in the department of mechanical engineering National Institute of Technology, Tiruchirappalli on September 2012. He has published papers in reputed International/National Journals and Conferences. His area of interest includes, polymer matrix composites, GTA welding, optimization techniques and numerical simulation. With his own interest practically he developed in-depth of knowledge in PAW, RSW, CMT and MMA welding. Related to his field, he was studied and finished level - II in Non-Destructive Testing (NDT) from ISNT Tiruchirappalli chapter in an association with BHEL- Tiruchirappalli. currently he was working in NIT-Patna as an assistant professor in mechanical engineering.

Anil Kumar Rout has completed his B. Tech in Mechanical Engineering from National Institute of Technology Agartala, Tripura, India and MME from Jadavpur University, Kolkata, India. Currently, he is pursuing his PhD. at Indian Institute of Technology, Guwahati, Assam, India.

Sandipan Roy received his PhD degree in Aerospace Engineering and Applied Mechanics in 2017 from Indian Institute of Engineering Science and Technology Shibpur India. Presently he is associated with SRM Institute of Science and Technology Tamil Nadu as Research Assistant Professor. His core areas of research include Orthopedic Biomechanics, Dental Biomechanics, Dental Biomaterials, Solid Mechanics.

Leo Kumar S. P. is presently working as an Associate Professor, PSG College of Technology, Coimbatore, Tamil Nadu, India.

Niranjan Sahoo is currently working as Professor in the Department of Mechanical Engineering, IIT Guwahati and has been affiliated to IIT Guwahati since 2004. He received his Ph. D Degree from IISC Bangalore, India. His broad research interest includes, Fluid Mechanics, Thermodynamics and Heat Transfer. He has undertaken various interdisciplinary courses within Department and NPTEL platform. Besides, he has published several research papers in peer-reviewed International Journals and Conferences. He guides several PhD scholars.

Jay Prakash Tripathi is presently Assistant Professor at Mechanical Engineering Department of Thapar Institute of Engineering & Technology, Patiala. He received his B. Tech. degree in mechanical engineering in 2012 from M. J. P. Rohilkhand University, Bareilly, India. After that, he graduated with Ph.D. in 2017 from Mechanical Engineering Department of IIT Dhanbad, India. He has around two years of post-Ph.D. research and teaching experience. He has published five research papers in SCI journals. He is GATE qualified, recipient of JRF and SRF fellowships and associate member of IEI India. His research focus includes Bond Graph modelling, condition monitoring, design and control of system dynamics, particle swarm optimization, and fluid power systems.

Index

M

N

O

P

R

S

T

V

W

www.ingramcontent.com/pod-product-compliance
Lightning Source LLC
LaVergne TN
LVHW080527300926
849162LV00007B/1554